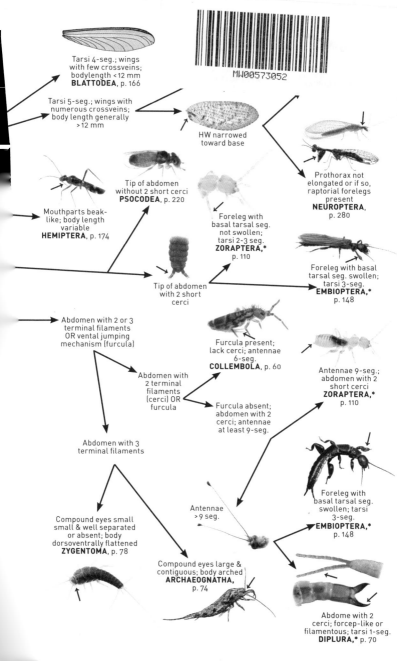

Tarsi 4-seg.; wings with few crossveins; bodylength <12 mm **BLATTODEA**, p. 166

Tarsi 5-seg.; wings with numerous crossveins; body length generally >12 mm

HW narrowed toward base

Prothorax not elongated or if so, raptorial forelegs present **NEUROPTERA**, p. 280

Mouthparts beak-like; body length variable **HEMIPTERA**, p. 174

Tip of abdomen without 2 short cerci **PSOCODEA**, p. 220

Foreleg with basal tarsal seg. not swollen; tarsi 2-3 seg. **ZORAPTERA,*** p. 110

Foreleg with basal tarsal seg. swollen; tarsi 3-seg. **EMBIOPTERA,*** p. 148

Tip of abdomen with 2 short cerci

Abdomen with 2 or 3 terminal filaments OR ventral jumping mechanism (furcula)

Furcula present; lack cerci; antennae 6-seg. **COLLEMBOLA**, p. 60

Antennae 9-seg.; abdomen with 2 short cerci **ZORAPTERA,*** p. 110

Abdomen with 2 terminal filaments (cerci) OR furcula

Furcula absent; abdomen with 2 cerci; antennae at least 9-seg.

Abdomen with 3 terminal filaments

Compound eyes small small & well separated or absent; body dorsoventrally flattened **ZYGENTOMA**, p. 78

Antennae >9 seg.

Foreleg with basal tarsal seg. swollen; tarsi 3-seg. **EMBIOPTERA,*** p. 148

Compound eyes large & contiguous; body arched **ARCHAEOGNATHA**, p. 74

Abdome with 2 cerci; forcep-like or filamentous; tarsi 1-seg. **DIPLURA,*** p. 70

*Not commonly encountered.

QUICK INDEX TO ORDERS

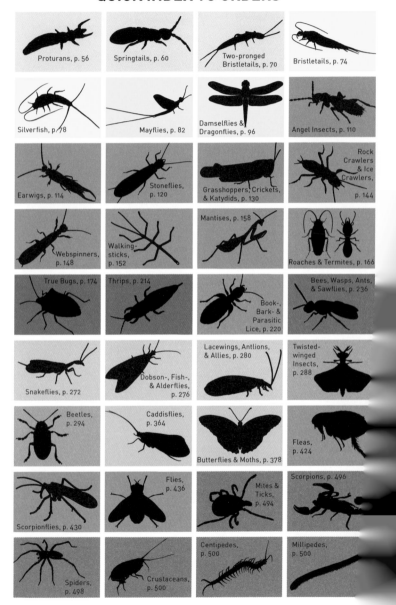

Proturans, p. 56

Springtails, p. 60

Two-pronged Bristletails, p. 70

Bristletails, p. 74

Silverfish, p. 78

Mayflies, p. 82

Damselflies & Dragonflies, p. 96

Angel Insects, p. 110

Earwigs, p. 114

Stoneflies, p. 120

Grasshoppers, Crickets, & Katydids, p. 130

Rock Crawlers & Ice Crawlers, p. 144

Webspinners, p. 148

Walking-sticks, p. 152

Mantises, p. 158

Roaches & Termites, p. 166

True Bugs, p. 174

Thrips, p. 214

Book-, Bark- & Parasitic Lice, p. 220

Bees, Wasps, Ants, & Sawflies, p. 236

Snakeflies, p. 272

Dobson-, Fish-, & Alderflies, p. 276

Lacewings, Antlions, & Allies, p. 280

Twisted-winged Insects, p. 288

Beetles, p. 294

Caddisflies, p. 364

Butterflies & Moths, p. 378

Fleas, p. 424

Scorpionflies, p. 430

Flies, p. 436

Mites & Ticks, p. 494

Scorpions, p. 496

Spiders, p. 498

Crustaceans, p. 500

Centipedes, p. 500

Millipedes, p. 500

Insects of North America

Insects of North America

John C. Abbott &
Kendra K. Abbott

Princeton University Press
Princeton and Oxford

Published by Princeton University Press
41 William Street, Princeton, New Jersey 08540
99 Banbury Road, Oxford OX2 6JX

press.princeton.edu

ISBN (pbk) 9780691232850
ISBN (e-book) 9780691232867
Library of Congress Control Number: 2022945058

British Library Cataloging-in-Publication Data is available

Editorial: Robert Kirk and Megan Mendonça
Production Editorial: Karen Carter
Jacket/Cover Design: Wanda España
Production: Steven Sears
Publicity: Caitlyn Robson and Matthew Taylor

Cover Credit: Photograph by Kendra Abbott /
Abbott Nature Photography

This book has been composed in DINOT.

The publisher would like to acknowledge the authors of this volume for providing the print-ready files from which this book was printed.

Printed on acid-free paper. ∞

Printed in Italy

10 9 8 7 6 5 4 3 2 1

CONTENTS

Acknowledgments vii

Introduction
What Is an Insect? 1
Classification and Nomenclature 6
Studying Insects 18
How to Use This Guide 50

Proturans (Protura) 56
Springtails (Collembola) 60
Two-pronged Bristletails (Diplura) 70
Bristletails (Archaeognatha) 74
Silverfish (Zygentoma) 78
Mayflies (Ephemeroptera) 82
Damselflies & Dragonflies (Odonata) 96
Angel Insects (Zoraptera) 110
Earwigs (Dermaptera) 114
Stoneflies (Plecoptera) 120
Grasshoppers, Crickets, & Katydids (Orthoptera) 130
Rock Crawlers & Ice Crawlers (Notoptera) 144
Webspinners (Embioptera) 148
Walkingsticks (Phasmatodea) 152
Mantises (Mantodea) 158
Cockroaches & Termites (Blattodea) 166
True Bugs (Hemiptera) 174
Thrips (Thysanoptera) 214
Booklice, Barklice, & Parasitic Lice (Psocodea) 220
Bees, Wasps, Ants, & Sawflies (Hymenoptera) 236
Snakeflies (Raphidioptera) 272
Dobsonflies, Fishflies, & Alderflies (Megaloptera) 276
Lacewings, Antlions, & Allies (Neuroptera) 280
Twisted-winged Insects (Strepsiptera) 288
Beetles (Coleoptera) 294
Caddisflies (Trichoptera) 364
Butterflies & Moths (Lepidoptera) 378
Fleas (Siphonaptera) 424
Hangingflies & Scorpionflies (Mecoptera) 430
Flies (Diptera) 436
Non-Insect Arthropods (spiders, centipedes, etc.) 492

Photographic Credits 504
Glossary 508
Literature 532
Index 538

ACKNOWLEDGMENTS

We have tried to create a unique guide to a diverse fauna that we hope will serve a wide audience. However, we did not want to duplicate the efforts of other North American insect field guides. Two very useful guides in this genre are the *Kaufman Field Guide to Insects of North America* by Eric Eaton and Kenn Kaufman and the *National Wildlife Federation Field Guide to Insects and Spiders of North America* by Arthur Evans. Both guides are very useful to anyone trying to identify many of the common species they may encounter. Alternatively, we set out to create a guide that would help the user identify insects to the level of family, a useful level of identification for amateurs and professionals. Many species-level identifications are provided throughout the book, but we wanted our guide to be an update to the *Peterson Field Guide to Insects* by Donald Borror and Richard White. We wanted to capture the spirit of that critical book while updating the taxonomy and offering our own creative aspects.

This endeavor was certainly not done alone, though. We have benefited from the teaching and mentorship of many colleagues and friends over the years. We would like to thank the following people for reviewing various sections of this book, providing valuable feedback and or providing content: James Adams, John Ascher, Thomas Atkinson, Ernest Bernard, Betsy Betros, Harry Blewitt, David Bolin, Pam Bolin, David Bowles, Matt Bowser, Valerie Bugh, Jerry Cook, Chris Difonzo, Mike Ferro, Luke Jacobus, Cheryl Johnson, Joshua Jones, Adem Keskin, Barrett Klein, Boris Kondratieff, Greg Lasley, Kelly Miller, Piotr Naskrecki, Liam O'Brien, Heath Ogden, Hans Pohl, Bill Ravlin, Brian Scholtens, Sean Schoville, Derek Sikes, Robert Sites, Graeme Smith, Vincent Smith, Manfred Ulitzka, John VanDyk, Milton Ward, Andrew Warren, Jim Weber, Lynne Weber, Alex Wild, Brandon Woo, Ian Wright, and Diane Young. Val Bugh, Arthur Evans and Stephen Marshall graciously read over the entire book for us and provided valuable comments and feedback. We extend a special thanks to Mike Ferro, who spent a great deal of time going through the book and providing valuable feedback and editorial suggestions that has resulted in a much improved resource.

We are thankful to individuals and organizations who provided specimens to photograph: Kim Bauer, James Berghdahl, David Bolin, Pam Bolin, Fossil Rim Wildlife Center, Holly Haefele, Lee Hoy, David Moellendorf, Alexander Nguyen, Dean Ryder, Anne Terry, Tracy Aviary, Riley Trapp, Westgate Pet and Bird Hospital, Diane Young, and Zookeeper Exotic Pets. Charlie Covell and Andy Warren (McGuire Center for Lepidoptera & Biodiversity) provided access to specimens to complete the book.

ACKNOWLEDGMENTS

We are grateful to the many outstanding photographers who let us use their photographs throughout the book. Rather than crediting each photo, to save space, we provide credits at the end of the book. Without these photographers, this book would not have been possible.

The pictorial keys in the book were inspired by those found in the Borror and White *Peterson Field Guide to Insects*. We used multiple sources to create these keys, including the amazing volumes by Stephen Marshall, *Insects, Beetles* and *Flies*, W. Patrick McCafferty's *Aquatic Entomology*, Gerald Fauske's key to superfamilies of Lepidoptera (www.ndsu.edu/ndmoths/ndmoths), and *Borror and DeLong's Introduction to the Study of Insects (7th edition)* by Charles Tripplehorn and Norman Johnson.

The BugGuide.net and iNaturalist.org communities are a tremendous resource and collectively provide an amazing wealth of knowledge. Without help from the members of these communities, this book would have certainly suffered. Many users on one or both sites assisted us in providing locations and identifications of species.

We thank the many individuals that we have spent time with chasing bugs. This book would not have been possible without the many friendships we have forged over the years. We continue to be grateful for all these friends who have shared their tremendous knowledge with us.

Finally, this field guide was originally meant to be an update to the *Peterson Field Guide to Insects of America North of Mexico* by Donald Borror and Richard White, published in 1970. However, shortly before completion of this book Houghton Mifflin Harcourt decided to end the historic Peterson Field Guide series. Thankfully, Robert Kirk and Princeton University Press decided to pick this title up. We are very thankful to Robert and his team. Finally, we appreciate Lisa White for providing us the opportunity to take on this project and her thorough and careful work as copy editor!

Seven-spotted Lady Beetle
(*Coccinella septempunctata*)
Coccinellidae

WHAT IS AN INSECT?

Insects are members of the phylum Arthropoda, a large, diverse group of organisms; representing over 80% of all animals on the planet, the overwhelming majority of arthropods are insects. Arthropods have an external skeleton (exoskeleton), jointed appendages (*arthropod* means "jointed foot"), a segmented body, and paired jointed appendages. They are bilaterally symmetrical, and the exterior portions of the body and appendages are made up of cuticle, containing chitin (pronounced with a hard K like kraken), a complex carbohydrate (the second-most common polysaccharide in nature behind cellulose, found in plants) that provides rigidity. The external skeleton means that insects, and all arthropods, must shed their cuticle in order to grow. This hormonally driven process is called *molting*, and the resulting shed exoskeleton is called an *exuviae*.

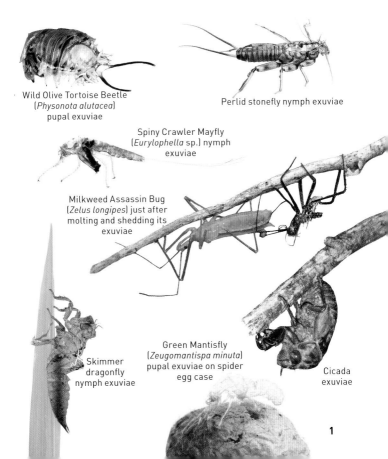

Wild Olive Tortoise Beetle
(*Physonota alutacea*)
pupal exuviae

Perlid stonefly nymph exuviae

Spiny Crawler Mayfly
(*Eurylophella* sp.) nymph
exuviae

Milkweed Assassin Bug
(*Zelus longipes*) just after
molting and shedding its
exuviae

Skimmer
dragonfly
nymph exuviae

Green Mantisfly
(*Zeugomantispa minuta*)
pupal exuviae on spider
egg case

Cicada
exuviae

1

INTRODUCTION

STRUCTURE

This field guide focuses on the Hexapoda, a subphylum largely made up of insects (see discussion on classification, p. 6). In addition to the features mentioned above that characterize the phylum, hexapods possess a 3-part body, 3 pairs of jointed legs, compound eyes and a pair of antennae. Most also bear a pair of wings.

Each body region of the insect (tagmata) is responsible for particular functions. The head, home to the mouthparts, eyes, and antennae, is responsible for sensing the environment and handling food. The legs and wings are found on the heavily muscular thorax, which is responsible for movement. The third body region of an insect is the abdomen, which usually doesn't have a lot of external structures present but contains and protects most of the insect's vital organs.

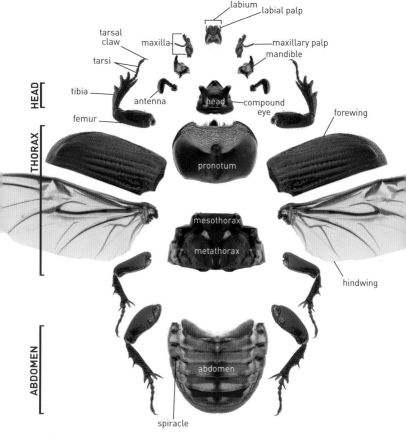

ANATOMY

Insects have an internal body cavity, the haemocoel, housing the organs. They have an open circulatory system (no blood vessels) where the organs are bathed in haemolymph (analogous to blood) circulated throughout the cavity by a "heart" or dorsal aorta (running along the top of the body). Insects don't have a single spinal cord that runs along their back; they have two ventral nerve cords on the underside of the body. These form a paired ganglia in each body segment. Insects don't have lungs. Air enters their body through paired external openings on their sides called spiracles, which usually connect to a tracheal trunk, or large tubules that then feed into a network of smaller trachea running throughout the body. The digestive system is comprised of a fore-, mid-, and hindgut. The tracheae and fore- and hindgut are lined in cuticle and therefore must be shed with each molt the insect undergoes. Most insects have excretory organs called malpighian tubules that act like kidneys. They are found in the abdominal cavity and empty into the digestive system where the mid- and hindgut meet.

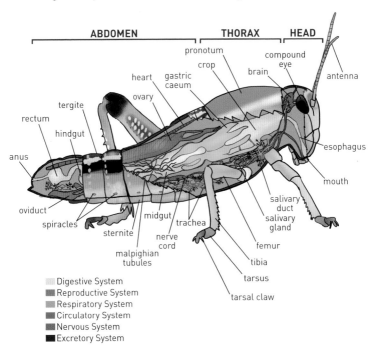

Digestive System
Reproductive System
Respiratory System
Circulatory System
Nervous System
Excretory System

GROWTH AND DEVELOPMENT

Insects grow by shedding their exoskeleton through a process called molting. With each molt they shed their exoskeleton as well as the linings of the fore- and hindgut and respiratory tubes. Molting is a precarious time that may take place nearly instantly or can take hours, depending on the species. Right after molting, the insect is usually soft-bodied, pale or even white in color, and very fragile. This stage is called *teneral*. Each molt an insect goes through is called an *instar*.

One of the keys to the success of insects is their ability to radically transform themselves, a process called *metamorphosis*. Insect metamorphosis is either simple (ametabolous and hemimetabolous) or complete (holometabolous). Ametabolous development occurs in primitively wingless hexapods (springtails, two-pronged bristletails, bristletails, and silverfish). Nymphs (the immature stage) hatch from an egg and look like a smaller version of the adult. Unlike all other insects, they continue to molt after reaching sexual maturity. The nymphs and adults are often found living together.

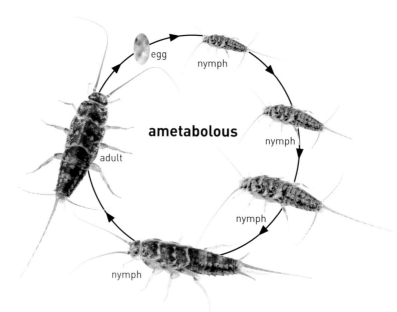

egg

nymph

ametabolous

nymph

nymph

adult

nymph

Insects that develop external wings as nymphs have hemimetabolous metamorphosis and include the dragonflies, mayflies, stoneflies, grasshoppers, mantids, walkingsticks, true bugs, and a few other related orders. The nymph hatches from an egg and goes through a series of molts (5 to 20 or more, depending upon the species) until reaching a winged (usually), sexually mature adult. With each molt, the external wing pads visible on the nymph get larger and larger. In some groups the nymphs and adults are found in the same habitat (e.g., true bugs and grasshoppers) and in others the nymphs may be aquatic while the adults are not (e.g., dragonflies and mayflies).

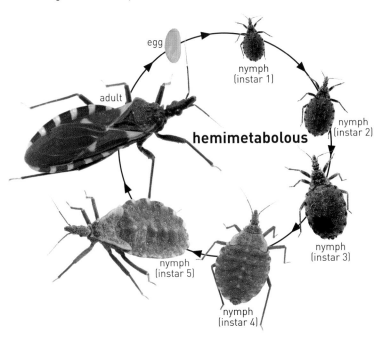

egg

nymph (instar 1)

adult

nymph (instar 2)

hemimetabolous

nymph (instar 3)

nymph (instar 5)

nymph (instar 4)

Complete metamorphosis (holometabolous) is the most familiar type of development: egg, larva, pupa, and adult. The pupa is an additional stage that occurs between the larva (immature) and adult stages. During this stage, the insect undergoes a complete tissue breakdown and rebuilding. Insects with complete metamorphosis have larvae that look remarkably different from the adults. Every beetle species has a grub, every fly species a maggot, every butterfly and moth a caterpillar. They do not develop external wing pads and often have different mouthparts and feeding habits. They have fewer instars (usually 3 to 5 molts) than insects with simple metamorphosis. The evolutionary development of the pupal stage is partially responsible for a massive species radiation in insects. Only one-third of insect orders are holometabolous, but but they contain 90% of all insect species! There is no doubt that holometaboly has been responsible for much of the success in insects.

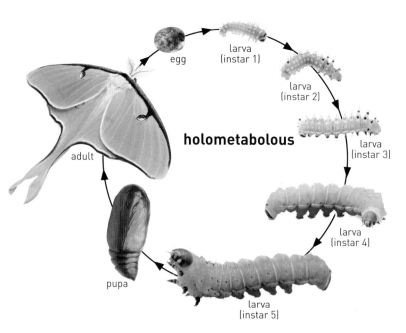

egg

larva
(instar 1)

larva
(instar 2)

holometabolous

larva
(instar 3)

adult

larva
(instar 4)

pupa

larva
(instar 5)

CLASSIFICATION AND NOMENCLATURE

Insects exhibit a dazzling array of feeding habits (fluids, solids, plants, animals) and can be found in nearly every habitat imaginable. The single exception is the open ocean, where only a few species have managed to invade. With such diversity, it can be difficult to categorize groups with specific characteristics; the takeaway is that there are almost always exceptions to generalities.

To study, communicate and better understand the relationships within such diversity, we use a classification based on a hierarchy. This system, first established by the eighteenth-century Swedish botanist Carl Linnaeus, groups organisms by similarity. The fundamental unit in this system is the species, which is given a binomial name made up of a genus and specific epithet, similar to a first and last name. A species is a population of organisms that are generally reproductively isolated, similar in appearance, genetics, and ecology and have newly evolved characters that set them apart from close relatives. One or more similar species are grouped together in a genus, multiple genera into a family, multiple families into an order and so on. This naming system is meant to follow the true relationships among species. As our understanding of the insect family tree changes, names are changed and updated as well. It can be frustrating, but the goal is precise communication and to show accurate relationships among species and groups.

Domain Eukarya
Kingdom Animalia
Phylum Arthropoda
Subphylum Hexapoda
Class Insecta
Order Coleoptera
Suborder Polyphaga
Superfamily Scarabaeoidea
Family Scarabaeidae
Subfamily Rutelinae
Tribe Rutelini
Genus *Chrysina*
Specific epithet *gloriosa*
Author & Year (LeConte, 1854)

The genus and specific epithet are combined to form a 2-word species name that is latinized in its spelling and italicized. Only the generic name is capitalized. Sometimes the person who described the species (author) and year the species was described are included. If the species is later moved to a different genus, the author and year are placed in parentheses. In the example above, John LeConte described *Plusiotus gloriosa* in 1854 (written *Plusiotus gloriosa* LeConte, 1854). A subsequent author determined the species was best placed in the genus *Chrysina*, so it is correctly written as *Chrysina gloriosa* (LeConte, 1854).

Depending upon the circumstances, intermediate groups may be used to help with classification. Some of these, such as subphylum and superfamily, are obvious in their placement. Others, like tribe (a designation between subfamily and genus), are not as obvious.

The names in the categories of superfamily through tribe have standardized endings (helpful, so you know the category, even if you don't know the insect):

Superfamilies end in the suffix "-oidea," pronounced "oi-dee-ə"
Families end in the suffix "-idae," pronounced "ih-dee"
Subfamilies end in the suffix "-inae," pronounced "ih-nee"
Tribes end in the suffix "-ini," pronounced "ī-nī"

The broadest group most insect enthusiasts are familiar with is order. There is no set suffix for orders within insects; most end in "-ptera," which means wing. Since there are only about 30, they are easy to memorize. Occasionally a trinomial is used to designate a subspecies or variety (e.g., Salt Creek Tiger Beetle, *Cicindela nevadica lincolniana*). Subspecies classification tends to occur in some groups more than others. A species that is unknown or not named may be designated by "sp." (single species) or "spp." (multiple species). For example, *Asterocampa* sp., would refer to a single, unnamed species of *Asterocampa*.

Scientific and common names come with costs and benefits. Common or English names are sometimes easier to remember, and many birds, mammals, and even reptiles and amphibians have them. Most insect species do not have common names, though more and more are being coined to facilitate the interest of enthusiasts, especially within readily identifiable groups (butterflies and dragonflies, for example). Even then, the names are usually not very well standardized. In this book, we have tried to use common names wherever possible (especially for large groups like orders, and occasionally families and species) because we feel it best serves our primary audience. We tried to use the most widely accepted common names, usually found on BugGuide.net or iNaturalist.org. An additional resource is the Entomological Society of America's Common Names of Insects Database (www.entsoc.org/common-names). We, however, encourage the use of scientific names (family, even genus and species) because they are more precise than common names and, while they do change, they are governed under a set of official rules and show relationships among the insects they are naming. In this guide, we capitalize common names of species.

The evolutionary history of arthropods has undergone considerable change and debate. The most closely related extant phyla are the onychophorans (velvet worms) and the microscopic tardigrades (water bears). The most recent evidence from molecular phylogenetics (see Giribet and Edgecombe, 2019) continues to support the idea that Arthropoda is monophyletic, a group with a single common ancestor. Extant arthropods fall into 2 additional

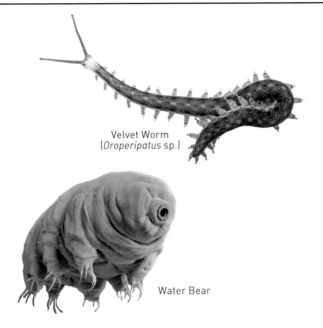

Velvet Worm
(*Oroperipatus* sp.)

Water Bear

large monophyletic groups, the Chelicerata (including sea spiders, horseshoe crabs, spiders, and their relatives) and the Mandibulata (including centipedes, millipedes, insects, and a number of lineages of crustaceans). Researchers continue to work on understanding the relationships among these groups, and it is beyond the scope of this book to delve into them further. Since our primary focus is to help with the identification of these groups, we have opted to use a 5 subphyla classification (including the extinct Trilobitomorpha) within the Arthropoda. Recent evidence suggests the Crustacea and Hexapoda may be considered a single subphylum called Pancrustacea.

The 5 subphyla recognized here include:

Subphylum **Trilobitomorpha** – trilobites (extinct)
Subphylum **Chelicerata** – spiders, scorpions, horseshoe crabs, & sea spiders
Subphylum **Myriapoda** – centipedes & millipedes
Subphylum **Crustacea** – shrimp, barnacles, lobsters, crabs, & woodlice
Subphylum **Hexapoda** – insects

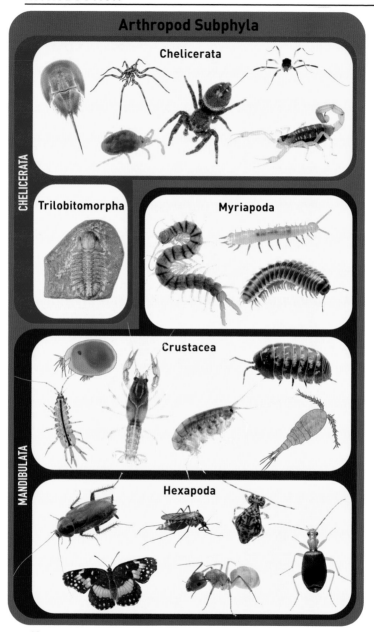

Within the Hexapoda, there are various subdivisions that are useful for understanding how insects are related to one another. These divisions continue to undergo change as our understanding improves. The phylogeny on the next page shows the group Entognatha, which have mouthparts that are retracted within the head (unlike insects) and are primitively wingless, meaning they and their ancestors have never developed wings. The Protura, Collembola, and Diplura used to be considered orders of insects, but are now thought of as their own orders or classes depending on the author. Entognatha is now recognized as a polyphyletic group that contains members that have similar characteristics, but lack a common ancestor.

The class Insecta includes 27 currently recognized orders, but some of these are disputed. These orders are often split into 2 informal groups, Palaeoptera and Neoptera. The Palaeoptera contains the extant orders Odonata and Ephemeroptera, which lack the ability to fold their wings flat over the abdomen. Recent evidence suggests the Palaeoptera is not monophyletic, meaning the orders may lack a common ancestor. The Neoptera is an informal group that contains the remaining orders, all of which can—or have an ancestor that could—fold their wings flat over the abdomen.

Within the Neoptera, the Exopterygota, also known as the Hemipterodea, are insects with simple metamorphosis: nymphs that develop external wing buds and do not go through a pupal stage. The Endopterygota, or Holometabola, have complete metamorphosis: internal wing development and the presence of a pupal stage.

A number of the historically recognized insect orders have undergone name changes, while others have been combined to form new orders. You may still see these names in older resources or simply because of a difference in opinion. We have provided a guide to some of these changes.

Name Used in This Guide	Name(s) Historically Used
Archaeognatha	Microcoryphia
Zygentoma	Thysanura
Notoptera	Mantophasmatodea + Grylloblattodea
Embioptera	Embiidina, Embiodea
Phasmatodea	Phasmida
Blattodea	Blattaria + Isoptera
Blattodea, Mantodea	Dictyoptera
Hemiptera	Heteroptera + Homoptera
Psocodea	Psocoptera + Pthiraptera (Anoplura + Mallophaga)

Phylogeny of the Hexapoda*

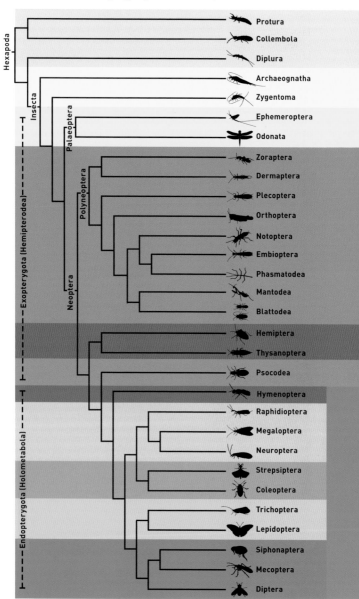

The **Hexapoda** is currently treated as a subphylum and contains arthropods that have a consolidated thorax and 3 pairs of legs.

The class **Entognatha** are wingless and have mouthparts retracted within their head.

The **Insecta** (groups from here down) have 3 pairs of legs, 3 main body parts, typically 2 pairs of wings, 1 pair of antennae and usually 1 pair of compound eyes.

The informal group **Palaeoptera** contains winged insects that lack the specialized anatomy to fold their wings flat over the abdomen.

The informal group **Neoptera** (this group and all below), contains winged insects with specialized anatomy to fold their wings flat over the abdomen. Even wingless insects in these groups are Neoptera because they had ancestors that could fold their wings.

The **Polyneoptera** (this group), sometimes called the Orthopteroids, represents one of the earliest insect radiations and contains the majority of hemimetabolous orders.

The **Condylognatha** is a superorder that contains the Hemiptera and Thysanoptera.

The **Panpsocoptera** contains the previously recognized Psocoptera (book and bark lice) and Pthiraptera (parasitic lice) orders.

The **Endopterygota** or **Holometabola**, are a monophyletic superorder of insects that go through complete metamorphosis: egg, larva, pupa, and adult. They undergo the most extreme metamorphosis, with larval and adult stages differing considerably both in form and behavior.

Endopterygota larvae have internal wing development during the pupal stage, while Exopterygota nymphs have external wing buds and don't go through a pupal stage.

This group contains the highest diversity, with about 850,000 known species divided among 11 orders. Coleoptera, Hymenoptera, Lepidoptera, and Diptera are the Big Four insect orders.

*Based on Misof et al. (2014).

INTRODUCTION

INSECT DIVERSITY

It is hard to say how many species of insects are known, much less how many actually exist. Digital databases have made lists easier and more accurate, but for large groups the numbers are still just estimates. And insect species are still being described on a daily basis. The numbers in the table to the right are taken from various sources and in most cases provide approximations. In total, 784 families are known to occur in North America (NA) north of Mexico and contain approximately 92,000 species, or about 10% of all insect species known on Earth.

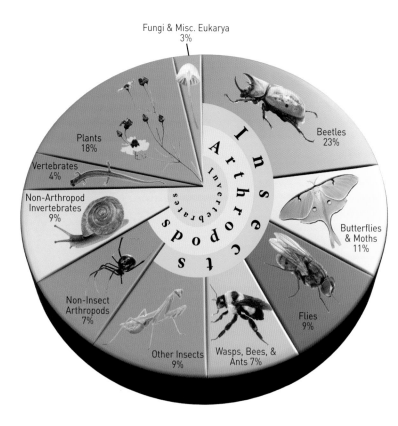

Fungi & Misc. Eukarya
3%

Plants
18%

Vertebrates
4%

Non-Arthropod
Invertebrates
9%

Non-Insect
Arthropods
7%

Other Insects
9%

Wasps, Bees, &
Ants 7%

Flies
9%

Butterflies
& Moths
11%

Beetles
23%

Arthropods

Insects

Invertebrates

Order	Families in NA*	Species in NA*	Species in the World
Proturans (Protura)	3	80	830
Springtails (Collembola)	21	1,000	9,000
Two-pronged Bristletails (Diplura)	5	173	1,000
Bristletails (Archaeognatha)	2	35	350
Silverfish (Zygentoma)	3	19	550
Mayflies (Ephemeroptera)	21	657	3,800
Damselflies & Dragonflies (Odonata)	11	471	6,350
Angel Insects (Zoraptera)	2	2	44
Earwigs (Dermaptera)	5	27	1,800
Stoneflies (Plecoptera)	9	670	3,400
Grasshoppers, Crickets, & Katydids (Orthoptera)	16	1,200	20,000
Rock & Ice Crawlers (Notoptera)	1	15	60
Webspinners (Embioptera)	3	11	380
Walkingsticks (Phasmatodea)	5	29	3,600
Mantises (Mantodea)	7	28	2,500
Roaches & Termites (Blattodea)	9	94	7,600
True Bugs (Hemiptera)	108	10,200	107,000
Thrips (Thysanoptera)	6	1,000	6,300
Book, Bark & Parasitic Lice (Psocodea)	44	1,220	11,000
Bees, Wasps, Ants, & Sawflies (Hymenoptera)	80	18,000	153,000
Snakeflies (Raphidioptera)	2	22	260
Dobsonflies, Fishflies, & Alderflies (Megaloptera)	2	46	400
Antlions, Lacewings, & Allies (Neuroptera)	11	400	4,700
Twisted-winged Insects (Strepsiptera)	5	84	600
Beetles (Coleoptera)	145	25,200	400,000
Caddisflies (Trichoptera)	28	1,500	14,000
Butterflies & Moths (Lepidoptera)	90	12,000	182,500
Fleas (Siphonaptera)	8	325	2,500
Scorpionflies (Mecoptera)	5	85	600
Flies (Diptera)	127	17,000	153,000
TOTALS	784	91,593	1,097,124

*North of Mexico.

INTRODUCTION

With over 1.7 million species currently described, the pie graph on page 14 shows the relative proportions of major groups. Seventy-five percent of described species are invertebrates; it is estimated that this number is probably closer to 95%. Sixty-five percent of all described species are arthropods, and nearly 60% are insects. The numbers used above are extracted from the International Union for Conservation of Nature (IUCN Red List of Threatened Species 2014.3. Summary Statistics for Globally Threatened Species. Table 1: Numbers of threatened species by major groups of organisms, 1996–2018).

GLOBAL THREATS TO INSECTS

Given insects make up a majority of the animals on Earth, it is critical to understand their role in the environment and whether they are in need of conservation. Insects are essential components of our ecosystem, and we rely on them to perform a myriad of ecosystem services without which humans couldn't survive. While 1.7 million insect species have been described, there are likely 5 million to 30 million on the planet. This just highlights how little we know about the most diverse group of animals on Earth.

Insect decline has been receiving a lot of attention. Hallmann et al. (2017) noted a 75% loss of flying insect biomass over 27 years in Germany. The overwhelming trend in research is that there is a decline in insects. However, there is also research indicating that insects are not declining and are increasing in certain parts of the world. One of the reasons it is so hard to pinpoint whether there is a decline and what the reason may be for the decline is that there are so many confounding factors. Wagner et al. (2021) refers to insect decline as death by a thousand cuts. Meaning that there are many factors working together to cause the decline. One of the challenges studying insect decline is to figure out which stressor may be the most important to a group.

Some of the stressors causing insect decline point back to climate change (top half of figure on the following page). Insect ranges tend to be moving toward the poles, as climate change predicts. When insects cannot move up a mountain or towards the poles, their options are limited. Some species are declining or going extinct while others may hybridize to compete for the same resource as related species. Other climate-pointing disruptions include increased fire risk and increased storm intensity. Storms are becoming stronger and more frequent, with damaging flooding. Another climatic extreme is droughts, with periods of decreased precipitation becoming longer and more frequent affecting all life. Finally, nitrification of the planet from fertilizer and products of fossil fuels are affecting groups that are adapted to low-nutrient conditions.

There are several factors outside of climate change affecting insect populations (bottom half of figure below). The most obvious and universal reason for species decline is human overpopulation. The more people we have, the more we need urbanization to accommodate the increased population size, which results in loss of habitat. Along with urbanization is increased global trade, which accelerates the movement of invasive species and pathogens to new ranges which can devastate populations. With increased populations, agricultural intensification increases. This leads to more deforestation and use of insecticides to produce more food. Various types of pollution, like light, air, chemical, and sound, can also have a negative effect on insects. Any one of these stressors may not cause a decline in insects, but all together they create a death by one thousand cuts.

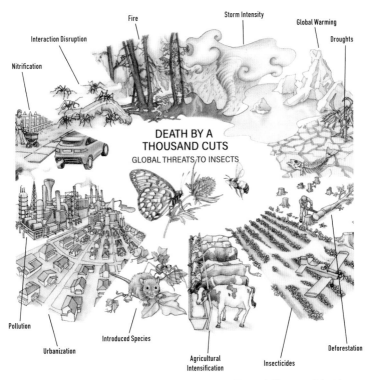

Artwork courtesy Virginia Wagner in Wagner et al. (2021).

STUDYING INSECTS

You can explore and learn about insects through observation only, photography, or making an insect collection. Each has its own benefits, and they are certainly not mutually exclusive. Historically, focusing on taxonomy and studying the diversity of insects meant making a collection. This is still the primary way by which most entomologists work because it is necessary to study all but the best known or largest insects under a microscope to identify them.

Making a collection is still a great way for many to learn about and enjoy insects. Before the age of digital photography, this was how every insect enthusiast and professional invariably got started. We don't encourage collecting and killing insects just for the sake of killing, but one of the reasons that insects are so popular is their accessibility. They are literally everywhere, in all terrestrial and freshwater habitats that can you imagine. This, along with their reproductive strategy, which typically results in dozens, to hundreds or even thousands of offspring per individual, makes them the perfect subject for every naturalist. Additionally, entomology is a field in which enthusiasts and non-professionals can make serious contributions. There are likely new species waiting to be discovered in your own yard or local park.

Physical collections are critical for the discovery and description of new species. They are also required for genetics studies that have proved so important in understanding the relationships of insects. If you do choose to make an insect collection, we encourage you to take it seriously; properly curated and labeled specimens are always welcome at a local university or museum collection.

Not everyone will want to make a physical insect collection, which is fine, because digital photography offers an amazing opportunity to collect insects in a different way. Nearly everyone has a great camera on their phone that they carry everywhere they go. If you want to take it to the next level, there are a host of more advanced digital cameras available. More importantly, many online resources let you share your images and get and give help identifying your insects. Social media platforms like FaceBook (are home to numerous groups that post and identify photos. Flickr (Flickr.com) is a photo-sharing site that is also great for finding, posting, and discussing identified photos. The best place to get help with identifying photos of insects from North America is BugGuide (BugGuide.net). This is a community of knowledgeable professionals and enthusiasts dedicated to making the most accurate identifications possible.

Many citizen science initiatives focused on insects. There are too many to list here, but try searching the internet for a specific group of interest and citizen science (e.g., "bumblebees citizen science" or "Odonata citizen science") and you will likely find opportunities to contribute. Most citizen science is done through the contribution of photos to a project, but not always; some solicit physical collections of specimens or other data. iNaturalist (iNaturalist.org) is by far the most popular citizen science website. This is because it has a beautiful, easy-to-use platform and doesn't focus on a particular group, but rather is one-stop shopping for submitting your natural history observations. It is a large and growing community of amateurs and professionals. It also uses some very impressive artificial intelligence to suggest possible IDs for photos submitted. Some of our other favorite online resources are OdonataCentral. com for dragonflies and damselflies, Moth Photographers Group (MothPhotographersGroup.msstate.edu) and e-butterfly.org for butterflies. Please keep in mind that photos, no matter how good they are, cannot replace specimens, and it simply may not be possible to make a confident identification. Nevertheless, all of these resources provide a fantastic opportunity for the insect collector and non-collector alike to learn.

Whether you are going to observe, make a physical collection, or photograph insects, you may enjoy using some basic collecting techniques. For example, the most standard piece of insect-collecting equipment, the aerial or butterfly net can be used as a catch-and-release piece of equipment to collect all flying insects, not just butterflies. We have tried to provide both general and specific guidance in the following sections on collecting, curating, and photographing insects and their relatives.

ETHICS & RESPONSIBLE COLLECTING

The first thing most people will ask when considering collecting insects and other arthropods is whether doing so will impact their populations. There are very few regulations on collecting insects, unlike hunting, fishing, or activities involving vertebrates. This is, in part, because these groups of animals have such different reproduction rates and numbers. An acre of land might hold one—or maybe a dozen—particular vertebrates, but can have thousands, tens of thousands, or even millions of a particular insect or arthropod.

There are some species that are found in smaller numbers, usually because they are found in particularly sensitive and locally restricted environments or because their life history is tied to a rare host plant or other limiting resource. The number of insects and arthropods killed through collecting activities will never come close to the number killed by their natural predators, such as birds, frogs, reptiles, and mammals. Make no mistake, though—humans do have an impact on insect populations through habitat destruction, light pollution, introduction of exotic species, even just driving on the highway. The IUCN (International Union for Conservation of Nature and Natural Resources) is trying to assess the conservation status and need for as many species as possible. They maintain a Red List (https://www.iucnredlist.org/) where this information can be viewed by species or group.

Collecting insects is essential to understanding, studying, and even conserving them. When an organism is properly curated and labeled, it becomes a specimen that can last for hundreds of years. It is utterly unique and irreplaceable and can be used by scientists and researchers for generations to come. You should always try to collect and curate specimens so they can be incorporated into a museum collection when you can't keep them anymore. For those who choose to collect, for themselves or academic collections, we encourage responsible collecting. Carolyn Trietsch and Andrew Dean came up with the Insect Collectors' Code (Trietsch & Dean, 2018) based on the Hippocratic oath to "facilitate awareness and discussion of the ethics of insect collecting, especially amongst entomologists-in-training (students)." The authors of the code received input and advice from members of the Entomological Collections Network (https://ecnweb.net) and hope that it *serves as a vehicle to instill ethical behavior in entomology.*

The Insect Collectors' Code

I strive to fulfill, to the best of my ability, the following ideals:

- I will respect the hard-won scientific gains of those entomologists in whose steps I walk and gladly share my scientific gains and knowledge with those who are to follow.

- I will aid in the dissemination of scientific knowledge, both to those who study insects and those who do not.
- I will not discriminate against others, and I will strive to create a safe working environment, whether in the field, the classroom, or the lab.
- I will treat insects humanely. As a collector, it is within my power to take insect life; I will not take insects that will not be deposited in a natural history collection or otherwise made available for research and education. While bycatch is often unavoidable, I will, to the best of my ability, attempt to reduce the unnecessary loss of insect life and find use for these specimens.
- I will consider the ecological impact of removing insects and their products (galls, nests, etc.) from the environment when collecting, whether the species are protected by law, known to be declining, or are considered to be of least concern. I will strive to avoid or minimize disturbance to the environment while collecting.
- I will secure appropriate permits prior to collecting insects, and I will honor and uphold the provisions stated by each permit. I will keep copies of all permits on my person while collecting and furnish them to authorized agents upon request. I will save all permits associated with specimens as proof that they were collected legally.
- I will keep detailed field notes of my collecting activities and will make these available to the greater scientific community.
- I will prepare and label specimens according to standards established by professional entomologists who work with collections.
- I will properly store all specimens under my care, and I will not allow specimens to become damaged or degraded through neglect.
- I will properly use and dispose of preservatives, killing agents, and other chemicals associated with specimen collection and preparation. I will never use these chemicals to harm myself or others.
- I will make arrangements for any personal specimens or collections in my possession to be deposited in a museum in the event of my untimely death.
- I will not create false data.

May I always act so as to preserve the finest traditions of natural history, and so long as I uphold these traditions and the stated ideals, may I long experience the joy of my contributions to the furthering of scientific knowledge.

PERMITS

In much of the United States and Canada you do not need a permit to collect insects. However, most state, provincial, and national parks, and some municipal parks, require permits. Be sure to talk with appropriate administrators before making collections. There are very few insects that are listed as threatened or endangered at the federal or state level. It is unlikely that you would encounter these during the course of normal insect hunting. It is important to collect legally and maintain all permits and paperwork associated with your collection. These documents should always be associated with your collection and will almost certainly be required by any institutions you may ultimately choose to donate your collection to.

COLLECTING METHODS

The basic tools needed for collecting insects depends on your groups of interested, but generally consist of a good hand lens or loupe, Pigma Micron pens, forceps, net, jars or containers for specimens, kill jar and/or ethanol vial(s), glassine or paper envelopes, and an aspirator. The most important tools, however, are patience and an enthusiasm for discovery!

The aerial or butterfly net is probably the first piece of equipment that will come to mind. It is an essential piece of equipment for many entomologists. Aerial nets come in all sorts of sizes and variations, and most users have their favorite variety. The basic idea is a net bag on a pole. A bag with a diameter of 15 or 18 inches will satisfy the needs of most collectors, but smaller or larger bags have their benefits. The bag color is usually white, great for seeing what is in the bag, but some people prefer green or a darker color so the insect is less likely to see the bag coming at it. The pole length is also variable. A length of 3 feet is fairly standard, but

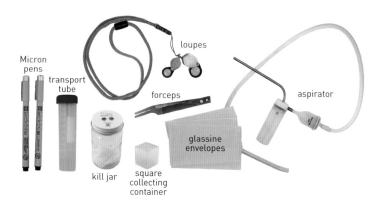

Micron pens

loupes

transport tube

forceps

aspirator

glassine envelopes

kill jar

square collecting container

there are much longer poles for catching canopy insects. There are also collapsible poles, which can offer you great flexibility. Whatever size and type of aerial net you employ, it is generally good practice to try and come from behind your subject, especially with fast-moving and visually acute subjects like dragonflies. Good follow-through is usually important when using this type of net. Swing through the insect and then flip the bag over the hoop so that it is captured at the tip. A rookie mistake is to immediately let the bag drop downward and look straight into the bag to see what you caught. Insects will quickly take advantage of the opportunity and fly straight up and out.

Another useful capture technique with an aerial net is to "pancake" your subject, especially for insects on the ground that could fly (tiger beetles and even butterflies puddling). Hold the tail end of the bag up so it's fully extended while slowly approaching the subject on the ground. When you think you can approach no closer, quickly slam the net over the insect. (Don't break your net!) Lift the tail of the net in the air so that your subject will go up and be captured in the tip. Note that many insects are adept at finding their way out if the net rim is not flush with the ground.

18"diameter, collapsible, white aerial net

canvas sweep net

D-frame dip net

scoops for collecting aquatic insects

rectangular dip net

18" diameter green aerial net

15" diameter short aerial net

modified Needham's scraper dredge

As we tell our students, aerial nets are designed to be swept through the air, not through vegetation or water. There are nets better designed for those habitats. A sweep net has a bag made of a heavy canvas material and a shorter, stouter handle. The robust design lets you collect insects off vegetation, even thorny and woody plants. Sweep using a left to right motion through the vegetation.

Aquatic dip nets are designed for capturing insects in the water. They have a heavy-duty net bag, usually with canvas on at least the bottom and or front area. Collect insects by scooping or dragging the net through the water. In streams with flowing water, place the bag against the bottom so the current flows into the bag and then kick up substrate in front of the net, allowing the current to take the insects into the net. Smaller scoops, made from old sports racquets or lacrosse sticks with fine-mesh hardware cloth work well, especially for undercut banks. There are also more heavy-duty dredges for collecting individuals that burrow into the substrate.

Your fingers are fat. To handle small bugs, you'll want forceps. There are many different kinds, what works best for you will depend on the insects you are studying. Soft-touch or featherweight forceps are best when working with live insects. Made of flexible stainless steel, they are designed to grab, but not crush your subject. Large or fast-moving arthropods (beetles and centipedes, for example) may be hard to grab with these. Spade-tip forceps are best for handling butterflies and moths so you don't damage their wings. Most forceps are hard and will vary by type of point and length. The sharper the tip, the smaller a subject you can manipulate.

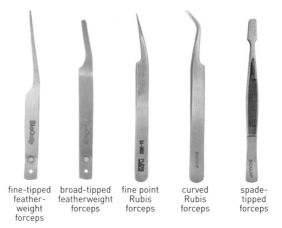

fine-tipped feather-weight forceps

broad-tipped featherweight forceps

fine point Rubis forceps

curved Rubis forceps

spade-tipped forceps

When the subject is too small or fast, an aspirator can be a useful tool. There are numerous homemade and commercially available designs out there, but the general idea is a container with 2 tubes. One tube is long and flexible, and you suck air through it. The other tube is short and stiff, and you suck the bug up through that one. Put some screen over the end of the tube you suck through—otherwise you might eat the bug! Many now have integrated filters to keep the user from inhaling noxious smells, dangerous mold spores, and other similar unwanted contaminants. The aspirator is a great tool for collecting ants, termites, small beetles, and similar insects.

In many cases insects collected in the field are too small to see any real detail without the aid of a magnifier. A hand lens or jeweler's loupe is an essential piece of equipment for these cases. They come in a variety of magnifications (usually 10x to 20x), and some even have integrated lighting. Higher magnification is not always better. It can take some practice to become proficient, but once you do, you won't leave home without it.

It is good to carry a variety of vial sizes, especially if your intent is to photograph live insects. Obviously avoid putting predaceous insects in with potential prey, but it is also a good idea to keep individuals (outside of social insects) separated if you are keeping them alive before preserving them or letting them go. We often use square containers to collect live insects because the flat edge makes it more difficult for the subject to evade the container when placed on or up against a flat surface.

If you are building a collection, you should kill insects quickly and humanely. There are a variety of ways, and as you might guess, it depends on the arthropod group. Soft-bodied insects, including immatures, and non-insect arthropods (spiders and other arachnids, millipedes, centipedes) should be preserved in alcohol, preferably ethanol, though isopropyl (rubbing alcohol) can work in a pinch and as a temporary solution. Place individuals directly in 70–80% alcohol to kill and preserve them. For preserving large specimens in alcohol, parboil them first. This kills bacteria that might damage the specimen as it's being preserved, "cooks" the tissues (think hard-boiled egg), and may result in better color fixing. Heat water to a boil, take it off the heat, and drop the live insect in. Remove the insect from the water after a couple of minutes, place on a paper towel to absorb excess water, then put it in the 70–80% ethanol for final preservation. Some groups benefit from the fixing of tissues with a specific chemical solution or process, and this is discussed within each order as appropriate.

INTRODUCTION

Most insects are hard-bodied as adults and can be killed using a kill jar. A kill jar uses a chemical, typically ethyl acetate or acetone, to kill the insect. Both chemicals are regular ingredients in fingernail polish remover or can be purchased in pure form from a biological supply company. Any jar with a good screw-cap lid will work. Place tightly wadded tissue or paper towel in the bottom of the jar and pour in enough chemical to moisten the paper, but not so much there is excess liquid in the bottom. Place additional tissue or paper towels in strips above the ethyl acetate. This gives insects something to hold onto and keeps them from coming into direct contact with the solvent which is not preferable for many groups (it may alter colors or mat hairs). Ethyl acetate and acetone will evaporate quickly when the jar is open, so you will need to recharge as necessary. A common kill jar design is to pour plaster of paris in the bottom of the jar to absorb the chemical. If you are using a glass jar, wrap tape around it so that if it breaks, it will not shatter. You should also label your kill jar to indicate it contains poison.

Instead of using a single large jar for everything, we encourage using multiple smaller jars or tubes. Centrifuge or transport tubes with a 50 ml volume work great.. You can separate things like tougher beetles from softer and more fragile insects that they could destroy. It is best to not put butterflies and moths in kill jars directly. They have scales that will get knocked off and end up on other insects. If available, place butterflies and moths in a glassine envelope first (restricting their movement) then place those envelopes in a larger kill jar (with no other insects). Envelopes are also good for collecting dragonflies and damselflies (see the chapter on Odonata for details). If you don't have access to ethyl acetate, a good way to kill most insects is to freeze them for 24 to 48 hours. Try not to put them in a container with a lot of air, as the moisture in the container will freeze and later saturate your insect(s) in water.

Just as insects are diverse, so are the ways to collect them. There are many specialized methods and tools. Most entomologists—even professionals—use homemade equipment. We mention a few of the most common collection techniques here. Collecting insects at lights is something anyone can do. Many insects are attracted to lights at night, probably for navigation. Simply turning on your porch light will attract insects (but don't use yellow or red light bulbs). If you want to up your game, try using a mercury vapor bulb (available from your local hardware store) or a "black light," but make sure to get a "bl" vs. "blb" type bulb. Mercury vapor and black lights put out a broad light spectrum, including ultraviolet light, which for some reason (no one really knows) attracts many different types of insects. If you can, place the light in front of a white sheet. This reflects the light, gives the insects a place to land, and makes it easy to collect or photograph them. Putting a sheet below the light is also a good idea as it will be easier to see the insects on the ground around the light. Check tree trunks and dark areas around the light for insects, particularly beetles, that landed before they made it to the sheet.

Malaise traps are screen tents designed to capture flying insects. The different styles function in the same basic way. In the center of the trap is a dark baffle, a screen above that is a lighter-colored peaked roof with a collecting head at the top. Collecting pans can be placed beneath the center screen. A flying insect doesn't see the screen and smacks into it. Some insects will drop down (typically beetles) and end up in the pans, and some will fly up (flies and wasps) and will hit the roof and eventually go into the collecting head. The collection head can have a wet preservative (propylene glycol, available as non-toxic antifreeze, works well as it won't evaporate) or be dry with a poison to kill the insects. The pans should have a wet preservative. These traps can be modified to be hung off the ground in a tree canopy.

A pitfall trap is designed to capture insects running along the ground. It is simply a container (ranging in size from a cup to a 5-gallon bucket) sunk so the top is flush with the ground. You can add one or more "fences" made of metal flashing to help direct the insects into the trap. Propylene glycol can be used as a preservative in the bottom, and it can be fitted with a cover wider than the container itself and elevated just above it to act as a rain shield.

A Lindgren funnel trap is a series of nested funnels spaced apart and designed to mimic a tree. It is good for capturing wood-boring insects and can be equipped with pheromones to attract certain species. At the bottom of the funnels is a collecting jar, again usually filled with propylene glycol. The idea is that the insect will be attracted to the trap and fall through the successive funnels into the collecting jar at the bottom.

A pan trap is an easy and readily available way to collect micro-Hymenoptera as well as other day-flying insects. All you need is a colored pan or container filled with soapy water. Different colors will attract different species. Colored plastic bowls used for picnics and outdoor food gatherings can work very well. The containers can simply be placed on the ground or they can be sunk even with the ground surface to double as pitfalls. Add some soapy water; the soap is used to break the surface tension of the water so the insect can't escape. Check every day or your specimens will mold.

For long-term preservation, replace propylene glycol or soapy water with 70–80% ethanol. Pour the sample through a fine-mesh

merucry vapor bulb

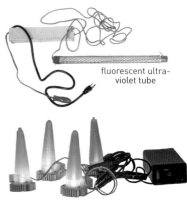

fluorescent ultra-violet tube

battery-operated LED ultraviolet lights

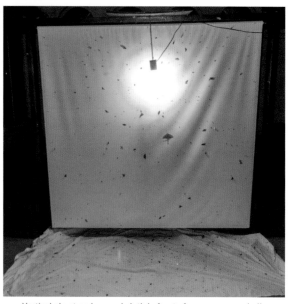

Vertical sheet and ground cloth in front of mercury vapor bulb.

Lindgren funnel trap

Townes-style malaise trap

sieve, coffee filter, or aquarium mesh net. Rinse specimens with water, then place them in ethanol. The ethanol will continue to preserve and will also break down any remaining propylene glycol or soap.

A beating sheet is a fun and productive way to see and capture insects hidden on trees and bushes. Hold a sheet under vegetation and whack it with a sturdy stick. A heavy canvas cloth suspended by 2 cross poles or even a white umbrella will make a great beating sheet. Many insects will be dislodged and fall onto the sheet. Angle the sheet up and insects will crawl up rather than fly away. Use an aspirator to quickly collect small insects.

A Berlese funnel is great for collecting small arthropods that are found in leaf litter, soil, and other similar substrates. They are pretty easy to engineer from items available at your local hardware store. You need a 2.5- to 5-gallon bucket with a large funnel that will rest inside. Place the smaller end of the funnel in a collecting jar (use a moist napkin if you want to keep the insects alive, or use ethanol to immediate kill and preserve). Rest a piece of fine hardware cloth in the funnel. Place the leaf litter or other substrate on the hardware cloth. Put a low-wattage light bulb or other heat source on top (do not let it touch anything flammable). Small arthropods will want to move away from the light, down the funnel and into the collecting jar. Always use a cloth bag (it breathes), never plastic, to hold a leaf-litter sample and keep the sample cool and moist until you put it in the funnel.

You don't need fancy traps, though. You can find many species by simply flipping rocks and fallen logs and looking at specific habitats that attract many insects, like fungi and flowers. A small pry bar and/ or trowel can be useful when looking in and under rotten logs or rocks. Always replace rocks and logs—they are very special long-term habitats for some species.

Using a beating sheet and aspirator.

Berlese funnel for collecting small arthropods from leaf litter and similar substrates.

Perhaps the most important aspect of making an insect collection is the careful documentation of when and where each insect was collected. Without this information the specimens have little to no scientific value. A field journal is the best way to document your outings and what you find. An exact location, including GPS coordinates, and date are the minimal requirements. It is also helpful to document any ecological or habitat information associated with the specimens. This information can be captured through a traditional written journal or digitally. Always make field labels for your specimens (often done in "lots," e.g., everything collected during a single event or place and time). You will need this information when you make permanent labels for your specimens.

CURATING INSECTS
Fluid Preservation

Immature and soft-bodied adult insects and all non-insect arthropods should be permanently preserved in 70–80% ethanol and kept in tightly sealed glass vials or jars (preferably with polyseal caps). Labels should be put in the vial, never outside. Many inks will run in alcohol, so use a pencil or a Pigma Micron pen and archival-quality heavier weight paper for handwritten labels. Printing black with a color laser printer on heavy archival paper is an alternative. When possible, place 2 or 3 copies of a label in a vial in case one is damaged, and wrap labels around the inside of the vial, as loose labels can chop up specimens.

Four-dram vial with polyseal screw cap and handwritten labels.

Pinning

Hard-bodied insects, other than dragonflies and damselflies, should be pinned soon after killing. A brittle, dry specimen can lose legs, and antennae or shatter if it's pinned, so it's best to pin fresh specimens. There are techniques for relaxing dried insects; see below. Once dry, a specimen shouldn't be touched. The pin provides a handle, a place to keep the label and keeps it from touching the drawer. Always use pins made specifically for insect mounting. They will be either stainless steel or double-lacquered and come in a variety of sizes from 000 to 7; the larger the number, the thicker the pin. Size 2 or 3 is a good all-purpose choice; sizes 000 to 0 should be used cautiously, as they are very thin, easy to bend, and insects will "boing" (break off and go flying) if they are even slightly bent when inserted into foam. The larger sizes, 4 and up, are needed only for the largest of insects (big moths, beetles, grasshoppers, etc.).

Each group of insects is traditionally pinned in a specific, standard spot meant to minimize damage to potentially valuable morphological characters. Most insects should be pinned on the right side. This leaves the entire left side and midline undamaged, so characters needed for identification are visible. Make sure the pin is straight and perpendicular to the body of the insect. Pinning techniques vary, but it is generally easiest to put the insect on a flat surface, steady it with a finger on each side, insert the pin in the correct spot, make sure the pin is straight and perpendicular to the insect, and then push all the way through. Once the pin is through, place the insect at the appropriate height. Always aim for consistency within a collection. A height of 8–10 mm between the top of the pin and the top of the insect is best. A pinning block is a useful tool for placing insects and labels at consistent heights. A pinning block is a block with small holes drilled at set heights that are used as a guide. Usually one hole is slightly larger than the others to accommodate the head of the pin. Insert the

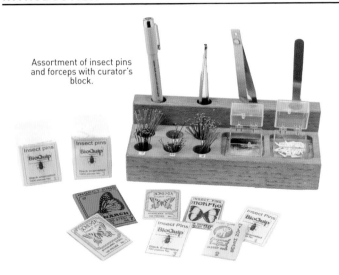

Assortment of insect pins and forceps with curator's block.

pin upside down with the insect on it to set the correct height of the insect on the pin.

An insect will stay in whatever position it dries in. If legs and antennae are sticking out they will probably get broken off, and they will certainly take up a lot of space. Insert the pin deep into the foam so the insect is resting on top. You can carefully use brace pins to arrange the legs and antennae close to the body as needed. Each order or group has a set of ideal standards, and you'll learn to keep structures needed for identification visible. Once the specimen is dry, you can remove the brace pins and put it in your collection. Pinning takes practice and patience but can be a fun part of the entomological experience.

Relaxing Specimens
If you are unable to pin your insects fresh, you can use a relaxing box to rehydrate them. Use an airtight plastic food storage container. Place absorbent material such as paper towels or sand in the bottom. Moisten the material until it is saturated, but so there is not standing water. Place a gridded frame made out of plastic or hardware cloth on top of the absorbent material to keep your specimen from coming in direct contact with the moist substate. Put some pinning foam below the grid to hold insects that have already been pinnned. Mold can quickly become a problem with relaxing boxes! Placing a moth ball (naphthalene) or a few moth ball crystals (paradichlorobenzene) in the chamber will act as a good mold inhibitor. Let the insect sit in the

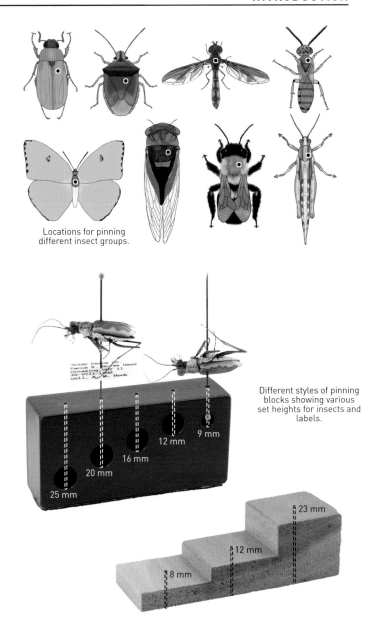

Locations for pinning different insect groups.

Different styles of pinning blocks showing various set heights for insects and labels.

9 mm
12 mm
16 mm
20 mm
25 mm

23 mm
12 mm
8 mm

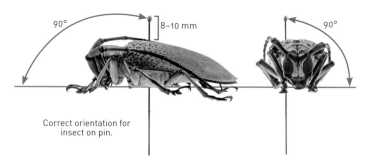

90° |8-10 mm 90°

Correct orientation for
insect on pin.

box for 8 to 24 hours depending upon the size and dryness. Once it's
relaxed, pin, arrange, and dry quickly.

Point Mounting
Some insects are too small to be pinned directly and are preserved
using a technique called pointing. The insect is glued to the end of a
triangular piece of card stock that is placed on a pin. Especially for
beginners, it can be confusing as to when you would pin versus point
small insects. A good general rule is, if in doubt, use a point. You can
always remove a point from an insect and pin it, but you can't remove a
pin without damaging the insect.

Use archival-quality Bristol board, 100 lb. weight, to make points. You
can cut your own, but it is best and faster to use a point punch so they
are consistent. Use heavy pins, #2–4, so they don't bend easily. Place
the point at the same height a pinned specimen would go, 8–10 mm
between the point and the top of the pin. You can use a pinning block or
you can fashion a block specifically for points. We use old phone books
that have been cut to 1 inch wide and wrapped tightly with tape. Place in
a petri dish or on some other hard surface and you have a handy point
pinning block. Make sure the surface of any point pinning block is firm,
otherwise points will bend when the pin goes through them.

When pointing there are 3 goals: 1) orient insect correctly, 2) firmly
attach insect, 3) obscure insect as little as possible. If you satisfy all
3, you did a good job. All this takes practice. Always attach the point
to the right side of the insect (leaving the left side unobstructed). For
many insects the tip doesn't need to be bent, and it could even stick to
the side of the insect, or just underneath. For other insects, it may be
useful to bend the tip of the point down with forceps or even create a
step on which the insect can rest. Make sure the point obstructs only
the right side of center when looking at the bottom of the insect. Use
Elmer's white glue (it is water soluble and can be easily removed) to
attach the insect.

Relaxing box with both a pinned beetle and a butterfly laying on the grid (lid removed).

Point "pin cushion" used for quickly setting points to the correct height.

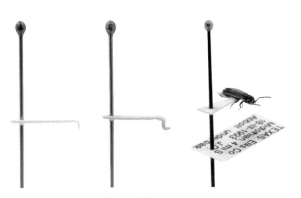

Points showing tips bent (if needed) and insect mounted.

We find it easiest to place an appropriate amount of glue on a hard or waxy surface and let it stand for a few minutes. It becomes tackier and grabs the specimens more easily. Place the insects you want to point on a paper towel or something similar. Make sure they are dry if you are removing them from ethanol. Then after preparing the point, if needed, touch the tip of the point to the glue, then to the right side of the insect. Position the insect so that it is straight and perpendicular to the pin. The process is very precise if done under a microscope.

Spreading Insects

Insects such as butterflies, moths, and some grasshoppers with wing characteristics that are important for identification are usually spread. A spreading board, or even foam with a groove cut in it, can be used. Pin the insect as described earlier, then insert the pin into the groove and set the body into the groove so the wings go straight from the body to the board. For butterflies and some moths, open the wings by gently placing a little pressure on to the sides of the thorax. This will open the wings a bit and allow you to insert the pin in the insect.

The top of most spreading boards slopes upwards. This compensates for the wings falling slightly after the insect is removed from the board. Everyone develops their own spreading technique. The general idea is to use wax strips or string to hold the wings in place while the insect dries. It sounds easy enough, but this definitely takes practice and is most easily done on fresh material. One technique is to use strips of wax paper, either the length of the entire board or cut to the size needed for each specimen, to hold the wings in place. Another technique is to use fine string or thread. Spade-tipped forceps and the fine point of an insect pin can be used to help maneuver the wings in place. The hind margin of the forewing should be at a 90° angle to the insect's body. The hindwing should be slightly tucked under the forewing. To set wings into position, begin by moving one forewing into position (or close to it), hold it in place with a strip of wax paper, then position the other forewing and hold it in place with wax paper. Move the hindwings into position and carefully, with small movements, adjust all the wings so they are symmetrical. Hold them in place with the wax strips (held down by pins). If the abdomen droops, use pins to elevate it so that it is at the same height as the thorax and head. Position antennae so they are parallel with the front margin of the forewing. This will provide some protection for these delicate structures. Leave the insect on the spreading board to dry for a week or so, then gently and carefully remove the strips of paper or string.

Steps to spread a butterfly. Start on one side, then the other, work back and forth to get the wings on each side in their proper position. Hold wings in place with glassine or wax paper. Never touch the wings with your fingers. Brace the abdomen and antennae, then allow to dry. Left; an alternative method to hold wings in place using string.

Enveloped Specimens

Dragonflies and damselflies are best preserved in clear envelopes with a data card behind them. If you collect a pair of mating odonates, keep them together to associate males and females of the same species. Grasp the dragonfly, hold the wings over the back, remove it from the net and place it directly in a glassine or paper envelope. Make sure the wings stay folded over the abdomen. Pairs caught in tandem or copula can usually be placed in the same envelope, but it is best to face them away from one another to avoid any unwanted predation. Label the envelope with the date, locality, your name, and species if known. Also be sure to note when you have pairs if they were taken in copula or tandem. It is helpful to have a 3 x 5 card file or similar box that you can store specimens in while collecting. Don't leave the specimens exposed to severe temperatures or sunlight. Ideally, the specimens will remain alive long enough to void their guts of any frass or excrement. This improves the preservation process by removing fats and oils.

Once you have returned home, it is time to acetone the specimens. We treat odonates with acetone because it removes fats and oils that cause the specimens to become greasy. It also fixes the colors to some degree, though, unfortunately, odonates don't retain their beautiful colors like beetles and butterflies do. Try not to let the acetone touch your skin; use forceps to move specimens, and always work in a well-ventilated area. Remove the specimens from the envelopes (do not lose the continuity of the label data and specimens themselves) and drop them in acetone to kill them. This will take only a minute. Remove a specimen and carefully straighten the abdomen, arrange the legs so they are pointed downward, and gently rotate the head so you are looking at the top, when the odonate is lying on its side (left side facing up). Finally, separate the forewings from the hindwings so they are not overlapping. This makes for a much easier examination of wing venation. Place the specimens back in their appropriate labeled glassine envelope and either punch holes (1/8 inch) or cut off the corners to allow the acetone to flow in and drain out. Make sure holes aren't so large that a small damselfly might slip out. Secure the flap with a paper clip and leave the specimens in acetone overnight.

Remove the envelopes from the acetone. Use forceps and drain off as much acetone as possible. Separate the envelopes and allow them to dry. Placing a fan on the envelopes forces the acetone to evaporate more quickly and results in better specimens. At this point the specimens are dehydrated and very brittle, so handle them carefully.

Gently remove the completely dried specimen(s) from the

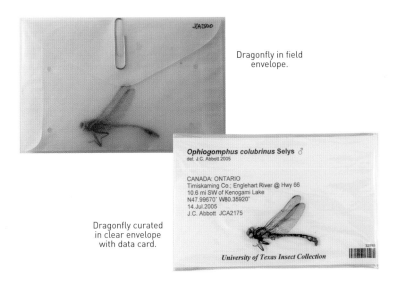

Dragonfly in field envelope.

Ophiogomphus colubrinus Selys ♂
det. J.C. Abbott 2005

CANADA: ONTARIO
Timiskaming Co.; Englehart River @ Hwy 66
10.6 mi SW of Kenogami Lake
N47.99670° W80.35920°
14.Jul.2005
J.C. Abbott JCA2175

University of Texas Insect Collection

Dragonfly curated in clear envelope with data card.

Drawer of damselflies organized in unit trays.

envelope and place it on the data card. Generally, a 3 x 5–inch card made of heavy archival paper is best. Write the collecting information and ID, if known, on the card. Special plastic envelopes are available that hold the card and the specimen. These enveloped specimens can then be stored in sealed containers (plastic shoe boxes), custom-made tight-fitting cardboard shoe boxes, or in drawers housed in cabinets.

Repairing Insects
Sometimes dried insects break. When this happens, it may be possible to repair them by simply gluing the parts back together. Always make sure to use a soluble glue, like Elmer's (water) or clear nail polish (acetone), never superglue. Be conservative with the amount of adhesive; you shouldn't obstruct important characters. For best results, use the tip of an insect pin to apply glue to a specimen while you look through a microscope or hand lens. Smaller parts can be placed in a gel capsule or glued to a card and pinned beneath the insect. This will keep everything in one place and associated.

Labeling
Once the insect has been pinned or curated, it must be labeled. A labeled insect is a specimen, valuable to science for a century and beyond; without a label, the insect is of little worth. Incorrect labels are worse than no label because they provide false information. Giant labels, labels that are set too low, and labels on paper that isn't acid-free are common mistakes. For pinned specimens use a laser printer, heavy archival paper, and a sans-serif font like Arial Narrow or Calibri size 4 (or 5). Work with "line spacing" to squeeze lines tightly together.

The **Locality Label** is the top label. It tells when, where, and how a specimen was collected. Use a pinning block to place this label under the specimen, high enough so that more labels can be added, but low enough so

> USA: ALABAMA: Tuscaloosa Co.
> Hurricane Cr.; 3.2 km S of
> Brookwood; N33.2150, W87.3384
> 22-Sep-2018, ex. fungi
> coll. J.C. Abbott #3109

that it can be read. Five lines are best. **Line 1**: country, state, and county the specimen was collected in. Abbreviate if necessary. **Line 2**: where the specimen was collected, a place found on a map (not "my backyard"), such as a town ("Iowa City") or relation to a place ("10 km E [east] Smallville"). **Line 3**: GPS coordinates, such as N 34.6797, W 82.8346 (only 4 numbers after the decimal are needed). Get these from Google Maps or metadata from your collecting site photo. **Line 4**: the date. Always write the month as a word or Roman numeral (11 May 1999, 11 V 1999). A date written as "11/5/99" could mean any of these possibilities: November 5, 1899; November 5, 1999; May 11, 1899; May 11, 1999. **Line 5**: information that would help someone else collect the insect, such as "sweepnet," "UV light," "ex. rotten log" (ex. is Latin for *from*), and the collector name (usually initials and last name). Wrap lines as needed to leave as little white space as possible and to make the information compact. Try to avoid using more than one label, but if you have more important information (host plant, etc.) you could add a second. Under the Locality Label may go a **Database Label** (both are married to the specimen and would never change), and finally the **Determination Label**, the name of the insect, can be set lowest of all because it may need to be turned upside down if the ID is incorrect or the insect's name changes. Try to standardize your labels as much as possible.

Collection Organization
If you become serious about insect collecting, it will not take long to amass a sizable collection. Make sure that your specimens are organized and accessible. Collections, from professional to personal, are kept in standardized insect drawers (the sizes are named for the place they were invented, CalAcademy, USNM, Cornell). The specimens are placed in unit trays, cardboard boxes with foam bottoms, that fill the drawer. Using unit trays, versus pinning directly into the bottom of a drawer, allows you to move, expand, and reorganize easily. Keep the collection organized phylogenetically so that species related to one another are kept together.

The Entomology Collections Network (ECN) is an active society of professional curators and collection managers who are always eager to share their experiences and techniques. Their website is https://ecnweb.net, and they have an active listserv where topics related collections topics are always being discussed. They will have good advice about where to purchase collection- and curation-related items.

Pest Prevention
There are bugs that eat dead bugs. The most likely cause of an insect collection being destroyed is from other insects, such as dermestid beetles and book lice. These small insects can find their way into the tiniest places and eat dried specimens very quickly. It is important to keep your dried specimens in tightly sealed containers. Insect drawers are expensive, but do the best job of keeping pests out. For years, entomologists used chemicals to deter pests, but it may not be healthy for humans. Now we use non-chemical control:

Collection of insects organized in unit trays within Cornell drawers.

keep specimens in sealed containers, freeze specimens before you add them to your collection, and check your specimens regularly— if you see dust under them, it may be a dermestid infestation. If you see evidence of an infestation, freeze the infected drawer for at least a week. Wrap the drawer in plastic before placing it in the freezer, and leave it wrapped once you remove it until the temperature equilibrates to prevent condensation. Beyond pests, light and mold can damage your collection. Keep the collection in a relatively dark, cool, low-humid environment.

OBSERVING LIVING INSECTS

A great way to learn more about insects and other arthropods is to keep them alive. In most cases, it doesn't take much space, and they are pretty easy to maintain and even breed. Keeping insects alive and observing behaviors such as feeding and mating is another great way to contribute to entomology. Many of these aspects are completely unknown for most arthropods. In most cases you don't need much—a small terrarium, aquarium, or critter keeper will work perfectly; even small jars and containers are perfect for many small species.

Some species have specific plant, animal, or fungal food requirements, so if you don't know what these are ahead of time, be sure to observe individuals in the field. For example, if you are

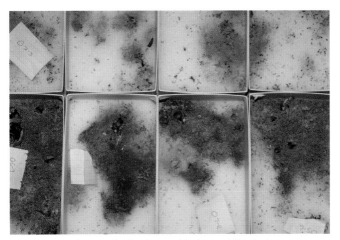

Insect collection destroyed by dermestid beetles!

bringing back caterpillars to rear, be sure to collect some of the food plant the caterpillar was eating, and pay attention to what it was so you can provide more. Even if something is known to eat a variety of plant species, many of them won't switch once they have started feeding.

While most arthropods are fairly "tough," you will likely need to provide a heat source and regularly spray the housing. Humidity is an important variable when keeping arthropods alive. It is important to provide cover in the container so that your critter(s) has options (basking vs. shade, for example) and you typically should not keep predators together. Many people are now keeping insects and arthropods as pets (especially cockroaches!). Many breeders are happy to help you start your own colony. You don't necessarily have to capture an insect to observe it—you can create a wildflower or pollen garden to attract them. Adding water features and letting fallen trees rot in place are just a couple of examples of easy things to do that will provide habitats for many insects and arthropods.

PHOTOGRAPHING INSECTS
Photographing insects is yet another way to enjoy them. Macrophotography used to be considered a highly specialized area of photography, which only the most serious and dedicated would venture into. The digital age has made photographing insects and

Examples of containers that can be used for observing or rearing live insects. Various food containers work well, as they are inexpensive and come in a variety of sizes. Puncture holes in the top, and add a mesh screen or cloth to allow air circulation. Insect Lore makes a variety of pop-up containers for keeping insects.

other arthropods much easier. Most importantly, almost everyone is carrying an exceptional camera everywhere they go—a phone! Most smartphones have multiple lenses, including a macro lens. The small sensors do a very good job of isolating the subject, even small ones.

Phone cameras are great for photos to post on social media and for photographing still subjects, but very high quality and moving subjects require different equipment—dSLR's (digital single-lens reflex) or mirrorless cameras. There are many options when it comes to a camera, and much about your decision will depend upon your personal preferences and ultimate objectives. Presently, mirrorless cameras are becoming more popular and have some real benefits for macrophotography, including an electronic viewfinder.

The biggest issue you will quickly confront when photographing small subjects like insects is depth of field—the amount of your subject that will be in focus. The closer you are to your subject, the shallower your depth of field will be, but because your subject is small, you need to be close to it to achieve the desirable magnification.

Focus stacking is a specialized technique for achieving greater depth of field. Multiple images of a subject are taken at different focal planes and combined using a special program so only the portions in focus are left. This can be done minimally in the field and extensively in studio situations. Many collections use focus stacking techniques to get stunning imagery with rich detail of their subjects.

In the field, the best way to achieve greater depth of field is to stop down your aperture. The smaller the aperture—or iris that light passes through—the greater the depth of field, but this comes at a cost as it results in softer images. Generally, it's best to photograph at apertures from f/11 to f/18 to achieve a good depth of field with minimal trade-off.

However, stopping down your aperture reduces the amount of light. To counter that, you can make the sensor more sensitive to light by increasing the ISO, but high ISOs result in noise. Try not to shoot at ISO's higher than 400 or 800, but cameras are improving remarkably in this area. Use a flash to add light. Flashes have many benefits, including stopping motion and allowing for creative illumination of your subject. Ring flashes, while convenient, are typically not the best option because the light is coming from a single plane, which results in a flat-looking image. Using one or more flashes that are off the camera and independent of one another is usually the best way to go. Many insects and arthropods are shiny, and a direct flash can result in harsh, unpleasant highlights. To overcome this, you need to diffuse the flash. This is done by placing a diffusion material between the flash and your subject. Many things can be used as a diffuser, for example frosted plastics, tracing paper, foam wrap, or anything that scatters light as it passes through. To achieve the best diffusion, position the diffusion material as close to your subject and as far away from your flash as possible.

Various lenses and options provide for magnification. There are techniques to reverse a lens to get greater magnification, but a macro lens will provide the best results. Macro lenses are specialized lenses that provide at least 1:1 magnification. There are

Photographing a subject on a fallen tree using a macro lens and diffused flash.

many macro lenses available and they usually fall into 3 focal lengths: 50 mm, 100 mm, and 200 mm. The longer the focal length, the farther you can be from your subject, but the larger and more expensive the lens. There are also several specialty macro lenses that provide as much as 5x magnification. Laowa provides many affordable options for different camera models. Extension tubes (spacing rings with no lens elements) can be added between the camera and lens to get even more magnification. These are a great, inexpensive addition to any macrophotographer's kit. Additionally, diopters, lenses that screw onto the filter thread of the mains lens, can be used to increase magnification. While they will result in some degradation in quality, they can be useful. Not all diopters are equal, and you generally get what you pay for.

Larger subjects like dragonflies, damselflies, grasshoppers, and many butterflies can be photographed using a close-focusing telephoto lens, without approaching the subject too closely. Many lenses are available, including ones specifically advertised for photographing bees and other insects. The main requirement is a relatively short minimal focusing distance; less than 2 m is ideal. Combining a telephoto with an extension tube and/or a teleconverter can create a powerful lens. Our favorite combination is a 300 mm f/4

Macrophotography setup with 100 mm macro lens, diffused twin flashes and modeling light.

telephoto coupled with a 1.4x teleconverter and a 25 mm extension tube placed between the camera and the lens. If we need more flexibility, a 100–400 mm f/4.5–5.6L telephoto with a 25 mm extension tube works well. Use a tripod to improve the sharpness of your shots when using this longer lens. Attach flashes to the camera's hot shoe and use a diffuser if you can. Now it is time to take photographs! One of the great benefits of digital photography is immediate feedback sand low cost, so you can take as many images as you want and toss the bad ones. Photograph everything and often!

There are many classes and opportunities to learn and improve your macrophotography. We offer courses directly through AbbottNature.com and BugShot.net.

Photographing a dragonfly using a telephoto lens, teleconverter and extension tube combination with a flash and tripod.

INSECTS AND PEOPLE

Insects are everywhere in our ecosystem; they are also everywhere in our culture. For many they are curiosities, for others inspirational. They are often an attraction at museum and other outreach events. Many invertebrates like mantids, beetles, millipedes, and spiders make great pets. In Japan, male stag beetles are kept and raised to participate in fights. They are also used as models in various forms of art, from jewelry and tattoos that we adorn ourselves with to plates, drapes, clothes, and more. They are favored subjects of many photographers. This fascination is not just something recent; they have been used throughout human culture, from engravings on walls and bones to models for prestigious jewelry.

One of the reasons we may be so fascinated with insects is their impact on humanity. At any one time, it is estimated that one-sixth of the human population is affected by a fly-borne illness. Other groups like crop pests can cause devastating indirect effects on our livelihood. In other cultures, these insect outbreaks are often taken advantage of, and the insects such as locusts are collected and used as food, a practice known as entomophagy. Insects are a far more sustainable source of protein than other animals. For instance, to grow one kilogram it takes 3.8 liters of water for crickets, 2,150 liters for chickens, 3,030 liters for pigs, and 9,500 liters for cows. This pattern is the same for the food required to sustain these different animals and the land use needed and greenhouse gas emitted. The requirements for insects are far less than for other animals. Chocolate chirp cookies are a good introduction to entomophagy. Use the recipe below if you would like to try them for yourself. Be cautious if you have a shellfish allergy.

Best Chocolate Chirp Cookies

1 cup butter, softened
1 cup white sugar
1 cup brown sugar
2 eggs
2 tsp. vanilla
1 tsp. baking soda
2 tsp. hot water
1/4 tsp. salt
2 cups chocolate chips
1/2 cup roasted crickets
3 cups flour

1) Preheat oven to 350° F.
2) Cream butter and sugars until smooth; beat in eggs and then vanilla.
3) Dissolve baking soda in hot water then add to batter along with salt.
4) Mix in 3 cups of flour.
5) Stir in chocolate chips; when combined, fold in crickets.
6) Bake for 10 minutes.

Environmental and Nutritional Benefits of Crickets

Water

Crickets require less water than other protein sources to produce 2.2 lbs. of meat.

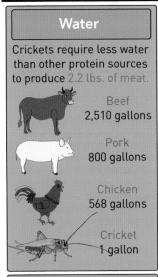

Beef
2,510 gallons

Pork
800 gallons

Chicken
568 gallons

Cricket
1 gallon

Food

Crickets require less feed than other sources to produce 1 lb. of meat.

Beef
17 lbs.

Pork
9 lbs.

Chicken
4 lbs.

Cricket
1 lb.

Greenhouse Gases

Crickets produce less greenhouse gases to make 2.2 lbs. of meat.

Beef
6.3 lbs.

Pork
2.5 lbs.

Chicken
0.7 lbs.

Cricket
0.002 lbs.

Nutrition

Cricket flour has more protein in 100 grams than other protein sources.

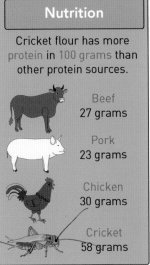

Beef
27 grams

Pork
23 grams

Chicken
30 grams

Cricket
58 grams

Modified from fix.com/blog/edible-insects.

HOW TO USE THIS GUIDE

Our intention with this book was not to provide another field guide to species of insects in North America, as several fantastic ones exist, including the Kaufman Field Guide by Eric Eaton and Kenn Kaufman and the National Wildlife Federation Guide by Arthur Evans. Rather, our intention is to fill the same niche as the *Peterson Field Guide to Insects of America North of Mexico* by Donald Borror and Richard White. That guide, like ours, was designed to help the user identify the families of insects that occur in North America north of Mexico. There are approximately 92,000 species known to occur in the United States and Canada, and no book can cover that. Instead, we present information on all 784 families that occur in the region.

This field guide differs from most guides to other organisms in North America in 2 significant ways. Most entomologists, from hobby to professional, learn insects at the level of family, then genera and species of their own groups. While family level identifications provide a more obtainable target for such a diverse group, large families have members that vary greatly in size, shape, and color. As a result, it is often necessary to focus on specific details rather than general appearance and color.

This guide focuses on the adult stage, though we occasionally provide photographs and characteristics of easily recognizable immatures. We have strived to use the most noticeable and easily distinguishable characters possible, but in many cases in-hand examination with magnification of at least 10x will be necessary; sometimes a dissecting microscope will be required. Once you have become familiar with the characters for a particular family, it will often be possible to identify the family from photographs, especially when the photographer has captured those characters.

If you are just starting out looking at insects, the first step is to recognize the different orders. This is generally straightforward, and we suggest using the key found on the endsheets at the front of the book. Within each order of insects, there is usually a suite of characters that you will want to focus on; becoming acquainted with those is the next step. From there, work on family level identification.

We have tried to provide common and/or representative example species for each family. We have provided a genus and/or species name for nearly all of these, but do keep in mind that they are meant to be representative and there may be other species that look very similar. Because our focus is primarily on the family level, we don't attempt to distinguish or recognize these similar-looking species.

Geographical Coverage

This guide covers families known to occur in the United States and Canada (or North America north of Mexico). Throughout the book, we abbreviate this as NA. We recognize our guide does not cover all of North America. The diversity increases considerably south of the United States border, and information is not as complete, so we have opted to constrain the geographical coverage to north of Mexico.

Taxonomic Coverage

We cover all 784 families of insects recognized in the United States and Canada at the time of printing. This is a moving target with our constantly changing understanding of taxonomic relationships. While every family is mentioned in the text, along with its characters, we don't include photographs for rare and less likely to be encountered families. **Uncommon families that you are unlikely to find, unless specifically looking for them, are indicated by an "*" in both the text and the keys.** There is disagreement among authorities as to what families in some groups to recognize. Don't get frustrated by this or the inconsistencies that you may recognize among this guide, others, and online resources. Rather, use it as an opportunity to learn more about the group and why the disagreement(s) may exist.

Terminology & Characteristics

We have tried to use clear and non-technical terms when possible, but much of the time, this would require a much lengthier description than space allows. To help with this, we have highlighted characteristics at the beginning of each order that are particularly useful, and sometimes specific, for that group. We use arrows to point to characters that are bolded in the text. We have also provided a pictorial glossary at the back of the book to help with learning characters. The endsheets at the back of the book are designed to serve as a quick reference for the most frequently used terms. We are not experts on each order and have therefore deliberately tried not to deviate from the terminology used by authorities in each group. This, however, may mean that terminology from order to order may seem inconsistent (wing venation is a good example of this), so we encourage you to pay careful attention. The terms like *forewing* and *hindwing* have been consistently abbreviated as **FW** and **HW** to save space. Measurements are given in millimeters and are given for the total body length (**TL**) or wingspan (**WS**). We use the symbols for male and female, ♂ and ♀ respectively, to distinguish between the sexes.

INTRODUCTION

Keys

In addition to the ordinal key at the front of the book, we have provided pictorial keys to common families and/or superfamilies for the larger orders. These keys were inspired by the pictorial ordinal key found in the *Peterson Field Guide to Insects*. These keys are largely modified from preexisting keys (see Acknowledgments). We have tried to make them as straightforward and clear as possible. Some groupings may be differentiated at the top of the key; from there, begin at the "Start Here" tab and follow a series of dichotomous (2) choices taking you to the order or family. We have tried to illustrate these keys as richly as possible. Characters that are bolded in the text are indicated with an arrow. Other characters may be highlighted in a color (typically green) and the corresponding text will also appear in that color.

If you have never used dichotomous keys, we encourage you to proceed with patience and caution. We have tried to simplify the keys for ease of use, but this means that they may work only for the more common members of a family, and rare species may not key out correctly. Use keys in combination with the text and photos within the chapter.

Layout

The layout is generally consistent within each order, and orders have been color coded for easy reference. These colors can be found in the Contents and Introduction sections. Scientific names for groups are listed in the color band on the left side of the page, and vernacular names are on the right.

Each order begins with a representative photograph of the adult on the left and generally an immature at the top on the right, and the following information is provided: **Derivation of Name, Identification, Classification, Range,** and **Similar Orders**. The next 2 pages contain additional characters for the order. On the left: **Food, Behavior, Life Cycle, Importance to Humans, Collecting and Preserving,** and **Resources**. The right page shows visual characteristics for the order. Under Collecting and Preserving, we have attempted to provide useful information on how to best collect and preserve members of the order. With the more diverse orders, this can be difficult to summarize in such a small space, so we encourage you to seek out additional information on these groups if you want to become a serious collector. Under Resources we have provided some of the most useful resources for the order. Full citations are in **Literature** on p. 532.

For larger orders, additional information in the form of a pictorial key and sometimes silhouettes of common families are provided. Within each order, families are listed on the left with corresponding photos on the right. Each family follows the same format: common or vernacular name, scientific name, general description, size measurements, and species numbers for the world and North America. Note that in most cases species numbers are approximate. Characteristics bolded in the text are indicated with arrows on the photos. Photos are usually enlarged to use the maximum space and are not relative in size to one another; refer to the size measurements in the description for details of total length (TL) or wingspan (WS).

The photos representing families are almost always of living individuals to best capture colors and postures. Most insects were photographed in a white box on a white background to capture the highest quality imagery possible. In some photos the background was removed. Families are differentiated by thin gray lines.

Vernacular Names
We recognize the importance of common or vernacular names for enthusiasts wanting to learn a group. Most insect species still lack these names, and even many families don't have vernacular names. For others, there can be more than one!. We tried to use the most widely accepted vernacular names. For those families lacking a vernacular name, we used the adjective form of the family, e.g., Bolbomyiidae is bolbomyiid. If there is no accepted vernacular name for a species, we did not create one.

Firefly
Photinus pyralis

Insects of
North America

Western Honey Bee
(*Apis mellifera*)
Apidae

Acerentomon sp.
Acerentomidae

Acerentomon sp.
Acerentomidae

PROTURANS
ORDER PROTURA

Derivation of Name
- Protura: *prot*, first; *ura*, tail. Referring to the lack of cerci or other structures at the end of the abdomen.

Identification
- 0.5–2.5 mm TL.
- Light brown or cream colored, soft-bodied and elongated.
- Head cone shaped, lacking antennae and eyes.
- Mouthparts are entognathous and sucking.
- Thorax wingless, bearing 3 pairs of legs. First pair modified for sensory function and held in front of body. Tarsi 1-segmented.
- Adults have 12 abdominal segments. S1–3 with rudimentary appendages that may have eversible vesicles at apices.
- Lack cerci; gonopore on S11.

Classification
- Classification and relationship to other Hexapoda somewhat unsettled. Sometimes treated as their own class.
- Identification of this group even to family level is difficult without slide mounting.
- Three or four families recognized in NA. We include the single species, *Hesperentomon macswaini*, sometimes placed in Hesperentomidae, within the Protentomidae, recognizing 3 families.
- World: 830 species; North America: 80 species.

Range
- Found worldwide except Antarctica.

Similar Orders
- Zygentoma (p. 78): larger size; prominent eyes; antennae present; 3 tails at tip of abdomen.
- Diplura (p. 70): antennae present; filamentous or forcep-like cerci present.
- Collembola (p. 60): antennae and usually eyes present; forked appendage (furcula) on underside of abdomen.
- Zoraptera (p. 110): antennae present; short 1-segmented cerci.
- Thysanoptera (p. 214): antennae and eyes present.

PROTURA

Food
- Adults and nymphs are believed to pierce and suck fluids of ectomycorrhizal fungi and free hyphae.

Behavior
- Have been found in densities of more than 90,000 per sq. meter.
- Found in damp soil, leaf litter, under logs and stones, almost always associated with fungi.
- Some found in subterranean mammal nests (probably coincidental).
- Unlike other hexapods, forelegs are not used for locomotion, but held in front of body, above the ground, and have sensory structures.

Life Cycle
- Anamorphosis with 4 juvenile stages: prelarva, larva I, larva II, and maturus junior—with 9, 9, 10, and 12 abdominal segments respectively. Adults have 12 abdominal segments and do not molt again.
- Males of the Acerentomidae have an extra stage (preimago), with partially developed genitalia, before becoming an adult.
- Species living near surface of soil have 1 generation per year.
- Very little is known about their life history; eggs have been observed in only a few species.

Importance to Humans
- Help incorporate fungal cytoplasm back into the soil. May aid in decomposition and the breaking down and recycling of leaf litter and other organic nutrients.

Collecting and Preserving
Immatures & Adults
- Collect from soil; use a Berlese funnel or soil flotation to extract individuals.
- Permanently store in 70% ethanol.

Resources
- "Catalog of the World Protura" by Szeptycki, 2007
- A Chaos of Delight—chaosofdelight.org
- *Soil Biology Guide* by Dindal, 1990
- *The Protura* by Tuxen, 1964

Acerentomon sp.
Acerentomidae

Styletoentomon sp.
Eosentomidae

Suborder Eosentomata
Lack striated band on S8. Tracheae and spiracles present.

EOSENTOMIDS Family Eosentomidae*
Mesothorax and metathorax with dorsal spiracles. Abdominal
appendages with 2 segments with terminal vesicle. Lack striate
band on S8. TL <1.5 mm; World 339, NA 45.

Suborder Acerentomata
Have striated band S8. Trachea and spiracles absent.

ACERENTOMIDS Family Acerentomidae*
Only the first pair of abdominal appendages with terminal vesicle.
Most abdominal segments with 2 transverse rows of dorsal setae.
Tracheae and spiracles absent. Striate band on S8. TL <1.5 mm;
World 335, NA 29.

PROTENTOMIDS Family Protentomidae*   Not illus.
At least 2 pairs of abdominal appendages with terminal vesicle. Most
abdominal segments with single transverse row of dorsal setae.
TL <1.5 mm; World 65, NA 5.

Acerentomon sp.
Acerentomidae

59

Golden Snowflea
(*Hymenaphorura cocklei*)
Onychiuridae

Folsomia candida
Isotomidae

SPRINGTAILS
SUBCLASS COLLEMBOLA

Derivation of Name
- Collembola: *coll*, glue; *embola*, wedge. Referring to the collophore (ventral tube) on underside of abdomen.

Identification
- 1–7 mm TL.
- Body elongate or globular; color can vary from white, to pink, purple or blue; primitively wingless.
- Eyes are a loose aggregation of no more than 8 facets (ommatidia) each. Some lack eyes entirely.
- Antennae 4-segmented and short; *Orchesella* and relatives appearing 5–6-segmented with terminal segments bearing a suture.
- Some have postantennal organ (PAO) between antenna and eyepatch.
- Mouthparts variable, from chewing to modified for sucking, but always concealed within the head.
- Abdomen with 6 segments; sometimes 2 or 3 terminal segments fused; in globular springtails, first 4 or 5 segments fused. S1 bears a ventral tube or collophore. This structure plays an important role in fluid and electrolyte balance. The eversible vesicles of this tube may also be used as a source of fluid for grooming or adhering to smooth surfaces.
- Most have a furcula arising ventrally from S4 allowing them to jump when disturbed. Furcula is held back at rest by the retinaculum on S3.

Classification
- One of the oldest hexapods, with fossils found in the Devonian.
- Previously recognized as a single order, recently split into 4 orders.
- Sometimes elevated to their own class.
- World: 9,000 species; North America: 1,000 species.

Range
- Can be found worldwide, including Antarctica.

Similar orders
- Archaeognatha (p. 74): larger; eyes large, meeting on top of head; 3 tail filaments.
- Zygentoma (p. 78): larger; 3 prominent tail filaments.
- Protura (p. 56): no antennae or tails.
- Diplura (p. 70): larger; 2 filamentous or forceps-like cerci.
- Zoraptera (p. 110): short 1-segment cerci; antennae 9-segmented.

Food
- Most feed on fungi, detritus, or fecal matter of other insects. A few groups feed on live vegetation, are carnivorous, or fluid-feeding.

Behavior
- Found in a diverse array of habitats, including damp soil, under rocks, in caves, and alongside fresh and marine waters. Some are associated with ant and termite nests.
- Have indirect sperm transfer. Stalked sperm packets (spermatophores) are deposited by the males on the substrate for wandering females to pick up or are placed directly on the female genital opening.
- Some species have been seen touching each other with antennae, and the male leads the female to the sperm packets. In others, a male will grab the female with his antennae to determine her receptiveness. If she is recedptive, he will lay sperm packets in a semicircle around her and then pull and push her towards them.

Life Cycle
- Ametabolous.
- Continuously molt throughout their lives; sexual maturity is reached after fifth instar.
- Some species exhibit diapause, including a regression of mouthparts and digestive system.
- Early instars with pattern less well-developed and less intense pigment compared to adults.

Importance to Humans
- Can be pests in gardens and greenhouses.

Collecting and Preserving
Immatures & Adults
- Collect from almost any habitat using a Berlese funnel, beat sheet, pitfall trap, sweeping vegetation, or by hand with an aspirator. A white enamel pan is helpful to see them.
- Collect directly in Von Torne's medium or 95% ethanol.
- Permanently store in 95% ethanol or cleared and mounted on microscope slides.

Resources
- Checklist of Collembola—collembola.org
- *The Collembola of North America North of the Rio Grande*, vols. 1-4 by Christiansen & Bellinger, 1998
- A Chaos of Delight—chaosofdelight.org

Smithurides sp. (Sminthuridae) exhibiting courtship dance with smaller male directing the larger female to a spermatophore.

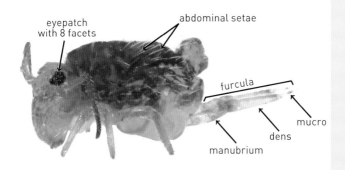

eyepatch with 8 facets

abdominal setae

furcula

mucro

dens

manubrium

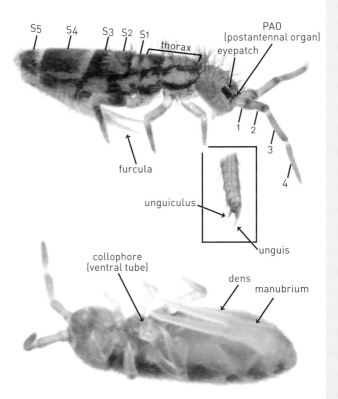

S5 S4 S3 S2 S1

thorax

PAO (postantennal organ)

eyepatch

1 2

3

4

furcula

unguiculus

unguis

collophore (ventral tube)

dens

manubrium

Order Poduromorpha

Body elongate. Setae dorsally on first thoracic segment. Last 2 antennal segments often fused. Relatively few, simple setae. Appendages short; Furculum often reduced or absent.

NEANURIDS Family Neanuridae

Last 2 antennal segments often fused dorsally. Unguiculus generally absent, but if present, setiform. Mouthparts lack molar plate and sometimes mandibles. Some found on marine beaches.
TL 0.5–2.5 mm; World 1,420, NA 60.

BRACHYSTOMELLIDS Family Brachystomellidae

Fourth antennal segment with well-developed apical bulb fused with third segment. Blunt-toothed maxilla; mandible absent. PAO with small number of lobes. Eight eyes on each side. Furcula small, but present and complete. TL 0.5 mm; World 130, NA 8.

ODONTELLIDS Family Odontellidae   Not illus.

Stout, oval body with short conical antennae. Well-developed furcula present. Five eyes per side. Lack apical bulb of fourth antennal segment. Unguiculus reduced or absent. Mandibles lacking.
TL 0.5–2.5 mm; World 130, NA 14.

PODURIDS Family Poduridae

Blackish-blue with reddish-brown appendages or entirely reddish-brown. **Antennae shorter than head.** S4 with 2 apical papillae, but lacking blunt setae. Furcula very long, reaching ventral tube when at rest. Head hypognathous. PAO and unguiculus absent. Found on the surface of freshwater ponds. TL <1.5 mm; World 4, NA 1.

HYPOGASTRURIDS Family Hypogastruridae

Furcula short, never reaching to tip of abdomen when extended, or absent and not ringed. Pseudocelli absent. Third antennal sense organ simple. Mouthparts variable. PAO, if present, rounded or oval usually with few tubercles. TL 1–2.8 mm; World 704, NA 62.

ONYCHIURIDS Family Onychiuridae

Nearly all species white, a few light yellow, one dark blue. Eyes absent (except *Lophognathella*). Pseudocelli present at least on antennal base or dorsum of S5. Furcula absent or relatively short and not ringed. Long, parallel-sided PAO with many tubercles in a groove.
TL 0.5–2 mm; World 600, NA 40.

TULLBERGIIDS Family Tullbergiidae*   Not illus.

Lack ocelli, pigment, and furcula. Found in soil.
TL 0.5–2.5 mm; World 231, NA 31.

PACHYTULLBERGIIDS Family Pachytullbergiidae*   Not illus.

Sensiphorura marshalli distinguished from similar species by fusion of last 2 antennomeres and dorsal/ventral division of sense organ on third antennomere. TL 0.5 mm; World 6, NA 1.

NEANURIDS

Sensillanura caecaa

Pseudachorutes sp.

PODURIDS

Podura aquatica

BRACHYSTOMELLIDS

HYPOGASTRURIDS

Brachystomella parvula

Hypogastrura sp.

ONYCHIURIDS

Golden Snow-flea
(*Hymenaphorura cocklei*)

Paronychiurus ramosus

ELONGATE SPRINGTAILS

65

ELONGATE-BODIED SPRINGTAILS
Order Entomobryomorpha
Body elongate. Antennae longer than head. First thoracic segment reduced dorsally and membranous. Furcula usually reaching ventral tube at rest. PAO generally present.

TOMOCERIDS Family Tomoceridae
Mucro hairy. Elongated antennal segments; fourth segment shorter than third. Body scaled coarsely. S4 shorter than or about equal to S3. Unguis with single ventral lamella. Retenaculum quadridentate. Found in soil litter and caves. TL 2–7 mm; World 160, NA 16.

PARONELLIDS Family Paronellidae
Eyes and pigment absent. Dens of furcula straight, not ringed, and lacking spines, but with dorsal, terminal bladder-like projection. Mucro generally blunt, bi-, or tridentate. S4 greatly elongate. TL 1–2.5 mm; World 500, NA 4.

ONCOPODURIDS Family Oncopoduridae   Not illus.
Body with glassy-looking scales. Row of blunt setae dorsally on fourth antennal segment. Dentes with dentate spines. Mucro subequal in length to dens or longer. Furcula densely scaled ventrally. Retenaculum quadridentate. TL 0.5–2 mm; World 55, NA 7.

ISOTOMIDS Family Isotomidae
Antennal segments distinct. Well-developed mouthparts. Pronotum reduced, lacking setae. **S3 and S4 similar in length.** Dental spines simple or absent, mucro of furcula usually shorter than dens. PAO present in most and setae at most unilaterally ciliate. Unguis with single inner margin.
TL 1–3.5 mm; World 1,400, NA 200.

ENTOMOBRYIDS Family Entomobryidae
Dens of furcula dorsally crenulate and curving upward basally in line with manubrium. **S4 long**, often twice as long as S3. Often numerous setae on head and thorax. Many species brightly colored with distinctive patterns. TL 2–5 mm; World 1,700, NA 162.

Order Neelipleona
Globular body. Furcula complete. At least S3–S5 fused. Furculum well-developed. Setae sparse and lacking scales.

NEELIDS Family Neelidae*   Not illus.
Body globular. Antennae shorter than head. Eyes absent. Dentes divided in 2 parts. Thorax relatively large and often, along with abdomen, exhibiting fusion. Furcula segmentation distinct. Can be common in forest soils and litter; also found in caves.
TL 0.5 mm; World 35, NA 8.

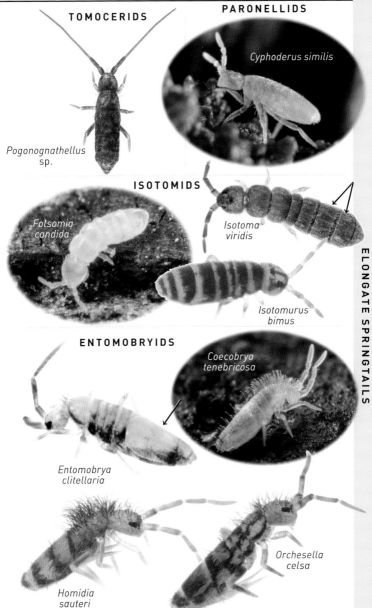

TOMOCERIDS

PARONELLIDS

Cyphoderus similis

Pogonognathellus sp.

ISOTOMIDS

Folsomia candida

Isotoma viridis

Isotomurus bimus

ENTOMOBRYIDS

Coecobrya tenebricosa

Entomobrya clitellaria

Homidia sauteri

Orchesella celsa

ELONGATE SPRINGTAILS

GLOBULAR SPRINGTAILS
Order Symphypleona
Body globular. PAO absent. Antennae longer than head or at least 1 ommatidium on each side. Thorax and S1–4 fused. S5–6 usually forming a small papilla. Furcula segmentation distinct.

KATIANNIDS Family Katiannidae
Large white raised areas between eye clusters. Eight eyes on each side. Numerous setae dorsally. Well-developed trochanteral organ on hind legs. Lack mucronal seta. *Vesicephalus* lack tubercles on filaments of ventral tube. TL 1–2 mm; World 200, NA 30.

SMINTHURIDIDS Family Sminthurididae*   Not illus.
Male antennae modified for grasping base of female antennae. Smaller males are often seen attached to female's head. She places her head on ground; male places spermatophores nearby, encouraging her to take them. TL 0.5–1.5 mm; World 150, NA 20.

ARRHOPALITIDS Family Arrhopalitidae*   Not illus.
Two or fewer eyes on each side. Trochanters with 3 setae; oval or round trochanteral organ. Apical unguicular filaments slender. Dens with inner basal projection. Mucro trough-shaped with both edges serrate. Most species known from caves, others in leaf-litter. TL 0.2–1.5 mm; World 140, NA 32.

SMINTHURIDS Family Sminthuridae
Antennae clearly longer than head; segment 4 with many subsegments. Antennal segments 1–4 fused; segment 5 may be distinct or fused with segment 4. Coxae and trochanters characteristically shaped with setae present. Furcula well-developed; mucro elongate and often serrate. Dens with many setae. Active jumpers. TL 0.7–3 mm; World 250, NA 140.

COLLOPHORIDS Family Collophoridae*   Not illus.
Fourth antennal segment with 3–5 weakly demarcated subsegments or none. Four eyes on each side. S5 fused with rest of abdomen. Dentes lack spines. Single species, *Collophora quadrioculata,* known from AZ. TL 0.5 mm; World 9, NA 1.

BOURLETIELLIDS Family Bourletiellidae
S4 with small number of subsegments. Linear bothrotricha pattern. Mucro with generally smooth margins. Posterior pretarsal setae always absent. Two to three heavy, clavate tenent hairs arranged parallel to long axis of tibiotarsus. TL 0.7–2.5 mm; World 250, NA 29.

DICYRTOMIDS Family Dicyrtomidae
Fourth antennal segment much shorter than third. Antennae elbowed between second and third segments. Eight eyes on each side. Mucro with both dorsal edges serrate and lacking setae. TL 1.3–3 mm; World 200, NA 25.

SYMPHYPLEONA

KATIANNIDS

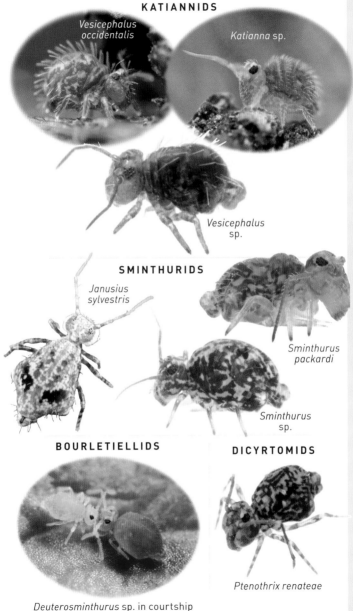

Vesicephalus occidentalis

Katianna sp.

Vesicephalus sp.

SMINTHURIDS

Janusius sylvestris

Sminthurus packardi

Sminthurus sp.

BOURLETIELLIDS

DICYRTOMIDS

Ptenothrix renateae

Deuterosminthurus sp. in courtship

Metajapyx sp.
Japygidae

Slender Dipluran
Campodeidae

TWO-PRONGED BRISTLETAILS
ORDER DIPLURA

Derivation of Name
- Diplura: *diplo*, 2; *ura*, tail. Referring to the 2 terminal cerci.

Identification
- 2–50 mm TL; most ~10 mm.
- Light brown to white, soft bodied, elongated and slightly flattened.
- Mouthparts chewing, but recessed in the head.
- Long antennae and 2 cerci, which may be modified as pincers.
- Eyes and ocelli absent.
- Wingless.
- Tarsi 1-segmented.
- Some species with darker abdominal tip.
- S1-7 or S2-7 have small paired finger-like appendages.

Classification
- Historically placed in Thysanura along with silverfish and bristletails.
- Sometimes elevated to their own class containing 2 orders.
- World: 1,000 species; North America: 173 species.

Range
- Found worldwide except Antarctica.

Similar orders
- Archaeognatha (p. 74): eyes large, meeting on top of head; thorax arched; with shorter, middle tail filament in addition to 2 cerci.
- Zygentoma (p. 78): eyes present; 3 tails of similar length at tip of abdomen; thicker body; mouthparts more exposed; tarsi 3- or 4-segmented.
- Protura (p. 56): no antennae or tails; minute size.
- Collembola (p. 60): no tails; forked appendage on underside of abdomen.
- Zoraptera (p. 110): short 1-segmented cerci; antennae 9-segmented.

Food
- Adults and nymphs eat a variety of food, including decaying vegetation and other insects.

Behavior
- Found in damp soil under logs and stones; some subterranean.
- Locate food or prey with their antennae and will stalk them.
- Species with forceps-like cerci sometimes use them to capture prey, but more often use them for defense.

Life Cycle
- Ametabolous.
- Eggs are deposited in clumps in soil, vegetation, cracks in wood, or on the soil surface.
- Some females guard eggs and immatures for several molts.
- In some carnivorous species, the young end up eating the parent as they grow.
- Males lay sperm droplets near a female; she then gathers the sperm to fertilize the eggs.

Importance to Humans
- Can be pests in gardens.
- May play key role in indicating soil quality and anthropogenic impact.

Collecting and Preserving
Immatures & Adults
- Collect in Berlese funnel or by hand.
- Permanently store in 70% ethanol.

Resources
- "Diplura of North America" by Allen, 2002
- "The families Projapygidae and Anajapygidae (Diplura) in North America" by Smith, 1960
- "A checklist and bibliography of the Japygoidea (Insecta: Diplura) of North America, Central America, and the West Indies" by Reddell, 1983
- A Chaos of Delight—chaosofdelight.org

Suborder Rhabdura
Cerci many-segmented; not forcep-like.

PROCAMPODEIDS Family Procampodeidae*   Not illus.
Cerci 8-segmented and shorter than antennae. Appendages on segments 2–7 and hairlike setae with spots on segments 3–7. Only 1 species in NA, *Procampodea macswaini*, found in CA.
TL ~ 5 mm; World 2, NA 1.

ANAJAPYGIDS Family Anajapygidae*   Not illus.
Cerci 8-segmented and shorter than antennae. Appendages on segments 1–7 and hairlike setae with spots on segments 5–7. Members of this family are scavengers. Short, stout setae discharge secretions. Only 1 species in NA, *Anajapyx hermosa*, found in CA.
TL ~5 mm; World 8, NA 1.

Suborder Rhabdura cont.

SLENDER DIPLURANS Family Campodeidae
Cerci about as long as antennae and many-segmented. Largest family of diplurans and most commonly encountered. TL 7–13 mm; World 200, NA 62.

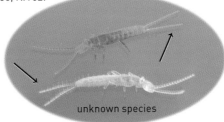
unknown species

Suborder Dicellurata
Cerci of adults 1-segmented and forcep-like.

FORCEPSTAILS Family Japygidae
Cerci 1-segmented and forcep-like. Labial palpi present and hairlike setae with spots on segments 4–6. Predatory, catching prey with cerci. TL 7–50 mm; World 408, NA 55.

Metajapyx sp.
Metajapyx sp. head with labial palps visible
Evallijapyx sp.

PARAJAPYGIDS Family Parajapygidae
Cerci 1-segmented and forcep-like. No labial palpi and antennal segment 4 without hairlike setae with spots. TL 2.5–13 mm; World 62, NA 6.

Parajapyx isabellae

Rock Bristletail
(*Machiloides banksi*)

Rock Bristletail
(*Machilinus* sp.)

BRISTLETAILS
ORDER ARCHAEOGNATHA

Derivation of Name
- Archaeognatha: *archae*, ancient; *gnath*, jaw. Referring to their primitive mouthparts.

Identification
- 7–20 mm TL.
- Small, soft bodied with elongated and cylindrical body.
- Thorax slightly arched.
- Compound eyes touching on top of head.
- Ocelli always present.
- Each mandible with single point of attachment.
- Hind coxae have styli, sometimes also on middle coxae.
- Abdomen with finger-like style on S2–9. S2–7 with 3 sclerites (coxopodites on each side of a median sternite) on the underside of abdomen.
- Two cerci and longer median filament.

Classification
- Historically placed in Thysanura along with silverfish, springtails, and diplurans.
- Most ancestral insect order; fossils date to the mid-Devonian.
- World: 350 species; North America: 35 species.

Range
- Found worldwide except Antarctica.

Similar orders
- Zygentoma (p. 78): eyes not contiguous on top of head; thorax not arched; 3 tails at tip of abdomen similar in length.
- Protura (p. 56): no antennae or tails; minute size.
- Collembola (p. 60): no tails; forked appendage on underside of abdomen.
- A variety of larvae and wingless adults from other orders: lack finger-like projections on underside of abdomen.

Food
- Immatures and adults eat algae and lichens. Use ancestral monocondylic mandibles as a pick when feeding.
- Will scavenge dead arthropods and even eat their own exuviae.

Behavior
- Will jump long distance, as far as 20 cm when disturbed.
- Several types of mating occur. A male will determine female's receptivity by tapping her with his maxillary palps. Then he secretes a line of silk thread to her ovipositor. Drops of sperm move down the thread and fertilization takes place internally. In other groups, like *Petrobius*, sperm is directly placed on the female ovipositor. In some, the male creates a stalk-like spermatophore that is retrieved by the female.

Life Cycle
- Ametabolous.
- Eggs laid in crevices. Females lay up to 30 eggs at a time.
- Immatures look like small versions of adults and can take up to 2 years to become sexually mature, depending on the species, temperature, and food.
- Adults continue to molt after sexual maturity and can live up to 4 years.
- During molting, they glue themselves to the substrate. The glue is likely made up of feces. If the glue fails to stick to a surface, they will not be able to molt and will die.

Importance to Humans
- Can be a food source for other insects.

Collecting and Preserving
Immatures & Adults
- Collect in Berlese funnel or by hand from rocks, leaf litter, tree trunks, and under logs.
- Permanently store in 70% ethanol.

Resources
- *Soil Animals* by Schaller, 1968
- *Soil Biology Guide* by Dindal, 1990

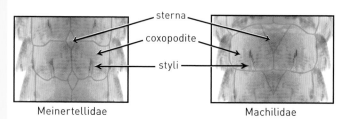

Meinertellidae   Machilidae

Ventral view of abdominal sternites.

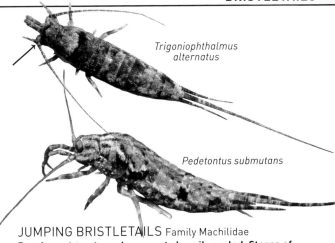

Trigoniophthalmus alternatus

Pedetontus submutans

JUMPING BRISTLETAILS Family Machilidae
Basal-most 2 antennal segments heavily scaled. Sterna of S2–7 large and triangular, extending at least halfway down the coxopodite. TL ~12 mm; World 250, NA 24.

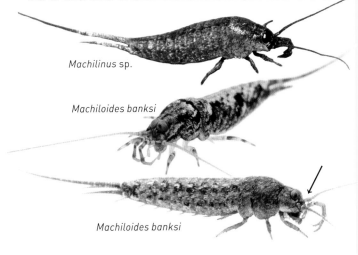

Machilinus sp.

Machiloides banksi

Machiloides banksi

ROCK BRISTLETAILS Family Meinertellidae
Base of antennae without scales. Sterna of S2–7 small and extending only slightly or not at all between coxopodite; never extending to half the length of the coxopodite.
TL 10–20 mm; World 170, NA 6.

Gray Silverfish
(*Ctenolepisma longicaudata*)
Lepismatidae

Blind Silverfish
(*Battigrassiella wheeleri*)

SILVERFISH
ORDER ZYGENTOMA

Derivation of Name
- Zygentoma: *Zyg*, a yolk; *entoma*, an insect. Referring to the placement of silverfish as one of the most ancestral insects.

Identification
- 5–20 mm TL.
- Small, softbodied, with an elongated and slightly flattened body.
- White to gray and brown, with or without scales.
- Ancestrally wingless.
- Antennae filamentous; short or almost as long as body.
- Lack eyes entirely or with compound eyes; with or without ocelli.
- Biting-chewing mouthparts.
- Abdomen with finger-like projections on underside.
- Two cerci and a median caudal filament only slightly longer.

Classification
- Historically placed in Thysanura along with springtails, diplurans, and bristletails or by themselves. To avoid confusion recent authors have created the name Zygentoma for the group.
- Recently the genus *Tricholepidion* was removed from Lepidotrichidae and placed in its own family, Tricholepidiidae. Lepidotrichidae now contains only extinct species known from Baltic amber.
- World: 550 species; North America: 19 species.

Range
- Found worldwide except Antarctica.

Similar orders
- Archaeognatha (p. 74): eyes large, meeting on top of head; thorax arched.
- Protura (p. 56): no antennae or tails; minute size.
- Collembola (p. 60): no tails; forked appendage on underside of abdomen.
- A variety of larvae and wingless adults from other orders: lack finger-like projections on underside of abdomen.

Food

- Nymphs and adults are generally omnivores. Subterranean species are generally vegetarian.
- Species found in houses feed on starches in books and other paper products. Within their guts natural chemical processes break down cellulose.

Behavior

- Cannot jump, but move quickly.
- Often found trapped in sinks or bathtubs in houses.
- Some species exhibit elaborate courtship behavior to ensure delivery of sperm. Males and females face each other and vibrate their antennae. The male then runs away and the female follows. The male then deposits a spermatophore or sperm packet that the female takes it up through her ovipositor to fertilize eggs.

Life Cycle

- Ametabolous.
- Eggs are white or yellow and oval-shaped; laid in crevices and cracks near a food source and hatch in 2 to 8 weeks.
- Immatures look like smaller adults and can molt up to 30 times a year. Take from 3 months to 3 years to develop to sexually mature adult.
- Adults live anywhere from 3 to 6 years.
- Females can lay up to 100 eggs in a lifetime.

Importance to Humans

- A few can be pests in houses, eating paper, books, glue from wallpaper, or fabric in clothes or tapestries.
- Currently being researched for use in the production of biofuel.

Collecting and Preserving

Immatures & Adults

- Collect in Berlese funnel or by hand.
- Permanently store in 70% ethanol.

Resources

- *Soil Animals* by Schaller, 1968
- *Ecology of Soil Animals* by Wallwork, 1970
- *Soil Biology Guide* by Dindal, 1990

FOREST SILVERFISH Family Tricholepidiidae*   Not illus.
With compound eyes and ocelli. Body brown, not covered in scales. Usually a reddish-brown color and has 5-segmented tarsi. Single species in this family, *Tricholepidion gertschi*. Found under redwood bark in northern CA and OR. TL 7–18 mm; World 1, NA 1.

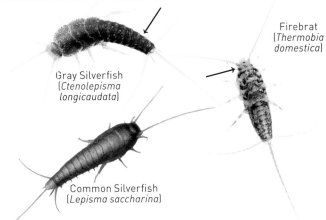

Gray Silverfish
(*Ctenolepisma
longicaudata*)

Firebrat
(*Thermobia
domestica*)

Common Silverfish
(*Lepisma saccharina*)

SILVERFISH AND FIREBRATS Family Lepismatidae
**Compound eyes well separated; lack ocelli. Body covered in
scales.** Most common group of silverfish. Some species, like the
Common Silverfish and Firebrat, can be found in homes. Will
readily feed on starchy foods, including papers and books. Other
species can be found under bark, leaf litter, and logs.
TL 8–12 mm; World 308, NA 14.

Nicoletia
wheeleri

Battigrassiella
wheeleri

SUBTERRANEAN SILVERFISH Family Nicoletiidae*
Lack compound eyes and ocelli. May or may not have scales.
Found in the southeast from FL to TX. Usually found underground
associated with ant and termite nests, mammal burrows, and in
caves. Can also be found under logs and leaf litter.
TL 5–20 mm; World 300, NA 4.

Common Burrower Mayfly
(*Hexagenia limbata*)
Ephemeridae

Isonychia sp. nymph
Isonychiidae

MAYFLIES
ORDER EPHEMEROPTERA

Derivation of Name
- Ephemeroptera: *ephemeros*, short-lived; *ptera*, wing. Many mayflies live as winged adults for only a short period (often a day or less). The common name is a reference to many species emerging in May.

Identification
- 2–31 mm TL (excluding cerci).
- Soft-bodied with long, cylindrical shape.
- Prominent triangular-shaped FW and much smaller, rounded, or sometimes absent HW; both with numerous veins and held together over body when at rest.
- Antennae short, bristle-like, and inconspicuous.
- Two to three long filaments (cerci) originating from tip of abdomen.

Classification
- Three suborders (Carapacea, Furcatergalia, and Pisciforma) have historically been recognized, but current molecular evidence suggests these are not monophyletic groupings, so we have not included them here. Twenty-two families in NA (plus Euthyplociidae found in Mexico) largely separated by wing venation.
- Wing venation and genitalic characters are important.
- World: 3,800 species; North America: 658 species.

Range
- Can be found worldwide except Antarctica. Nymphs found in various freshwater environments.

Similar orders
- Odonata (p. 96): HW smaller than FW; lack long filamentous tails off abdomen.
- Megaloptera (p. 276), Neuroptera (p. 280): antennae long; HW smaller than FW; lack long filamentous tails off abdomen.
- Hymenoptera (p. 236): antennae long; wings with fewer veins; generally lack long filamentous tails off abdomen.
- Diptera (p. 436): 1 pair of membranous wings; lack long filamentous tails off abdomen.
- Nymphs have 2 to 3 caudal filaments (tails) and gills along abdomen. May be confused with Odonata (may have caudal gills, but lack filaments) and Plecoptera (2 caudal filaments and lack gills along basal and middle portion of abdomen).

Food
- Nymphs (also referred to as larvae or naiads) are mostly detritivores and/or herbivores.
- Adults and subimagos have vestigial mouthparts and do not feed.

Behavior
- Many species have spectacularly large swarming flights when males rhythmically fly up and down. They grab females from below, and mating takes place in flight.
- Many species have synchronized emergences and a few (particularly lake species) in such large numbers they are characterized as pests.

Life Cycle
- Hemimetabolous.
- Eggs are either dropped over the water, laid directly on the surface, or attached to objects in the water; as many as 400 to 3,000.
- Nymphs with 15–25 instars on average, varying considerably even within a species. Most species have 1 to 2 generations per year and can take several weeks to 2 or more years to develop. Common in running water, but found in a wide range of other aquatic habitats.
- A winged subimago stage (characterized by dull, cloudy wings and a fringe of minute hairs) precedes the reproductively mature adult and rarely lasts beyond 24 hours; sometimes only a few minutes. Rarely, the subimago is sexually mature.
- Adults generally live only hours to a day.

Importance to Humans
- Grazing by nymphs has a significant impact on primary producers in streams and rivers.
- Both nymphs and adults are an important food source for birds, freshwater fish, and other insects, such as dragonflies and damselflies.
- Mayflies are the primary models for artificial flies in angling or fly fishing. Adults are called *spinners* and subimagos *duns* or *subs*.
- Larvae are useful indicators of water quality.

Collecting and Preserving
Immatures
- Collect with a dip net and place in 70% ethanol.

Adults
- Collected with net, then placed in 70% ethanol.
- Allow subimagos to molt before preserving. Use a small container with permanently fixed substrate (foam or similar) to house them while molting; crumpled paper in a vial will also work.
- Retain the subimago exuviae with the adult.
- Do not store many individuals together or with other insects such as beetles; mayflies are delicate and easily damaged.

Resources
- Mayfly Central—entm.purdue.edu/mayfly
- EphemeropteraGalactica—ephemeroptera-galactica.com
- *The Mayflies of North and Central America* by Edmunds, et al., 1976
- *Mayflies: Anglers Study of Trout Water Ephemeroptera* by Knopp and Cormier, 1997

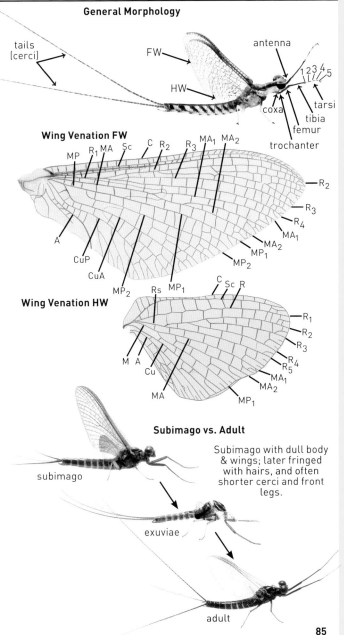

General Morphology

tails (cerci)

FW

HW

coxa

antenna

1 2 3 4 5

tarsi

tibia

femur

trochanter

Wing Venation FW

MP R₁ MA Sc C R₂ R₃ MA₁ MA₂

R₂

R₃

R₄

MA₁

MA₂

MP₁

MP₂

A

CuP

CuA

MP₂

Rs

MP₁

Wing Venation HW

C Sc R

R₁

R₂

R₃

R₄

R₅

MA₁

MA₂

MP₁

M A

Cu

MA

Subimago vs. Adult

Subimago with dull body & wings; later fringed with hairs, and often shorter cerci and front legs.

subimago

exuviae

adult

85

KEY TO MAYFLY FAMILIES

Rarely encountered families not included in the key. See text for descriptions.

Acanthametropodidae, p. 94
Ametropodidae, p. 94
Arthropleidae, p. 92
Behningiidae, p. 88

Neoephemeridae, p. 88
Oligoneuridae, p. 92
Pseudironidae, p. 94

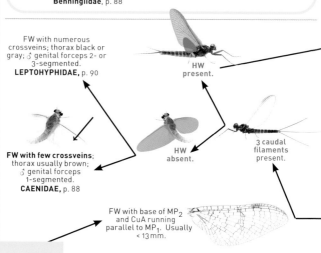

FW with numerous crossveins; thorax black or gray; ♂ genital forceps 2- or 3-segmented.
LEPTOHYPHIDAE, p. 90

HW present.

FW with few crossveins; thorax usually brown; ♂ genital forceps 1-segmented.
CAENIDAE, p. 88

HW absent.

3 caudal filaments present.

FW with base of MP_2 and CuA running parallel to MP_1. Usually < 13 mm.

START HERE

FW with base of MP_2 and CuA strongly divergent from MP_1; Large, usually >13 mm.

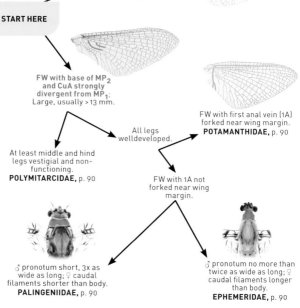

All legs well-developed.

FW with first anal vein (1A) forked near wing margin.
POTAMANTHIDAE, p. 90

At least middle and hind legs vestigial and non-functioning.
POLYMITARCIDAE, p. 90

FW with 1A not forked near wing margin.

♂ pronotum short, 3x as wide as long; ♀ caudal filaments shorter than body.
PALINGENIIDAE, p. 90

♂ pronotum no more than twice as wide as long; ♀ caudal filaments longer than body.
EPHEMERIDAE, p. 90

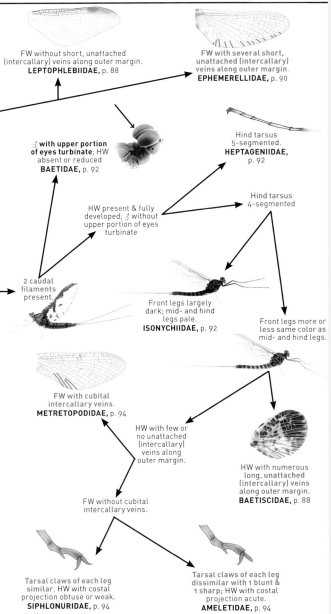

FW without short, unattached (intercallary) veins along outer margin.
LEPTOPHLEBIIDAE, p. 88

FW with several short, unattached (intercallary) veins along outer margin.
EPHEMERELLIDAE, p. 90

♂ with upper portion of eyes turbinate; HW absent or reduced
BAETIDAE, p. 92

Hind tarsus 5-segmented.
HEPTAGENIIDAE, p. 92

HW present & fully developed; ♂ without upper portion of eyes turbinate

Hind tarsus 4-segmented

2 caudal filaments present.

Front legs largely dark; mid- and hind legs pale.
ISONYCHIIDAE, p. 92

Front legs more or less same color as mid- and hind legs.

FW with cubital intercallary veins.
METRETOPODIDAE, p. 94

HW with few or no unattached (intercallary) veins along outer margin.

HW with numerous long, unattached (intercallary) veins along outer margin.
BAETISCIDAE, p. 88

FW without cubital intercallary veins.

Tarsal claws of each leg similar; HW with costal projection obtuse or weak.
SIPHLONURIDAE, p. 94

Tarsal claws of each leg dissimilar with 1 blunt & 1 sharp; HW with costal projection acute.
AMELETIDAE, p. 94

BAETISCADAE • LEPTOPHLEBIIDAE • BEHNINGIIDAE • NEOEPHEMERIDAE • CAENIDAE

ARMORED MAYFLIES Family Baetiscidae

Body distinctly robust with stout thorax and abdomen tapering posteriorly. Prosternal projection present between fore-coxae. Vein A_1 in FW ends at outer margin. **Two tails present.** Single genus, *Baetisca*, in North America. Nymphs found in small to moderate-size streams and edges of lakes having moderate wave action. TL 7–13 mm; World 12, NA 12.

PRONGGILLED MAYFLIES Family Leptophlebiidae

Often reddish and somewhat delicate in form. **Eyes of male distinctly divided with upper portion having larger facets.** Wings lack short free marginal intercalaries between longer veins; 2–4 long intercalaries between CuA and CuP; vein CuP generally strongly recurved; HW present. **Three tails present.** Nymphs found in crevices on underside of rocks, logs, or sticks anchored in streams especially along margins and in backwaters. TL 4–12 mm (usually < 8 mm); World 700, NA 67.

TUSKLESS BURROWING MAYFLIES

Family Behningiidae*   Not Illus.

Thorax is purplish-brown and dorsoventrally flattened with all legs feeble and twisted; abdomen pale. Antennae inserted on anterolateral projections of head. Wings with Rs and MA forked about one-tenth the distance from base to wing margin; Rs forked closer to base than MA; MP and CuA originate from CuP. Middle caudal filament poorly developed. Nymphs found in large rivers and tributaries with sandy substrate. Single species in North America, *Dolania americana*. TL 12–16 mm; World 7, NA 1.

LARGE SQUAREGILLED MAYFLIES

Family Neoephemeridae*   Not Illus.

Male with large eyes, separated by half diameter of eye. FW with basal costal crossveins weak; MP_2 curved toward CuA near base; A_1 with 1–3 crossveins attaching to margin. Three tails present. Nymphs occur on debris in slow to moderately rapid streams. Single genus, *Neoephemera*, in North America. TL 7–18 mm (usually < 10 mm); World 15, NA 5.

SMALL SQUAREGILLED MAYFLIES Family Caenidae

Small, pale, and robust with **wings held out from body giving them a rectangular shape.** FW MA_2 attached to MA_1 at right angles by a crossvein; MP_2 nearly as long as MP_1; HW absent. **Abdomen small and contracted.** Three tails present. Nymphs generally found in flowing streams. TL 2–7 mm; World 230, NA 37.

ARMORED MAYFLIES

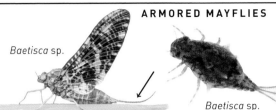

Baetisca sp.

Baetisca sp.

PRONGGILLED MAYFLIES

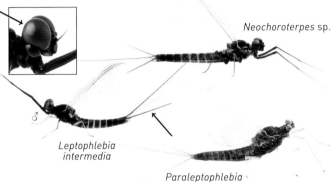

Neochoroterpes sp.

♂

Leptophlebia intermedia

Paraleptophlebia sp.

SMALL SQUAREGILLED MAYFLIES

Caenis sp.

Caenis sp.

Caenis sp.

SPINY CRAWLERS Family Ephemerellidae

Male with eyes nearly contiguous above. FW with costal crossveins absent or only a few weakly developed at apex; series of short intercalary veins along outer margin not attaching to other veins. **Two or three tails present.** Nymphs occur in a variety of lotic and lentic habitats. TL 5–19 mm; World 175, NA 75.

LITTLE STOUT CRAWLERS Family Leptohyphidae   Not Illus.

Eyes generally small and not touching in both sexes. General appearance resembles the Caenidae. FW fairly broad with cubito-anal area somewhat expanded; HW absent or greatly reduced. Three tails present. Nymphs found amid sand and gravel or on submerged rocks, sticks, and logs in small creeks to large rivers.
TL 2–8 mm; World 150, NA 24.

COMMON BURROWERS Family Ephemeridae

Large, often with **patterned wings.** MP_2 and CuA in FW strongly divergent from MP_1 basally; A_1 unforked and attached to hind margin by 2 to many veinlets. Two or three tails present. Nymphs found in streams and lakes where they burrow in soft substrate.
TL 10–30 mm; World 80, NA 12.

SPINY-HEADED BURROWERS Family Palingeniidae

Wings with numerous crossveins. A1 in FW with no more than 3 veinlets. Pronotum in male relatively short. Three tails with middle 1 shorter than body. In NA, nymphs are found burrowing in clay banks of large rivers. Single species, *Pentagenia vittigera*, found in eastern and central NA (including Plains).
TL 20–26 mm; World 32, NA 1.

WHITE BURROWERS Family Polymitarcidae   Not Illus.

Large with pale body and translucent wings with brownish veins along anterior margin. MP_2 and CuA strongly divergent from MP_1 at base. Middle and hind legs in both sexes and forelegs in female are vestigial. Male with 2 tails; 2 or 3 in female. Nymphs found burrowing in silt, clay bottoms and banks of streams, rivers, and lakes.
TL 15–20 mm; World 90, NA 6.

HACKLEGILLED BURROWERS

Family Potamanthidae   Not Illus.

Pale with top of head and thorax light reddish brown. FW with MP_2 and CuA arched at base away from MP_1; A_1 forked; HW with distinct costal projection. Three tails present. Nymphs found sprawling on rocks and woody debris in moderately-sized rivers and streams. Single genus, *Anthopotamys*, occuring in central and eastern NA, rarely on west coast. TL 7–13 mm; World 24, NA 4.

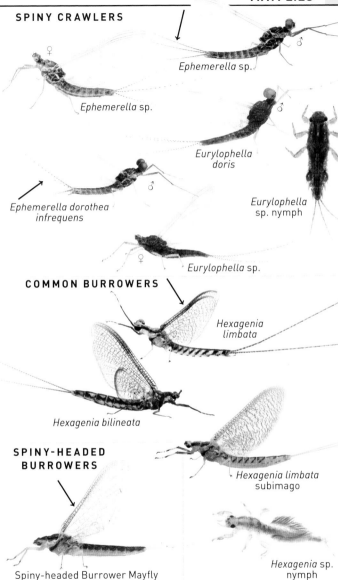

SPINY CRAWLERS

Ephemerella sp. ♂

Ephemerella sp. ♀

Eurylophella doris ♂

Eurylophella sp. nymph

Ephemerella dorothea infrequens ♂

Eurylophella sp. ♀

COMMON BURROWERS

Hexagenia limbata

Hexagenia bilineata

SPINY-HEADED BURROWERS

Hexagenia limbata subimago

Spiny-headed Burrower Mayfly
(*Pentagenia vittigera*)

Hexagenia sp. nymph

SMALL MINNOW MAYFLIES Family Baetidae

Males with prominent, turbinate eyes. FW with IMA, MA$_2$, IMP, and MP2 not attached basally; HW reduced, without veins, or absent. **Two tails present.** Nymphs found in a wide variety of lentic and lotic habitats. TL 2–15 mm; World 960, NA 133.

HOWDY BRUSHLEGGED MAYFLIES Family Isonychiidae

Body usually reddish or purplish-brown. Foreleg femora and tibia dark, contrasting with remainder of leg. Male with eyes large and contiguous dorsally. **Two tails present.** Nymphs are strong swimmers found in fast-flowing waters of creeks and rivers. Single widespread genus, *Isonychia*, in NA. TL 9–16 mm; World 34, NA 16.

OLIGONEURIID BRUSHLEGGED MAYFLIES

Family Oligoneuriidae*   Not Illus.

Forelegs shorter than middle pair. Wing venation reduced, but R$_3$ and IR$_3$ present in FW; MP in HW is forked near outer margin. Two or three tails are present. Nymphs generally found in leaf packs in swift-moving areas of steams and rivers with some sand. TL 5–10 mm; World 54, NA 9.

PALPHEADED MAYFLIES Family Arthropleidae*   Not Illus.

Male with well-separated eyes. Basal tarsal segment of forelegs slightly shorter than second segment. MA simple and unforked in HW; Rs forms a regular fork. Two tails present. Nymphs found in ponds or channels with little or no flow. Single species, *Arthroplea bipunctata*, found in northeastern NA. Some authorities place this species within the Heptageniidae. TL 7–8 mm; World 2, NA 1.

FLATHEADED MAYFLIES Family Heptageniidae

FW with 2 pairs of cubital intercalary veins; MP$_1$ and MP$_2$ forming a more or less symmetrical fork. **Two tails present.** Nymphs common in a variety of lotic habitats and the shallow littoral areas of lakes. TL 5–21 mm; World 600, NA 126.

HOWDY BRUSHLEGGED MAYFLIES

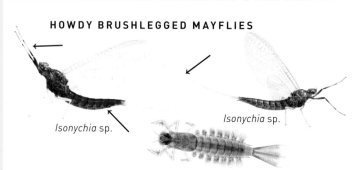

Isonychia sp.

Isonychia sp.

Isonychia sp. nymph

SMALL MINNOW MAYFLIES

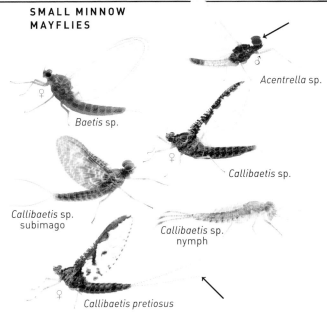

♀
Baetis sp.

♂
Acentrella sp.

♀
Callibaetis sp.

Callibaetis sp.
subimago

Callibaetis sp.
nymph

♀
Callibaetis pretiosus

FLATHEADED MAYFLIES

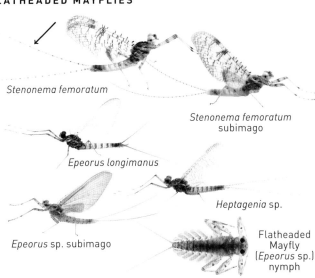

Stenonema femoratum

Stenonema femoratum
subimago

Epeorus longimanus

Heptagenia sp.

Epeorus sp. subimago

Flatheaded
Mayfly
(*Epeorus* sp.)
nymph

93

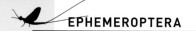

BOW-LEGGED MINNOW MAYFLIES
Family Acanthametropodidae*   Not Illus.
Rare and poorly known. HW large, half or more as long as FW and lacking a costal projection. Three tails present. Nymphs found in rapid, shallow rivers with sandy substrate.
TL 12–20 mm; World 4, NA 2.

COMBMOUTHED MINNOWS Family Ameletidae   Not Illus.
Body generally brown or yellow-brown. Male with large, contiguous eyes dorsally and forelegs nearly as long as body; basal tarsal segment one-third to one-half length of second segment. MP in HW simple and unforked or forked near base; HW with acute costal projection. Claws on each tarsus dissimilar; 1 is blunt, the other sharp. Two tails and a short median stub present. Nymphs usually found in small, rapidly flowing streams among cobble. Single widespread genus, *Ameletus*, in NA. TL 7–16 mm; World 60, NA 35.

SAND MINNOWS Family Ametropodidae*   Not Illus.
Males with larger facets in top half of eye. Foretarsi in male 5 times as long as tibiae. FW with 2 pairs of cubital intercalaries; A_1 attached to hind margin by several veinlets; HW with sharp costal projection. Three tails. Nymphs found in large rivers with clean silt and sandy substrate. Single genus, *Ametropus*, found in western and central NA. TL 15–25 mm; World 3, NA 2.

CLEFTFOOTED MINNOWS Family Metretopodidae   Not Illus.
Body typically brownish. Male with large compound eyes, nearly contiguous above. FW with MP_2 somewhat arched from MP_1 at base; 2–4 cubital intercalary veins present. Two tails present. Nymphs found on snags and submerged broken tree branches during winter months, emerging in spring.
TL 7–16 mm; World 11, NA 9.

PRIMITIVE MINNOWS Family Siphlonuridae   Not Illus.
Head and thorax generally brown with conspicuously light-and-dark-patterned abdomen. Male with large eyes, usually contiguous above and forelegs as long as or slightly longer than body. Two tails present. Nymphs generally found along edges of shallow pools and ponds where they cling to the vegetation, but also in slow-moving waters and backwaters of springs. TL 6–21 mm; World 50, NA 26.

CRABWALKERS Family Pseudironidae*   Not Illus.
Male with large, well-separated eyes. Male forelegs with tarsi twice length of tibiae. Two tails present. Nymphs found in medium-sized to large rivers with sandy substrate. Family represented by a single species, *Pseudiron centralis*, endemic to NA. Some authorities place this species within the Heptageniidae. TL 10–15 mm; World 1, NA 1.

Leptophlebia intermedia
Leptophlebiidae

Eastern Amberwing
(*Perithemis tenera*)
Libellulidae

Dancer nymph
(*Argia* sp.)
Coenagrionidae

DAMSELFLIES & DRAGONFLIES
ORDER ODONATA

Derivation of Name
- Odonata: *odont*, tooth. Referring to teeth on large mandibles in the adults.

Identification
- 20–115 mm TL; 25–140 mm WS.
- Long slender body with prominent wings, variably colored.
- Only insect order in which all adults are winged.
- Small hair-like (setaceous) antennae; distinct head with large compound eyes; thorax with small prothorax and large pterothorax; 2 pairs of densely veined membranous wings; 3-segmented tarsi; long slender abdomen with 10 segments.
- Male has secondary copulatory organs on ventral side of second and third abdominal segment.

Classification
- Two suborders and eleven families in NA. Suborders are separated by wing shape, venation, and number of appendages at end of abdomen.
- World: 6,350 species; North America: 471 species.

Range
- Can be found worldwide except Antarctica. Includes Wandering Glider (*Pantala flavescens*), longest known migratory insect, traveling 11,000 miles round trip across the Indian Ocean.

Similar orders
- Megaloptera (p. 276), Neuroptera (p. 280): antennae long; tarsi 5-segmented; wing venation different.
- Hymenoptera (p. 236): antennae long; 5-segmented tarsi; HW smaller than FW; wings with fewer veins.
- Diptera (p. 436): 1 pair of membranous wings.
- Ephemeroptera (p. 82): 2–3 long tails; HW smaller than FW; very soft-bodied.

Powdered Dancer
(*Argia moesta*)
Coenagrionidae

Food

- Nymphs (also referred to as larvae or naiads) are all predators, feeding on worms, small crustaceans, mosquito larvae, tiny fish, and even other dragonfly and damselfly nymphs with a modified hinged labium for grabbing prey.
- Adults feed primarily on other insects, mostly small biting flies and their relatives, but eat other odonates. Large dragonflies have been recorded feeding on hummingbirds.

Behavior

- Generally found in or near aquatic habitats.
- Perch on or atop vegetation, directly on the ground, or atop rocks.

Life Cycle

- Hemimetabolous.
- Eggs are laid in plants (endophytically) or in water (exophytically).
- Short-lived "prolarva" hatches from egg and is specially modified to free itself from egg and find its way to the proper microhabitat.
- Nymphs with 8–15 instars, lasting from 1 month to several years. Common in ponds, marshes, lakes, and streams, but can also occupy a wide range of other aquatic habitats.
- Adults generally live 4–6 weeks, with migratory species living several months.
- Some species, especially those living in ephemeral habitats, have multiple generations per year; others take several years to develop.

Importance to Humans

- Considered beneficial because they feed primarily on mosquitos and biting flies as both nymphs and adults.
- Embraced by the public similar to birds. Second only to butterflies in popularity.

Collecting and Preserving

Immatures

- Collect nymphs with a dip net and place them live in vial with moist vegetation; boil water, remove from heat, and drop live nymph in water; remove nymph from water after 1 minute.
- Permanently store in 70% ethanol.

Adults

- Collect individuals with net, then place in glassine envelope.
- Best preserved by initially dropping live individual in acetone, then after 2–3 minutes, remove and position legs and abdomen in glassine envelope with both wings arranged over abdomen. Resubmerge in acetone for 8–24 hours. Remove envelope from acetone and dry thoroughly.
- Store in clear cellophane or polypropylene envelope.

Resources

- OdonataCentral—OdonataCentral.org—checklists and maps
- *Dragonflies and Damselflies of the East* by Paulson, 2011
- *Dragonflies and Damselflies of the West* by Paulson, 2009
- *Dragonflies: Behavior and Ecology of Odonata* by Corbet, 1999
- *Dragonflies and Damselflies: A Natural History* by Paulson, 2019

Dragonfly Wings

Damselfly Wings

HW is broader basally than FW

HW and FW are the same size and shape

Dragonfly Nymph

Damselfly Nymph

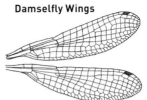

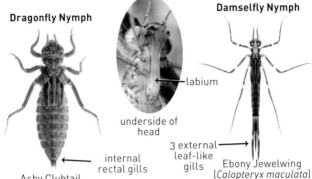

labium

underside of head

internal rectal gills

Ashy Clubtail
(*Phanogomphus lividus*)
Gomphidae

3 external leaf-like gills

Ebony Jewelwing
(*Calopteryx maculata*)
Calopterygidae

Wheel position

Male dragonfly grabs female by back of eyes to mate. The female then brings her abdomen up to male's secondary genitalia, forming the wheel position.

Male secondary genitalia

Halloween Pennant
(*Celithemis eponina*)
Libellulidae

Tandem position

Male may guard female by either staying in tandem position or hovering near female until eggs are laid. Smoky Rubyspot (*Hetaerina titia*) Calopterygidae.

KEY TO DAMSELFLY & DRAGONFLY FAMILIES

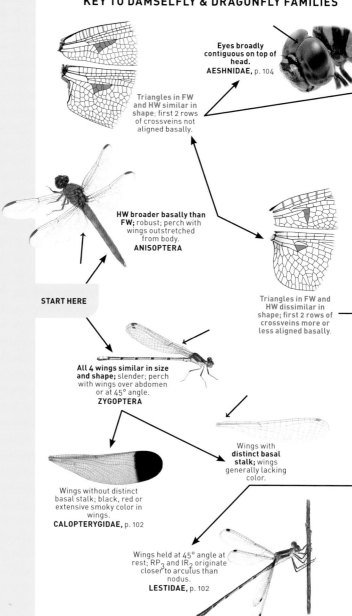

Eyes broadly contiguous on top of head.
AESHNIDAE, p. 104

Triangles in FW and HW similar in shape; first 2 rows of crossveins not aligned basally.

HW broader basally than FW; robust; perch with wings outstretched from body.
ANISOPTERA

START HERE

Triangles in FW and HW dissimilar in shape; first 2 rows of crossveins more or less aligned basally.

All 4 wings similar in size and shape; slender; perch with wings over abdomen or at 45° angle.
ZYGOPTERA

Wings with distinct basal stalk; wings generally lacking color.

Wings without distinct basal stalk; black, red or extensive smoky color in wings.
CALOPTERYGIDAE, p. 102

Wings held at 45° angle at rest; RP_3 and IR_2 originate closer to arculus than nodus.
LESTIDAE, p. 102

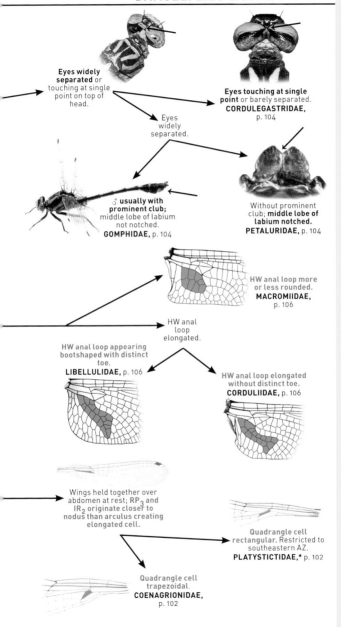

Eyes widely **separated** or touching at single point on top of head.

Eyes touching at single **point** or barely separated. **CORDULEGASTRIDAE,** p. 104

Eyes widely separated.

♂ usually with **prominent club;** middle lobe of labium not notched. **GOMPHIDAE,** p. 104

Without prominent club; **middle lobe of labium notched. PETALURIDAE,** p. 104

HW anal loop more or less rounded. **MACROMIIDAE,** p. 106

HW anal loop elongated.

HW anal loop appearing bootshaped with distinct toe. **LIBELLULIDAE,** p. 106

HW anal loop elongated without distinct toe. **CORDULIIDAE,** p. 106

Wings held together over abdomen at rest; RP_3 and IR_2 originate closer to nodus than arculus creating elongated cell.

Quadrangle cell rectangular. Restricted to southeastern AZ. **PLATYSTICTIDAE,*** p. 102

Quadrangle cell trapezoidal. **COENAGRIONIDAE,** p. 102

Suborder Zygoptera

Eyes transversely elongate, widely separated and never touching. Wings typically held together over abdomen; FW and HW similarly shaped. Discoidal cell (=quadrangle) not divided. Male with 2 pairs of terminal appendages (dorsal cerci and ventral paraprocts).

BROAD-WINGED DAMSELS Family Calopterygidae

Wings broad in middle, with color, and **lacking petiolate stalk.** Numerous antenodal crossveins. Large damselflies and only group in NA that engage in courtship. Found near streams. TL 35–66 mm; World 176, NA 8.

SHADOWDAMSELS Family Platystictidae*

Wings stalked; quandrangle rectangular (see key). Two antenodal crossveins and numerous postnodal crossveins. RP_3 arising closer to nodus than arculus. One species, *Palaemnema domina*, found in southeastern Arizona. TL 33–45 mm; World 264, NA 1.

SPREADWINGS Family Lestidae

Eyes and head in male often blue. Wings spread at **45° angle** when perched; stalked with 2 antenodal crossveins. RP_3 arising closer to arculus than nodus. Abdomen relatively large and stocky. Thorax and abdomen often metallic with varying degrees of pruinosity. Found at standing bodies of water. TL 31–51 mm; World 151, NA 19.

POND DAMSELS Family Coenagrionidae

Most diverse and common group of damselflies. **Wings at rest slightly above or overlapping abdomen.** RP_3 arising closer to nodus than arculus, with 2 antenodal crossveins. Quadrangle cell trapezoidal. Many small and dainty compared to other families. Males, in particular, can have red, orange, yellow, blue or green bodies. TL 20–51 mm; World 1,082, NA 103.

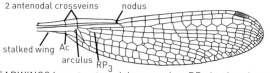

SPREADWINGS have 2 antenodal crossveins. RP_3 begins closer to arculus than nodus.

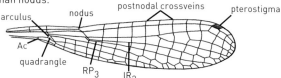

POND DAMSELS have 2 antenodal crossveins. RP_3 is closer to nodus than arculus.

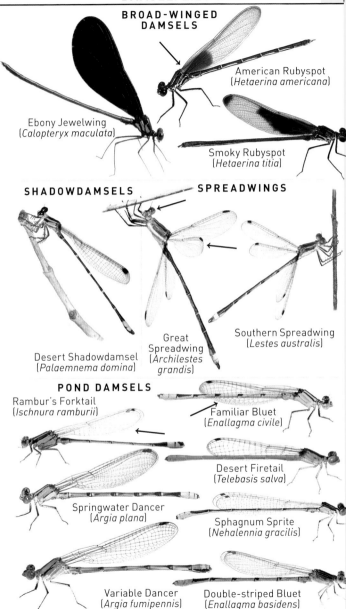

BROAD-WINGED DAMSELS

American Rubyspot
(*Hetaerina americana*)

Ebony Jewelwing
(*Calopteryx maculata*)

Smoky Rubyspot
(*Hetaerina titia*)

SHADOWDAMSELS

SPREADWINGS

Desert Shadowdamsel
(*Palaemnema domina*)

Great
Spreadwing
(*Archilestes grandis*)

Southern Spreadwing
(*Lestes australis*)

DAMSELFLIES

POND DAMSELS

Rambur's Forktail
(*Ischnura ramburii*)

Familiar Bluet
(*Enallagma civile*)

Desert Firetail
(*Telebasis salva*)

Springwater Dancer
(*Argia plana*)

Sphagnum Sprite
(*Nehalennia gracilis*)

Variable Dancer
(*Argia fumipennis*)

Double-striped Bluet
(*Enallagma basidens*)

Suborder Anisoptera

Eyes large and often touching at least at a single point, but never separated by more than their own width. HW broader basally than FW. At rest wings held away from the body. Discoidal cell divided into triangle and supertriangle. Male with pair of superior caudal appendages (cerci) and single inferior appendage (epiproct).

PETALTAILS Family Petaluridae

Gray or black in general color. Median lobe of labium with **triangular-shaped notch** on underside of head. **Eyes separated on top of head.** Wings clear with **pterostigma 0.3 in** or longer. Thorax and abdomen compact and stout, giving a more ancient appearance. Nymphs semi-terrestrial. TL 53–84 mm; World 11, NA 2.

DARNERS Family Aeshnidae

Among largest dragonflies. Often brilliantly colored blue, green, and brown. **Eyes meet broadly** at top of head. Triangles in FW and HW similar in shape. **Long slender abdomen, slightly swollen** (especially in male) where it attaches to thorax. Nymphs found in both standing and flowing water. TL 50–117 mm; World 426, NA 41.

CLUBTAILS Family Gomphidae

Greenish-yellow with brown or black stripes. **Eyes widely separated.** Triangles in FW and HW similar in shape. Male with **distinct club** or widening of posterior abdomen. Most often found resting on ground near streams and rivers. TL 40–60 mm; World 952, NA 101.

SPIKETAILS Family Cordulegastridae

Yellow stripes on brownish-black body. Brilliant green or blue **eyes meet at single point on top of head.** Thorax with dark lateral stripes. Legs relatively short. Named for **thin pointed ovipositor** found in some females. Nymphs generally found in small streams. TL 55–85 mm; World 51, NA 10.

Clearlake Clubtail
(*Phanogomphus australis*)

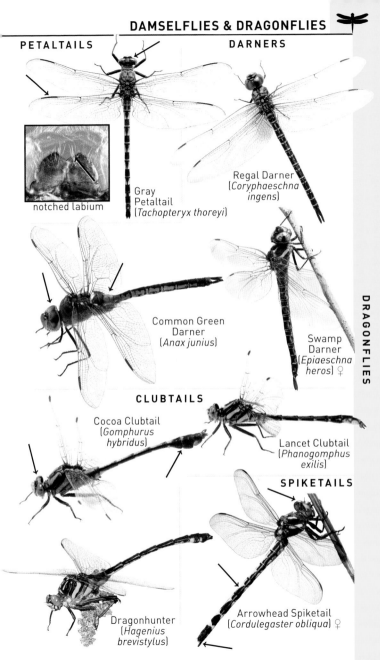

PETALTAILS

DARNERS

notched labium

Gray Petaltail (*Tachopteryx thoreyi*)

Regal Darner (*Coryphaeschna ingens*)

Common Green Darner (*Anax junius*)

Swamp Darner (*Epiaeschna heros*) ♀

CLUBTAILS

Cocoa Clubtail (*Gomphurus hybridus*)

Lancet Clubtail (*Phanogomphus exilis*)

SPIKETAILS

Dragonhunter (*Hagenius brevistylus*)

Arrowhead Spiketail (*Cordulegaster obliqua*) ♀

DRAGONFLIES

ANISOPTERA

CRUISERS Family Macromiidae

Strong fliers found along edges of rivers, streams, and lakeshores, where they perch vertically. Have large eyes (often **brilliant green**) and legs relatively long and spined. Body brown or black with a **yellow band around the thorax like a belt**. Anal loop circular. Abdomen in males often clubbed. Nymphs found in streams and rivers. TL 66–92 mm; World 121, NA 9.

EMERALDS Family Corduliidae

Mature individuals with brilliant **iridescent-green eyes.** Body dark, usually with pale ring basally on abdomen and noticeably hairy thorax. Anal loop usually a **weakly formed foot or boot,** lacking well-developed toe region (outlined in Slender Baskettail). Males with small auricle (ear-like lobe) on sides of S2. Breed in streams and ponds with high oxygen content. Many restricted to characteristic northern habitats such as bogs and fens. TL 27–77 mm; World 240, NA 50.

SKIMMERS Family Libellulidae

Largest, most diverse and common family of dragonflies. Often found at ponds, lakes, and marshes, where many perch readily. Often vividly colored with distinctive wing markings. Some species are sexually dimorphic. Anal loop forming a **foot or boot with well-developed toe** (see Eastern Pondhawk). TL 17–64 mm; World 968, NA 108.

Elfin Skimmer
(*Nannothemis bella*)

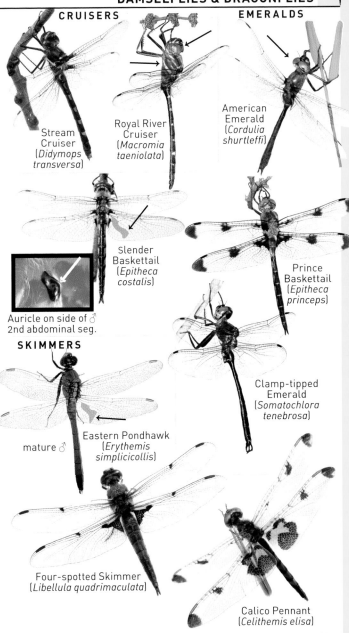

CRUISERS

EMERALDS

Stream Cruiser (*Didymops transversa*)

Royal River Cruiser (*Macromia taeniolata*)

American Emerald (*Cordulia shurtleffi*)

Slender Baskettail (*Epitheca costalis*)

Auricle on side of ♂ 2nd abdominal seg.

Prince Baskettail (*Epitheca princeps*)

SKIMMERS

Clamp-tipped Emerald (*Somatochlora tenebrosa*)

mature ♂

Eastern Pondhawk (*Erythemis simplicicollis*)

Four-spotted Skimmer (*Libellula quadrimaculata*)

Calico Pennant (*Celithemis elisa*)

DRAGONFLIES

107

ANISOPTERA

SKIMMERS CONT.

Slaty Skimmer
(*Libellula incesta*)

Widow Skimmer
(*Libellula luctuosa*)

Chalk-fronted
Corporal
(*Ladona julia*)

Flame Skimmer
(*Libellula saturata*)

Wandering Glider
(*Pantala flavescens*)

Variegated
Meadowhawk
(*Sympetrum corruptum*)

Black Saddlebags
(*Tramea lacerata*)

Dot-tailed
Whiteface
(*Leucorrhinia intacta*)

108

REPRESENTATIVE NYMPHS

Ebony Jewelwing
(*Calopteryx maculata*)
Calopterygidae

Dusky Dancer
(*Argia translata*)
Coenagrionidae

Northern Spreadwing
(*Lestes disjunctus*)
Lestidae

Gray Petaltail
(*Tachopteryx thoreyi*)
Petaluridae

Common Sanddragon
(*Progomphus borealis*)
Gomphidae

Broad-striped
Forceptail
(*Aphylla
angustifolia*)
Gomphidae

Common Green Darner
(*Anax junius*)
Aeshnidae

Arrowhead Spiketail
(*Cordulegaster obliqua*)
Cordulegastridae

Bronzed River
Cruiser
(*Macromia annulata*)
Macromiidae

Wandering Glider
(*Pantala flavescens*)
Libellulidae

Yellow-sided Skimmer
(*Libellula flavida*)
Libellulidae

Kennedy's Emerald
(*Somatochlora
kennedyi*)
Corduliidae

DRAGONFLIES

Hubbard's Angel Insect
(*Usazoros hubbardi*)

Hubbard's Angel Insect nymph [*Usazoros hubbardi*]

ANGEL INSECTS
ORDER ZORAPTERA

Derivation of Name
- Zoraptera: *Zor*, pure; *aptera*, wingless. Meaning purely wingless. When they were first described, only wingless forms were known.

Identification
- 1.5–3 mm TL.
- Resemble termites. Wingless forms are soft, with pale to brown-colored body, and lack eyes. Winged forms are dark, have compound eyes and ocelli.
- FW slightly larger than HW with few veins and no crossveins.
- Adult antennae moniliform 9-segmented; juveniles with fewer segments.
- Legs without modification; tarsi 2-segmented; cerci short and 1-segmented.

Classification
- First discovered in 1913, their relationship to other orders is still debated. Most recent evidence suggests they are closely related to Embioptera or Dictyoptera (roaches and termites).
- Sometimes given the vernacular name of Angel Insects.
- Historically 1 family, Zorotypidae; 1 genus, *Zorotypus*, in the world, but Kocárek et al. (2020) have recently proposed 2 families and 9 genera following their molecular phylogeny and differences in the male genitalia and reproductive strategies.
- Found throughout tropical areas.
- World: ~44 species; North America: 2 species.

Range
- Southeastern NA; MD to OK south to FL and TX.

Similar orders
- Embioptera (p. 148): tarsi 3-segmented, basal tarsal segment of foreleg enlarged; cerci 1–2-segmented; size 5–20 mm, antennae 16–32 segments.
- Blattodea (termites) (p. 166): tarsi 4-segmented; antennae usually short and moniliform; FW and HW similar in size.
- Psocodea (p. 220): antennae filiform; 2- or 3-segmented tarsi and lack cerci.

Food

- Feed on several parts of fungi, nematodes, mites, and other minute arthropods.

Behavior

- Live in small subsocial colonies in rotting wood or sawdust; only known insects with a male-dominated society.
- As colony grows and resources become scarce, winged individuals are produced and individuals fly to a new location. The number of winged females is generally greater than males. It is thought that the female mates prior to dispersal, which would account for the small number of winged males.
- When a new location has been found, wings are shed, leaving stubs on the thorax. These individuals are common in young colonies.
- Grooming within colonies is very elaborate and common and thought to be involved with removal of fungal pathogens.

Life Cycle

- Hemimetabolous.
- Courtship takes place by male stroking female with antennae; if the female is receptive, she will reciprocate.
- Male has a hook on S10 and female has a groove where the hook is placed. This hook is seen in fossils.
- Males of some species have been found to produce 1 large sperm the same length as the female. Sperm is deposited on the abdomen of the female, where external fertilization takes place.
- The size of sperm is thought to help prevent competition from other males, leaving very little space for more sperm.
- Nymphs similar to adults; just smaller and paler with less than 9-segmented antennae.
- Nymphs likely have 4–5 instars, which takes several months.

Importance to Humans

- Rare insects that are never pests because of their small numbers.

Collecting and Preserving

Immatures & Adults

- Hand collect in rotting wood and sawdust or with a Berlese funnel. An aspirator is useful in collecting without injuring.
- Preserve in 70% ethanol.

Resources

- *How to Know the Grasshoppers, Crickets, Cockroaches and Their Allies* by Helfer, 1987

Typical habitat where Zoraptera may be found.

Hubbard's Angel Insect
(*Usazoros hubbardi*)

ZOROTYPID ANGEL INSECTS Family Zorotypidae*
Minute soft-bodied insects, generally smaller and faster moving
than a termite. Wingless forms are pale, and winged forms are
generally darker. **Antennae are beaded** and 9-segmented as
adults; antennomere I is 2 times longer than antennomere III and
antennomere II as long as antennomere III. Wingless forms lack
eyes; winged forms with compound eyes and 3 ocelli. Male genitalia
asymmetrical. Single species, *Usazoros hubbardi*, found from the
Great Plains eastward in the US. TL 1.8–2.4 mm; World 21, NA 1.

SPIRALIZORID ANGEL INSECTS Family Spiralizoridae*
Not Illus.

Minute soft-bodied insects. Wingless forms are pale, and 0winged
forms are generally darker. Antennae are beaded and 9-segmented
as adults; antennomere I is 1.0 to 1.5 times times longer than
antennomere III and antennomere II one half length of antennomere
III. Male genitalia symmetrical. Single species, *Centrozoros snyderi*,
from FL through South America. TL 2.3–3.6 mm; World 22, NA 1.

Hubbard's Angel Insect (*Usazoros hubbardi*) on rotting wood.

European Earwig
(*Forficula auricularia*)

Ring-legged Earwig nymph
(*Euborellia annulipes*)

EARWIGS
ORDER DERMAPTERA

Derivation of Name
- Dermaptera: *derma*, skin; *ptera*, wings. Describing the FW as skin-like.

Identification
- 6–35 mm TL.
- Small, slender, long flattened insects usually with 4 wings. FW (tegmina) are thickened, leathery, and short; HW membranous and folded under FW at rest; large and fan-like shape.
- Mouthparts chewing; antennae threadlike, long and about half the length of the body; tarsi 3-segmented.
- Males and females have forcep-like cerci. Cerci usually larger and more curved in males. Males have 10 abdominal segments and females have 8.

Classification
- Classification has undergone recent changes and currently a single extant suborder (Neodermaptera) and 2 extinct suborders (Archidermaptera and Eodermaptera) are recognized.
- Oldest fossils date from the late Triassic–early Jurassic in England and Australia.
- Currently 5 families in NA (plus 1 in Hawaii).
- Half our fauna are introduced species.
- World: 1,800 species; North America: 27 species.

Range
- Found throughout the world except Antarctica and extreme Arctic.

Similar orders
- Embioptera (p. 148): tarsi 3-segmented, basal segment enlarged; cerci 1–2 segments; small 5–15 mm.
- Coleoptera (p. 294): with elytra usually extending over length of abdomen, but reduced in Staphylinoidea; no forcep-like cerci.
- Plecoptera (p. 120): flattened body; most with 4 membranous wings; filamentous cerci; found near streams.
- Phasmatodea (p. 152): *Timema* have 3-segmented tarsi; antennae more than half the length of body.
- Blattodea (p. 166): Antennae more than half body length; 5-segmented tarsi; head concealed by pronotum; cerci thin, short, not forcep-like.

Food
- Adults and nymphs usually omnivorous; few species are strictly herbivorous or carnivorous.

Behavior
- Most found in moist areas under debris. In dry areas in the west, they can be found in dead or dying cacti to avoid dessication.
- Most are found in southern regions, however, a few species are widespread in NA.
- Generally nocturnal, hiding during the day.

Life Cycle
- Hemimetabolous.
- Mate and lay eggs in soil, usually in the fall.
- Female exhibits parental care (subsocial) tending eggs to prevent fungal growth.
- Eggs hatch in the spring.
- Nymphs are similar to adults but smaller and paler.
- Nymphs are fed by the mother for the first 2 instars, then leave and fend for themselves.

Importance to Humans
- Considered beneficial, eating things like termites and other pest insects. A few species may be considered pests on crops and flowers or because of abundance in households.

Collecting and Preserving
Immatures
- Keep nymphs in 70% alcohol.

Adults
- Found in leaf litter, debris, grasses, roots, and soils. Some come to lights. Often collected in pitfall traps.
- Pin in right tegmina with hind legs pulled in close to body. If too small for a pin, place on a point.

Resources
- *How to Know the Grasshoppers, Crickets, Cockroaches and Their Allies* by Helfer, 1987

Lesser Earwig ♂
(*Labia minor*)

Male Forceps

Anisolabididae

Forficulidae

Labiduridae

Spongiphoridae

Lined Earwig
(*Doru taeniatum*)
with wings out
Forficulidae

RING-LEGGED EARWIGS Family Anisolabididae
Antennae 14–24 segments. Right forcep of male is more strongly curved than the left. Pronotum uniformly colored. **Tegmina rounded flaps, not meeting at basal margin** or absent. Some in this group are predaceous while others eat detritus.
TL 10–26 mm; World 400, NA 8.

COMMON EARWIGS Family Forficulidae
Antennae 12–16 segments. Second tarsal segment lobed and extending past third segment. Eats vegetation, including crops; sometimes considered a pest. Widely distributed; most common species is the European Earwig. TL 10–21 mm; World 460, NA 6.

STRIPED EARWIGS Family Labiduridae
Pronotum with 2 dark longitudinal stripes. Antennae 25–30 segments. Antennal segments 4–6 with combined length shorter than segment 1. Single introduced species in family found in southern US from FL to CA. Preys on invertebrates and emits foul odor when disturbed. TL 17–31 mm; World 70, NA 1.

LITTLE EARWIGS Family Spongiphoridae
Antennae 10–16 segments. Pronotum uniformly colored. Tegmina normally developed and wings usually present. Male forceps symmetrical. TL 2.5–11 mm; World 420, NA 10.

PYGIDICRANID EARWIGS Family Pygidicranidae*   Not illus.
Large flexible pad between tarsal claws. Male forceps curved strongly inward. Body with dense covering of hairs. Single introduced species in US, *Pyragropsis buscki*, found in Miami, FL. TL 10–16 mm; World ~200, NA 1.

LITTLE EARWIGS

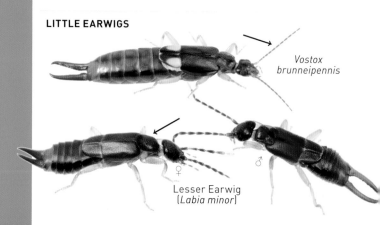

Vostox brunneipennis

♀

♂

Lesser Earwig
(*Labia minor*)

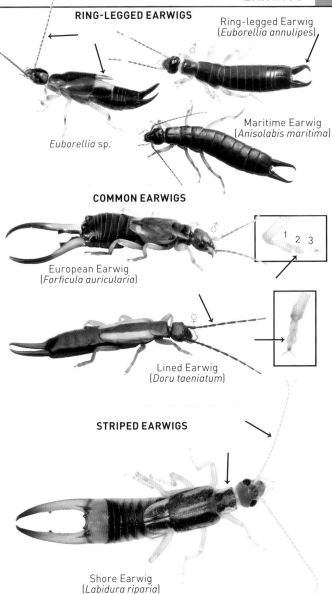

RING-LEGGED EARWIGS

Ring-legged Earwig
(*Euborellia annulipes*)

Euborellia sp.

Maritime Earwig
(*Anisolabis maritima*)

COMMON EARWIGS

1 2 3

European Earwig
(*Forficula auricularia*) ♂

Lined Earwig
(*Doru taeniatum*) ♀

STRIPED EARWIGS

Shore Earwig
(*Labidura riparia*)

Clio Stripetail Stonefly
(*Clioperla clio*)
Perlodidae

STONEFLIES
ORDER PLECOPTERA

Derivation of Name
- Plecoptera: *pleco*, folded, or twisted; *ptera*, wings. Referring to the anal region of HW that is folded when wings are at rest. Stoneflies refers to the presence of these insects on stones and rocks.

Identification
- 2–64 mm TL
- Small to medium-sized, somewhat flattened, soft-bodied insects.
- Usually with 4 wings held flat over abdomen; HW often a little shorter than FW and with enlarged anal area. Some species brachypterous, others apterous.
- Mouthparts chewing, but often not functional in adult.
- Antennae filamentous, long and threadlike.
- Usually have 3 ocelli, but sometimes 2.
- Tarsi 3-segments.
- Males and females have cerci of variable length.

Classification
- Classification has undergone recent changes, with some disagreement on the number of families. Generally, 2 suborders recognized, Euholognatha and Systellognatha.
- Oldest fossils are from the early Permian. The 2 suborders likely became differentiated during the breakup of Pangaea.
- Currently, 9 families recognized in NA.
- World: 3,400 species; North America: 670 species.

Range
- Found throughout the world except Antarctica and extreme Arctic; most abundant in cooler areas.

Similar orders
- Trichoptera (p. 364): tarsi 5-segments; basal area of HW without expanded anal lobe.
- Neuroptera (p. 280): tarsi 5-segments; basal area of HW without expanded anal lobe.
- Ephemeroptera (p. 82): HW small or absent; wings held together above body at rest; antennae short and bristle-like.
- Embioptera (p. 148): tarsi 3-segments; basal segment of foreleg enlarged; cerci 1–2 segments; small (5–15 mm).

121

Food
- Adults usually do not feed, but those that do are herbivores.
- Nymphs are usually detritivores, herbivores, or omnivores, but some are strictly carnivores.

Behavior
- Nymphs exhibit periodic drifting behavior in streams. At night, nymphs will let go of the substrate and let the current take them downstream, likely looking for new habitat and food.
- Males of some species drum on the substrate to attract virgin females with species-specific vibrational communications. Some males are capable of triangulating in on stationary, drumming, females.

Life Cycle
- Hemimetabolous.
- Eggs are various shapes and textures depending on where they are laid. Some are flared at the tips to fit into crevices.
- Nymphs are aquatic with filamentous external gills on the thorax near legs, around the head, and/or basally on the abdomen.
- Nymphs take 1–3 years to develop and go through up to 22 instars.
- Adults live a few months, mating and laying eggs; mating occurs on the ground or low in vegetation.
- Females can drop up to 6,000 eggs in the water.

Importance to Humans
- Nymphs and some adults are important fish food.
- The emergence of the salmonfly in the mountainous west attracts anglers from all over and is important to the economy.
- Used as indicators of good water quality.

Collecting and Preserving
Immatures & Adults
- Hand collect adults or use aspirator.
- Nymphs can be collected with a dip net.
- Keep both nymphs and adults in 70% alcohol.

Resources
- *American Stoneflies: A Photographic Guide to the Plecoptera* by Stark et al., 1998
- *Nymphs of North American Stonefly Genera (Plecoptera)* by Stewart and Stark, 2002

Isoperla sp.
(Perlodidae)

Wing Venation

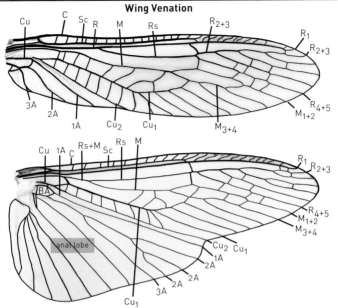

FW & HW of **Perlidae**. HW anal area with 5 or more veins. BA is basal anal cell.

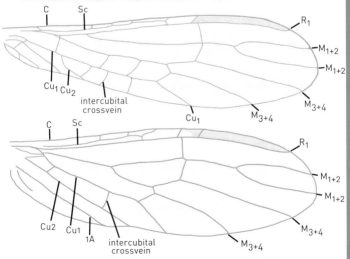

FW & HW of **Chloroperlidae**. FW anal area narrow with one or no crossveins. HW anal area reduced, with 1–4 veins.

KEY TO STONEFLY FAMILIES

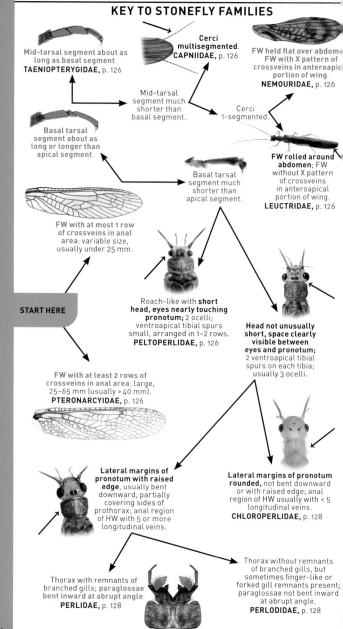

Mid-tarsal segment about as long as basal segment. **TAENIOPTERYGIDAE,** p. 126

Cerci **multisegmented. CAPNIIDAE,** p. 126

Mid-tarsal segment much shorter than basal segment.

FW held flat over abdomen; FW with X pattern of crossveins in anteroapical portion of wing. **NEMOURIDAE,** p. 126

Basal tarsal segment about as long or longer than apical segment.

Cerci 1-segmented.

FW rolled around abdomen; FW without X pattern of crossveins in anteroapical portion of wing. **LEUCTRIDAE,** p. 126

Basal tarsal segment much shorter than apical segment.

FW with at most 1 row of crossveins in anal area; variable size, usually under 25 mm.

START HERE

Roach-like with **short head, eyes nearly touching pronotum;** 2 ocelli; ventroapical tibial spurs small, arranged in 1–2 rows. **PELTOPERLIDAE,** p. 126

Head not unusually short, space clearly visible between eyes and pronotum; 2 ventroapical tibial spurs on each tibia; usually 3 ocelli.

FW with at least 2 rows of crossveins in anal area; large, 25–65 mm (usually > 40 mm). **PTERONARCYIDAE,** p. 126

Lateral margins of pronotum with raised edge, usually bent downward, partially covering sides of prothorax; anal region of HW with 5 or more longitudinal veins.

Lateral margins of pronotum rounded, not bent downward or with raised edge; anal region of HW usually with < 5 longitudinal veins. **CHLOROPERLIDAE,** p. 128

Thorax with remnants of branched gills; paraglossae bent inward at abrupt angle. **PERLIDAE,** p. 128

Thorax without remnants of branched gills, but sometimes finger-like or forked gill remnants present; paraglossae not bent inward at abrupt angle. **PERLODIDAE,** p. 128

Gravid female
Perlesta sp.
in flight
(Perlidae)

EUHOLOGNATHA • SYSTELLOGNATHA

Suborder Euholognatha
Basal tarsal segment similar in size or longer than apical segment.

apical — mid — basal

SMALL WINTER STONEFLIES Family Capniidae
Black bodied; emerge during winter months. Found in northern NA.
Wings are short or rudimentary in some species. Mid-tarsal segment
small and wedge shaped. Cerci multisegmented. Most diverse
stonefly family in NA. TL 3–16 mm; World 300, NA 160.

ROLLED-WINGED STONEFLIES Family Leuctridae
Black or brown bodies; most commonly encountered along small
streams in hilly or mountainous regions. **Wings rolled around body
at rest. FW with no X pattern.** Mid-tarsal segment small and wedge
shaped. Cerci 1-segmented. TL 5–15 mm; World 300, NA 130.

SPRING STONEFLIES Family Nemouridae
Brown to black bodies; emerge April to June. Wings held flat at rest;
anteroapical FW veins form an X pattern. Midtarsal segment small
and wedge shaped. Cerci 1-segmented. Nymphs found in small
streams with rocky bottoms. TL 5–15 mm; World 666, NA 73.

WINTER STONEFLIES Family Taeniopterygidae
Dark brown to black, emerge January to June. Antennae long. Basal
tarsal segments similar in size to midtarsal segment. Nymphs occur
in large rivers and streams. Some adults feed on flowers.
TL 7–15 mm; World 106, NA 35.

Suborder Systellognatha
Basal tarsal segment much shorter than apical segment.

apical — mid — basal

GIANT STONEFLIES Family Pteronarcyidae
Largest members of the order are in this family and are generally
gray, brown, or black, emerge April to July. FW with numerous
crossveins; anal area with 2 or more rows of crossveins. Labium
with 4 subequal short lobes. Nymphs are plant feeders and found in
medium to large rivers. Adults are nocturnal and come to lights.
TL 25–65 mm; World 12, NA 10.

ROACH-LIKE STONEFLIES Family Peltoperlidae
Pale yellow to dark brown, emerging April to July. Labial palps not
visible from above. FW anal area with 1 or no crossveins. Two ocelli.
Basal tarsal segment much shorter than apical segment. Found in
mountain streams in western and northern NA. Nymphs cockroach-
like in appearance. TL 10–20 mm; World 70, NA 23.

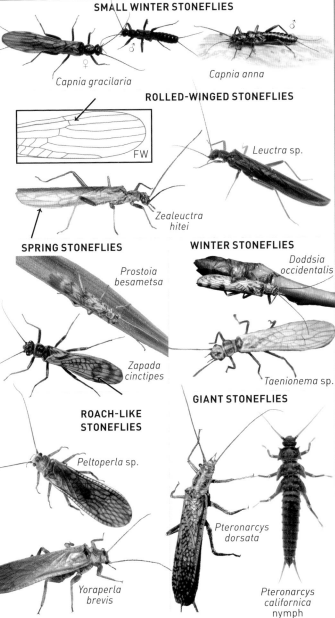

SMALL WINTER STONEFLIES

Capnia gracilaria ♀

♂

Capnia anna ♂

ROLLED-WINGED STONEFLIES

FW

Leuctra sp.

Zealeuctra hitei

SPRING STONEFLIES

Prostoia besametsa

Zapada cinctipes

WINTER STONEFLIES

Doddsia occidentalis

Taenionema sp.

ROACH-LIKE STONEFLIES

Peltoperla sp.

Yoraperla brevis

GIANT STONEFLIES

Pteronarcys dorsata

Pteronarcys californica nymph

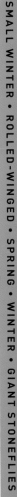

127

SYSTELLOGNATHA

Suborder Systellognatha cont.

GREEN STONEFLIES Family Chloroperlidae
Yellow or green; emerge during spring from creeks. FW anal area narrow with 1 or no crossveins. HW anal area reduced with 1–4 veins; sometimes lacking anal lobe. Pronotum without lateral ridges. TL 5–26 mm; World 400, NA 8.

COMMON STONEFLIES Family Perlidae
Most commonly encountered stoneflies. Emerge during spring and summer. HW anal area with 5 or more veins. Sternacostal sutures on metathorax not reaching furcal pits, vestigial thoracic gills with many branches. Nymphs live on the bottom substrate of rivers and are predaceous. TL 7–46 mm; World 1,000, NA 84.

PREDATORY STONEFLIES Family Perlodidae
Brown or yellowish; emerge during spring and summer. HW anal area with 5 or more veins. Sternacostal sutures on metathorax going through furcal pits. Some adults feed on pollen. Nymphs can be predaceous and are usually found in slow-moving rivers. TL 5–41 mm; World 300, NA 130.

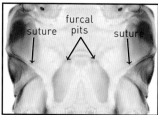

Ventral view of Common Stonefly metathorax.

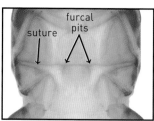

Ventral view of Predatory Stonefly metathorax.

Leuctra sp.

GREEN STONEFLIES

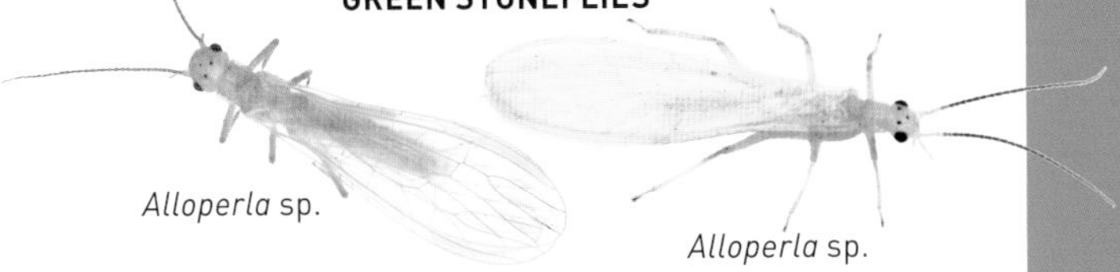

Alloperla sp.

Alloperla sp.

COMMON STONEFLIES

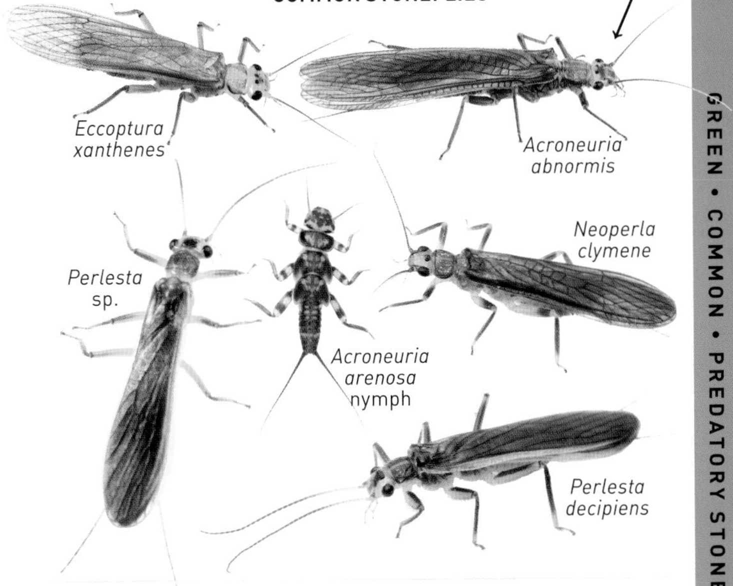

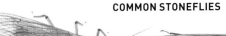

Eccoptura xanthenes

Acroneuria abnormis

Perlesta sp.

Acroneuria arenosa nymph

Neoperla clymene

Perlesta decipiens

PREDATORY STONEFLIES

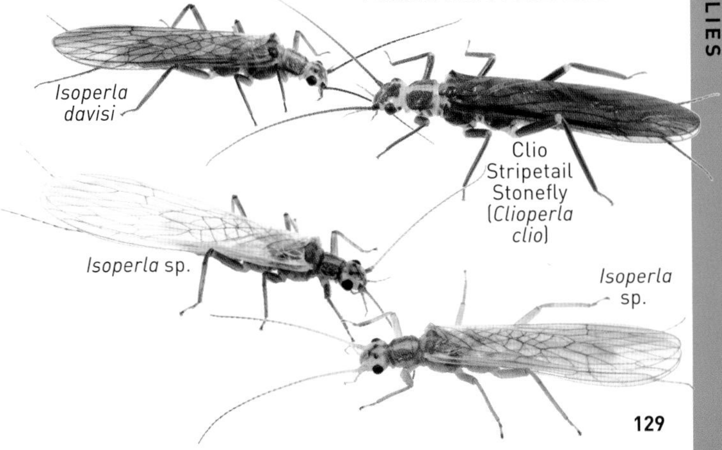

Isoperla davisi

Clio Stripetail Stonefly (*Clioperla clio*)

Isoperla sp.

Isoperla sp.

GREEN • COMMON • PREDATORY STONEFLIES

129

Cattail Toothpick Grasshopper
(*Leptysma marginicollis*)
Acrididae

Bush Katydid
(*Scudderia* sp.) nymph

GRASSHOPPERS, CRICKETS, & KATYDIDS
ORDER ORTHOPTERA

Derivation of Name
- Orthoptera: *ortho*, straight; *ptera*, wing. Referring to the straight FW.

Identification
- 5–155 mm TL.
- Generally, large, robust, brown or green with long leathery FW (tegmina).
- Antennae filamentous with numerous segments.
- Chewing mouthparts.
- Most with 2 pairs of wings: leathery FW with veins; broad, membranous HW with veins and folded fan-wise at rest under FW.
- One or both wings can be absent.
- Cerci generally present; ovipositor long and slender or short.
- Tarsi generally 3- to 5-segmented.

Classification
- Two suborders and sixteen families; suborders, in general, separated by antennae length and ovipositor shape.
- World: 20,000 species; North America: 1,200 species.

Range
- Found worldwide except Antarctica. Most in tropical grasslands.

Similar orders
- Coleoptera (p. 294): FW thickened, lacking veins; lack cerci; antennae usually less than 11-segmented; HW longer than FW.
- Hemiptera (p. 174): mouthparts piercing-sucking; FW usually with base thickened and tip membranous; antennae with 5 or fewer segments.
- Dermaptera (p. 114): FW thickened but short; cerci forceps-like.

Camel Cricket
(*Gammarotettix* sp.) feeding
on a spermatophylax
Rhaphidophoridae

Food
- Most adults and nymphs in the suborder Caelifera eat a variety of plant material, including roots.
- Members of the suborder Ensifera tend to be more omnivorous, eating plant material as well as other insects.

Behavior
- Sing by stridulation; rubbing 1 body part against another.
- Long-horned grasshoppers and crickets rub a sharp edge (scraper) on the top side of 1 forewing over a file-like ridge (file) on the underside of the other front wing.
- Short-horned grasshoppers rub hind legs against the tegmina to producing a clacking sound.
- Males generally do the singing; females of a few species produce soft noises. Songs are species specific to attract the opposite sex.

Life Cycle
- Hemimetabolous.
- All stages are terrestrial, though some species are associated with riparian habitats.
- Eggs laid on or in the ground or in vegetation.
- Nymphs similar in appearance to adults, but wings short or absent.
- Most are sexually mature when wings are fully developed.

Importance to Humans
- Many orthopterans consume vegetation and can be destructive to crops and other plants.
- Some species (locusts) dramatically increase their numbers when resources are scarce, migrating long distances and completely destroying large areas of crops along the way.
- A few species are predaceous, and a few are omnivorous.
- Commonly used as a food source for both humans and animals.
- Many species are kept as pets.

Collecting and Preserving
Adults
- Find by sweeping grass or using a beat sheet. Some come to lights.
- Should be pinned. If there is color in the hindwing, typically spread the left wing. Pull in hind legs close to the body.
- Large-bodied individuals should have the abdomen eviscerated and stuffed with cotton to prevent rot; an alternative method is to inject abdomen with alcohol.

Resources
- *Field Guide to Grasshoppers, Katydids, and Crickets of the United States* by Capinera et al., 2004
- *How to Know the Grasshoppers, Crickets, Cockroaches and Their Allies* by Helfer, 1987
- *The North American Grasshoppers—Volume I,* by Otte, 1981
- *The North American Grasshoppers—Volume II,* by Otte, 1984
- *The Songs of Insects* by Hershberger and Elliott, 2007
- Orthoptera Species File Online—orthoptera.speciesfile.org
- Singing Insects of North America—orthsoc.org/sina

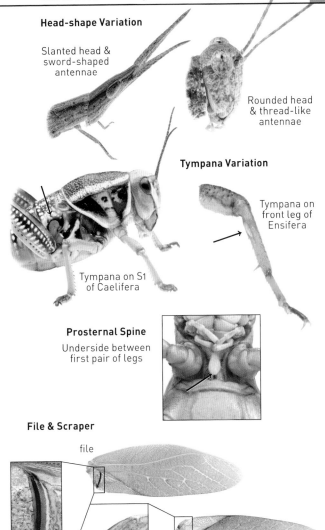

Head-shape Variation

Slanted head & sword-shaped antennae

Rounded head & thread-like antennae

Tympana Variation

Tympana on front leg of Ensifera

Tympana on S1 of Caelifera

Prosternal Spine

Underside between first pair of legs

File & Scraper

file

scraper

Stridulating file & scraper found in many Ensifera; top is upper side of FW, bottom is underside of FW

KEY TO ORTHOPTERA FAMILIES

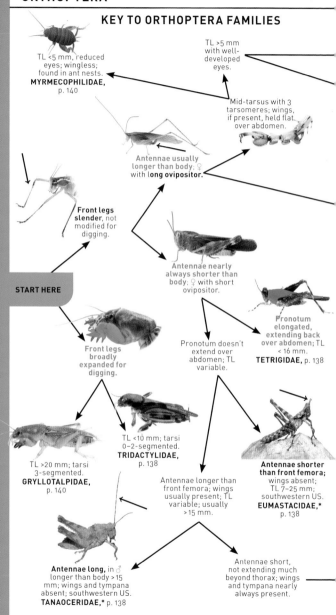

TL <5 mm, reduced eyes; wingless; found in ant nests.
MYRMECOPHILIDAE, p. 140

TL >5 mm with well-developed eyes.

Mid-tarsus with 3 tarsomeres; wings, if present, held flat over abdomen.

Antennae usually longer than body; ♀ with long ovipositor.

Front legs slender, not modified for digging.

Antennae nearly always shorter than body; ♀ with short ovipositor.

START HERE

Pronotum elongated, extending back over abdomen; TL < 16 mm.
TETRIGIDAE, p. 138

Pronotum doesn't extend over abdomen; TL variable.

Front legs broadly expanded for digging.

TL <10 mm; tarsi 0–2-segmented.
TRIDACTYLIDAE, p. 138

TL >20 mm; tarsi 3-segmented.
GRYLLOTALPIDAE, p. 140

Antennae longer than front femora; wings usually present; TL variable; usually >15 mm.

Antennae shorter than front femora; wings absent; TL 7–25 mm; southwestern US.
EUMASTACIDAE,* p. 138

Antennae long, in ♂ longer than body >15 mm; wings and tympana absent; southwestern US.
TANAOCERIDAE,* p. 138

Antennae short, not extending much beyond thorax; wings and tympana nearly always present.

GRASSHOPPERS, CRICKETS, & KATYDIDS

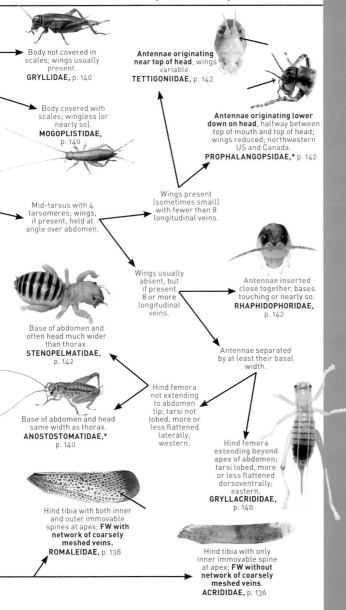

Body not covered in scales; wings usually present.
GRYLLIDAE, p. 140

Antennae originating near top of head; wings variable.
TETTIGONIIDAE, p. 142

Body covered with scales; wingless (or nearly so).
MOGOPLISTIDAE, p. 140

Antennae originating lower down on head, halfway between top of mouth and top of head; wings reduced; northwestern US and Canada.
PROPHALANGOPSIDAE,* p. 142

Mid-tarsus with 4 tarsomeres; wings, if present, held at angle over abdomen.

Wings present (sometimes small) with fewer than 8 longitudinal veins.

Wings usually absent, but if present 8 or more longitudinal veins.

Antennae inserted close together, bases touching or nearly so.
RHAPHIDOPHORIDAE, p. 142

Base of abdomen and often head much wider than thorax.
STENOPELMATIDAE, p. 142

Antennae separated by at least their basal width.

Base of abdomen and head same width as thorax.
ANOSTOSTOMATIDAE,* p. 140

Hind femora not extending to abdomen tip; tarsi not lobed, more or less flattened laterally; western.

Hind femora extending beyond apex of abdomen; tarsi lobed, more or less flattened dorsoventrally; eastern.
GRYLLACRIDIDAE, p. 140

Hind tibia with both inner and outer immovable spines at apex; **FW with network of coarsely meshed veins.**
ROMALEIDAE, p. 138

Hind tibia with only inner immovable spine at apex; **FW without network of coarsely meshed veins.**
ACRIDIDAE, p. 136

135

Suborder Caelifera

Grasshopper-like jumping insects with hind femora more or less enlarged. Tarsi 3 or fewer segments. Antennae relatively short. Tympana usually present on sides of S1. Ovipositor short. Feed almost exclusively on plants.

SHORT-HORNED GRASSHOPPERS Family Acrididae

Relatively short antennae. Head may be slanted, flat, or rounded. Pronotum not prolonged back over abdomen. Cyrtacanthacridinae, have a spine between front legs. Wings usually well developed, and Oedipodinae often have HW variously banded and colored. Tarsi 3-segmented. This group contains our most common grasshoppers. Many are important pests of cultivated plants. Most oviposit in the ground and overwinter in the egg stage. Six subfamilies generally recognized. TL 9–80 mm; World 8,000, NA 611.

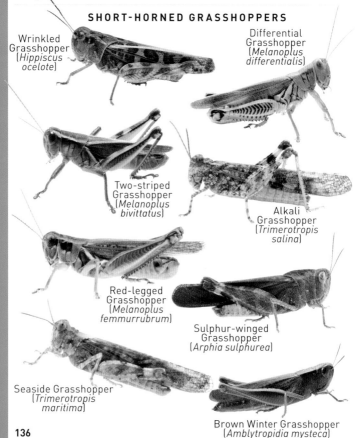

SHORT-HORNED GRASSHOPPERS

Wrinkled Grasshopper (*Hippiscus ocelote*)

Differential Grasshopper (*Melanoplus differentialis*)

Two-striped Grasshopper (*Melanoplus bivittatus*)

Alkali Grasshopper (*Trimerotropis salina*)

Red-legged Grasshopper (*Melanoplus femmurrubrum*)

Sulphur-winged Grasshopper (*Arphia sulphurea*)

Seaside Grasshopper (*Trimerotropis maritima*)

Brown Winter Grasshopper (*Amblytropidia mysteca*)

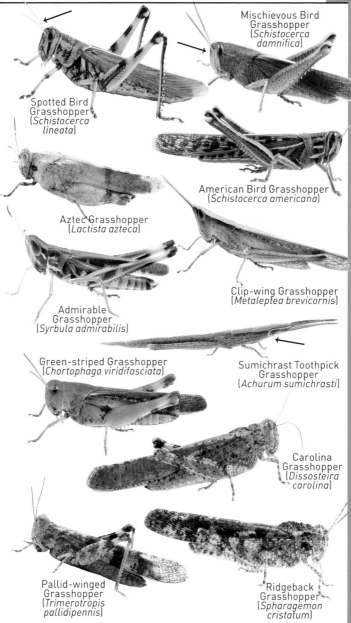

Mischievous Bird Grasshopper
(*Schistocerca damnifica*)

Spotted Bird Grasshopper
(*Schistocerca lineata*)

American Bird Grasshopper
(*Schistocerca americana*)

Aztec Grasshopper
(*Lactista azteca*)

Clip-wing Grasshopper
(*Metaleptea brevicornis*)

Admirable Grasshopper
(*Syrbula admirabilis*)

Green-striped Grasshopper
(*Chortophaga viridifasciata*)

Sumichrast Toothpick Grasshopper
(*Achurum sumichrasti*)

Carolina Grasshopper
(*Dissosteira carolina*)

Pallid-winged Grasshopper
(*Trimerotropis pallidipennis*)

Ridgeback Grasshopper
(*Spharagemon cristatum*)

SHORT-HORNED GRASSHOPPERS

137

LUBBER GRASSHOPPERS Family Romaleidae

Large, robust grasshoppers found mostly in the southwestern US. **FW often with prominent veins appearing as course mesh-like appearance.** Some have reduced HW and while often brightly colored, can't fly. **Two rows of immovable spines on hind tibia.** Some have roughened body surfaces and may be cryptically colored. Hind femur relatively slender compared to other Caelifera. Often with pronounced prosternal spine. One species, *Romalea microptera,* occurs in the east and is commonly used for student dissections. TL 18–145 mm; World 470, NA 9.

MONKEY GRASSHOPPERS Family Eumastacidae*

Similar to Acrididae but wingless. Medium-sized to small, usually brownish. **Antennae shorter than front femur. Often hold hind legs perpendicular to body.** Tympana generally absent. Common name refers to their ability to maneuver through trees and shrubs in chaparral. This mostly tropical group is restricted to the southwest in the US. TL 7–25 mm; World 247, NA 13.

PYGMY GRASSHOPPERS Family Tetrigidae

Pronotum elongated back over abdomen and points posteriorly. Hind tarsi 3-segmented, other tarsi 2-segmented. Tegmina small, pad-like, or absent. Overwinter as adults and are most often encountered in spring and early summer along shores and riparian zones. TL 6–16 mm; World 1,600, NA 30.

PYGMY MOLE CRICKETS Family Tridactylidae

Resembling crickets, but with **front tibiae enlarged and modified for digging.** Front and middle tarsi 2-segmented, hind tarsi 1-segmented or lacking. Antennae 11-segmented. Occur chiefly in moist sandy situations along shores of ponds and streams, where they burrow underground. TL < 16 mm; World 147, NA 7.

DESERT LONG-HORNED GRASSHOPPERS

Family Tanaoceridae* **Adults wingless** and active in the desert of the southwest, from NV to CA. Face less slanted than Monkey Grasshopper's and **top of head is rounded.** Antennae longer than body in males and shorter than body in females. Tympana and wings absent. File on S3. TL 7–26 mm; World 3, NA 2.

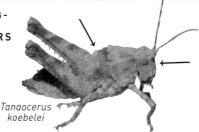

DESERT LONG-HORNED GRASSHOPPERS

Tanaocerus koebelei

GRASSHOPPERS, CRICKETS, & KATYDIDS

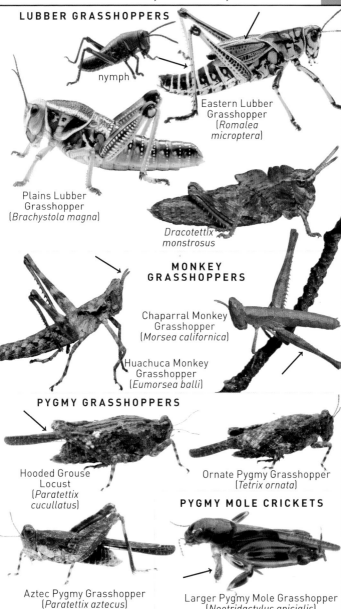

LUBBER GRASSHOPPERS

nymph

Eastern Lubber
Grasshopper
(*Romalea
microptera*)

Plains Lubber
Grasshopper
(*Brachystola magna*)

*Dracotettix
monstrosus*

MONKEY
GRASSHOPPERS

Chaparral Monkey
Grasshopper
(*Morsea californica*)

Huachuca Monkey
Grasshopper
(*Eumorsea balli*)

PYGMY GRASSHOPPERS

Hooded Grouse
Locust
(*Paratettix
cucullatus*)

Ornate Pygmy Grasshopper
(*Tetrix ornata*)

PYGMY MOLE CRICKETS

Aztec Pygmy Grasshopper
(*Paratettix aztecus*)

Larger Pygmy Mole Grasshopper
(*Neotridactylus apicialis*)

Suborder Ensifera
Includes katydids and crickets. With enlarged femurs. Antennae thin and usually longer than body. Tarsi 3- or 4-segmented. Tympana located on the upper ends of the front tibia. Sing by rubbing the file-like structures on FW together. Includes both predators, herbivores and omnivores.

MOLE CRICKETS Family Gryllotalpidae
Brownish body. **Forelegs spade shaped** for digging underground. **Hind legs not modified for jumping.** All tarsi 3-segmented. Body covered in hairs. Females lack ovipositor. Found digging in turf and sandy riparian areas, occasionally at lights. Can damage crops and turf in southern states. TL 20–30 mm; World 92, NA 7.

TRUE CRICKETS Family Gryllidae
Resemble katydids, but have 3 tarsi instead of 4. **Hind tibia with long spines. Ovipositor cylindrical or needle-like** instead of flattened. Right FW lays over the left. Usually overwinter as eggs. Eight subfamilies recognized. TL 5–36 mm; World 900, NA 112.

SCALY CRICKETS Family Mogoplistidae
Small and flattened. Found in sandy localities near water, often under debris. Body covered with scales that are easily knocked off. **Hind tibia lacks spines, but does have apical spurs.** Can be wingless or have very short wings. May resemble silverfish. TL 5–13 mm; World 364, NA 20.

ANT CRICKETS Family Myrmecophilidae*
Small, oval shaped, and wingless. Found in ant nests. Lack tympana. **Hind femur width twice the length.** Primarily feed on secretions from the ants. Sometimes considered a subfamily of Gryllidae. TL 2–5 mm; World 100, NA 4.

WETAS & KING CRICKETS Family Anostostomatidae*
Flightless and ground dwelling, resembling Camel Crickets, but antennae are more separated at base and body is less hump shaped. **Base of abdomen and head same width as thorax.** Can spin silk from maxillary glands. Found in southern CA and the Channel Islands off the coast of CA. TL 15–20 mm; World 251, NA 5.

RASPY CRICKETS Family Gryllacrididae
Antennae 5 times length of body. Tarsi lobed and flattened. Ovipositor reaches above abdomen. Nocturnal, feeding chiefly on aphids. Hide during the day in rolled-up leaves. Single Nearctic species found from CT to FL and west to south TX. TL 12–16 mm; World 600, NA 1.

GRASSHOPPERS, CRICKETS, & KATYDIDS

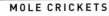

MOLE CRICKETS

Southern Mole Cricket
(*Neoscapteriscus borellii*)

TRUE CRICKETS

♀

Field Cricket
(*Gryllus* sp.)

Brown Trig
(*Anaxipha* sp.)

Two-spotted Tree Cricket
(*Neoxabea bipunctata*)

Western Tree Cricket
(*Oecanthus californicus*)

Red-headed Bush Cricket
(*Phyllopalpus pulchellus*)

SCALY CRICKETS

Forest Scaly Cricket
(*Cycloptilum trigonipalpum*)

ANT CRICKETS

Eastern Ant Cricket
(*Myrmecophilus pergandei*)

WETAS & KING CRICKETS

Silk-spinning
Cricket
(*Cnemotettix
bifasciatus*)

RASPY CRICKETS

Carolina Leaf-roller
(*Camptonotus carolinensis*)

CRICKETS

141

Suborder Ensifera cont.

HUMP-WINGED CRICKETS Family Prophalangopsidae*
Robust with humpbacked appearance. **Brown with black and pale yellowish markings. Antennal sockets halfway between head and suture between frons and clypeus.** Old group known mostly from fossils. Single genus in North America restricted to the northwest. TL 17–30 mm; World 8, NA 3.

CAMEL CRICKETS Family Rhaphidophoridae
Light tan to dark brown. Humpbacked appearance and wingless. **Very long antennae almost touch at their base.** Long legs. Tarsi laterally compressed without pads. Typically found in cool damp places like caves and tree holes. TL 10–36 mm; World 1,100, NA 150.

JERUSALEM CRICKETS Family Stenopelmatidae
Large with head and abdomen enlarged. Usually brown, with **black bands on abdomen**. Strong mandibles capable of biting. Found under stones and most common in the west. Females lay eggs several inches underground. TL 20–75 mm; World 200, NA 50.

KATYDIDS Family Tettigoniidae
Long, thin antennae usually longer than body. Antennae sockets located near top of head. Four tarsal segments. Overwinter as eggs. Most feed on plants, but some have powerful jaws and are predatory. Tympana on front tibiae.
TL 10–60 mm; World 6,400, NA 243.

ENSIFERA

KATYDIDS

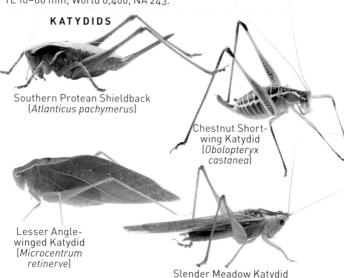

Southern Protean Shieldback
(*Atlanticus pachymerus*)

Chestnut Short-
wing Katydid
(*Obolopteryx
castanea*)

Lesser Angle-
winged Katydid
(*Microcentrum
retinerve*)

Slender Meadow Katydid
(*Conocephalus fasciatus*)

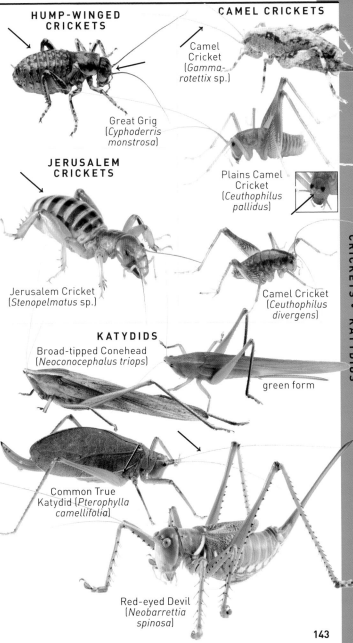

HUMP-WINGED CRICKETS

CAMEL CRICKETS

Camel Cricket (*Gammarotettix* sp.)

Great Grig (*Cyphoderris monstrosa*)

JERUSALEM CRICKETS

Plains Camel Cricket (*Ceuthophilus pallidus*)

Jerusalem Cricket (*Stenopelmatus* sp.)

Camel Cricket (*Ceuthophilus divergens*)

KATYDIDS

Broad-tipped Conehead (*Neoconocephalus triops*)

green form

Common True Katydid (*Pterophylla camellifolia*)

Red-eyed Devil (*Neobarrettia spinosa*)

CRICKETS • KATYDIDS

143

Grylloblatta campodeiformis
Grylloblattidae

Grylloblatta sp.

ROCK CRAWLERS & ICE CRAWLERS
ORDER NOTOPTERA

Derivation of Name
- Notoptera: *notos*, back; *ptera*, wing. Unfortunate name, as both groups within this order are completely wingless!

Identification
- 15–30 mm TL.
- Slender, long wingless body ranging from pale yellow to orange to brown among the species and life stages; all with thin short layer of hair over body.
- Antennae filiform, half as long as body.
- Small eyes and lacking ocelli.
- Tarsi 5-segmented.
- Cerci 8- or 9-segmented.
- Females have a long ovipositor between 2 cerci.

Classification
- Contains 2 groups, Grylloblattodea and Mantophasmatodea, originally each considered their own order. Only the Grylloblattodea occur in NA.
- First discovered in 1913 in Banff, Alberta, Canada.
- Xenonomia sometimes used instead of Notoptera.
- Two extant families: Grylloblattidae (Ice Crawlers), worldwide; 5 genera, only *Grylloblatta* occur in NA; Mantophasmatidae (Rock Crawlers or Gladiators) found in southern Africa.
- World: 60 species; North America: 15 species.

Range
- Northwestern US and western Canada, Japan, China, North Korea, South Korea, and Russia (Grylloblattodea) and South Africa and Namibia (Mantophasmatodea).

Similar orders
- Embioptera (p. 148): tarsi 3-segmented, basal segment enlarged; cerci 1–2 segments; some have wings; small, 5–16 mm.
- Dermaptera (p. 114): FW thick and short; cerci forcep-like.
- Plecoptera (p. 120): flattened body; most with 4 membranous wings; found near streams.
- Phasmatodea (p. 152): *Timema* have 3-segmented tarsi; forcep-like cerci; antennae more than half body length.
- Blattodea (p. 166): antennae more than half body length; wings usually present; body flattened; head concealed by pronotum.

Food
- Primarily scavengers of dead insects, moss, and other vegetation; sometimes cannibalistic and perhaps predaceous.
- Normally feed all year long.

Behavior
- Live on edges of glaciers under rocks, edges of snow fields and talus slopes, slopes, rotting wood, and in moss in rocky soils. Occur from sea level to more than 11,500 ft. elevation. Most abundant in alpine habitats.
- Found in areas where temperatures are at or below freezing. In summer they go underground to find colder temperatures.
- Distribution can be attributed to the last Pleistocene glaciation.
- Nocturnal, hiding during the day, avoiding daylight.

Life Cycle
- Hemimetabolous.
- Nymphs similar to adults, just smaller and paler.
- Reach adulthood after 6 or more years.
- Mate and lay eggs in summer and fall.
- Likely live up to 10 years, but little is known about their life cycle.

Importance to Humans
- They are decomposers that eat insects found under snow and ice.
- Ability to remain active just before they freeze solid makes them important research subjects.

Collecting and Preserving
Immatures & Adults
- Hand collect on snow, under rocks/debris or in pitfall traps.
- Preserve in 70% ethanol.

Resources
- *Biology of the Notoptera* edited by Ando, 1982
- "Ice Crawlers (Grylloblattodea)—the history of the investigation of a highly unusual group of insects" by Wipfler et al. 2014

Grylloblatta campodeiformis

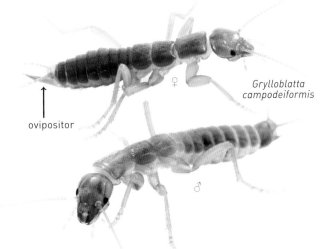

Grylloblatta campodeiformis

♀

↑ ovipositor

♂

ICE CRAWLERS Family Grylloblattidae*

Wingless, slender with elongated abdomen. Eyes are small or absent; lack ocelli. Antennae are long, filiform, and about half the length of body. Body is pale with fine hairs. Found under rocks and logs near the edge of glaciers at high elevations in the mountains of CA, MT, OR, WA, ID, and western Canada. Single genus, *Grylloblatta*. TL 15–30 mm; World 33, NA 15.

Male (left) and female (right) *Grylloblatta campodeiformis* in moss and ice. Female ovipositor is visible between cerci.

Webspinner in crevice of tree.

Oligotomid
Webspinner

WEBSPINNERS
ORDER EMBIOPTERA

Derivation of Name
- Embioptera: *embi*, lively; *ptera*, meaning wing. Presumably referring to their small size and quick movements rather than their flight capabilities.

Identification
- 4–10 mm TL.
- Elongated and slender; males usually with more robust head and often winged, while females and juveniles are wingless.
- Filiform antennae, 16–31 segments; ocelli lacking; chewing mouthparts;
- Tarsi 3-segmented. Front basal tarsus enlarged, with a large field of silk glands, each with a tubular silk ejector; hind femur is also enlarged to move backwards.
- Live under silk galleries on trees, bark, leaf litter, stones, and soil.

Classification
- Recent evidence suggests they are closely related to Phasmatodea.
- Some authorities use the ordinal name Embiidina or Embiodea.
- World: 380 species; North America: 11 species.

Range
- Found in southern NA and tropical areas throughout the world.

Similar orders
- Blattodea (p. 166): basal tarsus not enlarged; 4-segmented tarsi; galleries not silk lined.
- Zoraptera (p. 110): basal tarsus not enlarged; 2-segmented tarsi; 9-segmented antennae.
- Psocodea (p. 220): basal tarsus not enlarged; 2- or 3-segmented tarsi; lack cerci.

Webspinner webbing on tree.

Food
- Adults and nymphs eat various dead leaf material, bark, mosses and lichens.

Behavior
- Spend most of their life under the silk galleries.
- Can live communally, but females care only for their own young.
- Silk is spun from enlarged basal tarsus of foreleg.
- Avoid predators by moving quickly backwards in silk galleries using enlarged hind femur.
- Adult males often with wings; has the ability to fold them forward over its head so it can move backward through the webbing.
- Males fly during the day to find females.
- Can be reared in jars with leaf material and bark from where they are collected. Useful to get adult males for identification.

Life Cycle
- Hemimetabolous.
- Females lay eggs under silk in a single layer and cover the eggs with leaf material and fecal pellets.
- Females guard eggs until they hatch.
- Nymphs look similar to adults and live under silk galleries.
- Adults live about 6 months.
- Usually, only 1 generation per year.

Importance to Humans
- Sometimes webs frighten homeowners, but not generally thought of as pests.
- Ability to move forward and backward with same efficiency has been studied to improve robotic movement.

Collecting and Preserving
Immatures & Adults
- Hand collect from various habitats, looking for silk galleries.
- Some males will come to lights.
- Preserve in 70% alcohol.

Resources
- *How to Know the Grasshoppers, Crickets, Cockroaches and their Allies* by Helfer, 1987
- Embioptera Species File Online—embioptera.speciesfile.org

Black Webspinner
(*Oligotoma nigra*)

Anisembia texana

ANISEMBIID WEBSPINNER Family Anisembiidae*

Mandibles in adult males lack teeth at tip. Adult males with or without wings; females and immatures wingless. MA not forked in FW. **Left cercus 1-segmented.** Adults and immatures with 1 bladderlike papilla on underside of basal hind tarsus. TL ~10 mm; World 110, NA 3.

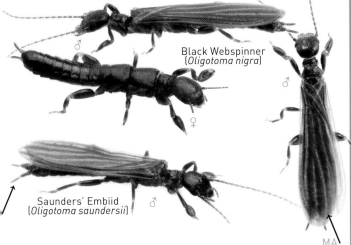

Black Webspinner
(*Oligotoma nigra*)

♂

♀

Saunders' Embiid   ♂
(*Oligotoma saundersii*)

MA

OLIGOTOMID WEBSPINNER Family Oligotomidae

Mandibles in adult males with teeth at tip. Adult males usually with wings, females and immatures wingless. **Left cercus 2 segments.** Adults and immatures of *Haploembia* with 2 bladder-like papillae on underside of basal hind tarsus; others with 1. **MA not forked in FW.** TL 5-10 mm; World 45, NA 3.

TERATEMBIID WEBSPINNER Family Teratembiidae* Not illus.

Mandibles in adult males with teeth apically. Left cercus 2 segments; MA forked in FW. A single species, *Diradius vandykei*, found in the southeast. TL 2.5-5 mm; World 50, NA 1.

Timema californicum
Timematidae

Prairie Walkingstick nymph
(*Diapheromera velii*)
Diapheromeridae

WALKINGSTICKS
ORDER PHASMATODEA

Derivation of Name
- Phasmatodea: *phasm*, phantom. Referring to the cryptic nature of walkingsticks, allowing them to disappear in their habitat.

Identification
- 7–205 mm TL.
- Often large, elongated, slender, and cylindrical; green or brown body.
- Usually long antennae; wingless except 1 species; tarsi usually 5-segmented; cerci 1-segmented; found on vegetation.

Classification
- Some authorities use the term Phasmida for the order.
- Historically included within Orthoptera as a suborder.
- Currently considered its own order; most closely related to Embioptera and Notoptera.
- World: 3,630 species; North America: 29 species.

Range
- Found worldwide in temperate and tropical habitats. Most species found in Indo-Malayan region.

Similar orders
- Mantodea (p. 158): prothorax and coxae greatly lengthened; front legs modified for grabbing prey; tarsi 5-segmented.
- Hemiptera (p. 174): some Reduviidae, but mouthparts piercing-sucking; FW usually with base thickened and tip membranous; antennae less than 5-segmented.
- Dermaptera (p.130): FW thickened but short; cerci forceps-like.

Gray Walkingstick
(*Pseudosermyle straminea*)
Diapheromeridae

PHASMATODEA

Food
- Adults and nymphs eat various plant species. Some are generalists, while others are species specific.
- Oak (*Quercus*) species are a common food source.

Behavior
- Similar in appearance to twigs.
- Will sway back and forth when disturbed.
- Some species have glands that emit a toxic substance used as a defense against predators.
- Nymphs can regenerate lost legs through molting.
- Some species are parthenogenetic (females can reproduce without a mate).
- Eggs are generally randomly distributed on the ground.
- Some eggs resemble the capitulum of a seed and have nutrients that attract ants to take the egg underground to their colony, where they are protected from parasites and other predators.
- Young sometimes resemble ants.

Life Cycle
- Hemimetabolous.
- Overwinter as eggs dispersed on the ground.
- Nymphs similar in appearance to adult, but have fewer antennal segments.
- A single generation per year.

Importance to Humans
- Generally not considered pest species, though they may occasionally occur in numbers sufficient to cause damage and kill vegetation.

Collecting and Preserving
Adults
- Can be shaken from trees using a beat sheet and collected with a sweep net.
- Some species come to lights at night.
- Should be pinned with body supported while drying.

Resources
- *How to Know the Grasshoppers, Crickets, Cockroaches and Their Allies* by Helfer, 1987
- *Stick Insects of the Continental United States and Canada* by Arment, 2006
- Phasmida Species File Online, phasmida.speciesfile.org
- *Common Insects of Texas and the Surrounding States* by Abbott and Abbott, 2020 (*Diapheromera* species identification)

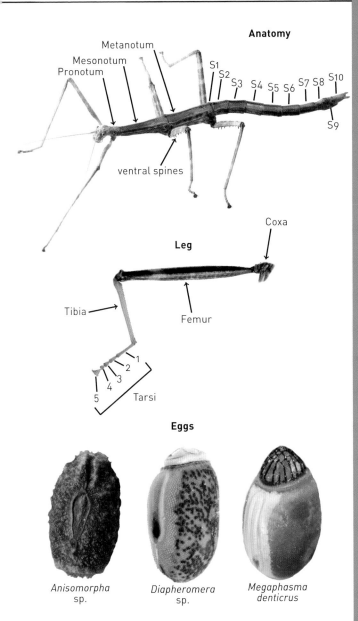

Anatomy

Pronotum
Mesonotum
Metanotum
S1 S2 S3 S4 S5 S6 S7 S8 S10
S9
ventral spines

Leg

Coxa
Tibia
Femur
1
2
3
4
5
Tarsi

Eggs

Anisomorpha sp.

Diapheromera sp.

Megaphasma denticrus

SHORT-HORN WALKINGSTICKS Family Heteronemiidae
Antennae shorter than front femur. Found from TX to southern CA.
Slender species ranging in color from red to brown. The 2 species
occurring in NA are *Parabacillus coloradus* and *P. hesperus*. TL
60–90 mm; World 71, NA 2.

WINGED WALKINGSTICKS Family Phasmatidae
One native species found in FL (*Haplopus mayeri*) that is large,
green to brown with small wings. **Head, pronotum, and mesonotum
with spines.** *Carausius morosus* has also been introduced in CA.
Size 80–130 mm; World 160, NA 2.

STRIPED WALKINGSTICKS Family Pseudophasmatidae
Stocky body, brownish-yellow to red, with or without **stripes along
body. Mesonotum less than 3 times as long as pronotum.** Two
similar species found mostly in the south are difficult to tell apart.
Smaller males often found riding on larger females.
TL 38–80 mm; World 61, NA 2.

TIMEMA WALKINGSTICKS Family Timematidae
Small, less than 1.25" long. Broad body with legs close together,
resembling earwigs. Green to pink or brown body color. **Tarsi
3-segmented.** Coxae under body not visible from above. Found in
the southwest. TL 10–25 mm; World 21, NA 10.

COMMON WALKINGSTICKS Family Diapheromeridae
Long thin body, either green or brown. **Antennae longer than
front femur.** Legs far apart and slender with coxae visible from
above. Tarsi 5-segmented. **Mesonotum 4 or more times as long as
pronotum.** Wings are absent. TL 38–112 mm; World 870, NA 19.

COMMON WALKINGSTICKS

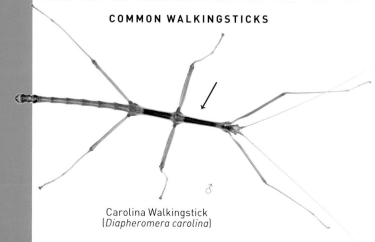

Carolina Walkingstick
(*Diapheromera carolina*)

SHORT-HORN WALKINGSTICK

Parabacillus sp.

WINGED WALKINGSTICKS

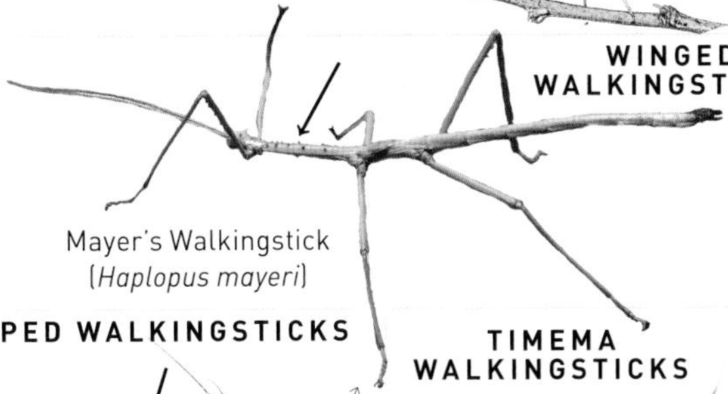

Mayer's Walkingstick
(*Haplopus mayeri*)

STRIPED WALKINGSTICKS

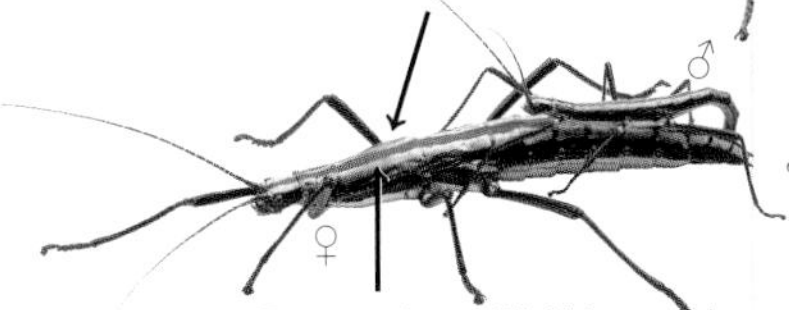

♂

♀

Southern Two-striped Walkingstick
(*Anisomorpha buprestoides*)

TIMEMA WALKINGSTICKS

Timema californicum

COMMON WALKINGSTICKS

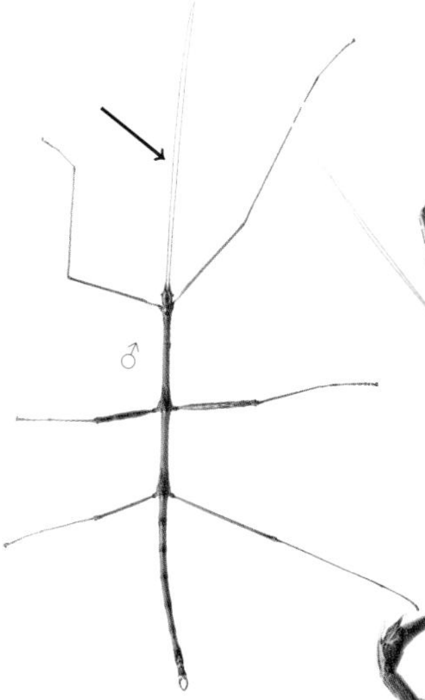

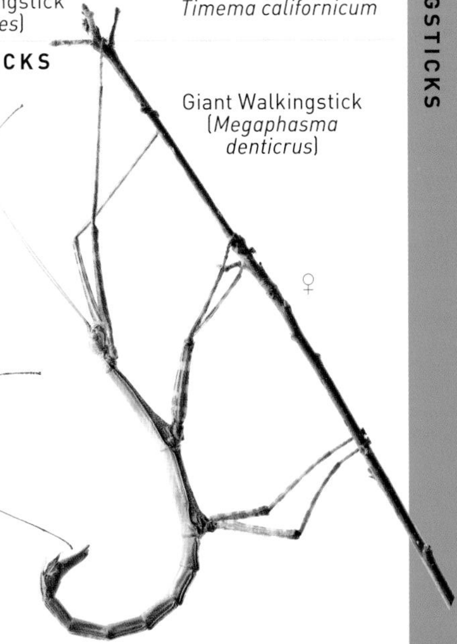

Giant Walkingstick
(*Megaphasma denticrus*)

♂

♀

Northern Walkingstick
(*Diapheromera femorata*)

157

Carolina Mantis
(*Stagmomantis carolina*)
Mantidae

Brunner's Mantis nymph
(*Brunneria borealis*)
Mantidae

MANTISES
ORDER MANTODEA

Derivation of Name
- Mantodea: *manti*, soothsayer or prophet; *ode*, like. Both the scientific and common names of praying mantid refers to their prayer-like posture.

Identification
- 10–170 mm TL.
- Large elongated insects with long pronotum and large raptorial front legs used for grabbing prey.
- Large bulging eyes; mouthparts chewing.
- Usually green to brown or gray; antennae long and slender; FW thick and leathery protecting membranous HW; legs long and thin; abdomen may be enlarged.

Classification
- Considered predatory roaches, generally placed in their own order.
- Historically placed within Orthoptera; sometimes placed in Dictyoptera along with roaches.
- World: 2,500 species; North America: 28 species.

Range
- Found worldwide except Antarctica; most diverse in the tropics.

Similar orders
- Neuroptera (p. 280): Mantispidae with membranous FW and front legs attached closer to head on prothorax.
- Phasmatodea (p. 152): do not have predatory, raptorial front legs.
- Hemiptera (p. 174): Reduviidae and Nepidae have piercing-sucking mouthparts.

Chinese Mantis
(*Tenodera sinensis*)
Mantidae

MANTODEA

Food
- Predators as adults and nymphs, eating all types of insects. Some larger species even eat small lizards and birds.

Behavior
- Found in various habitats from tree trunks to grass. Often found on flowers attacking flying insects coming in for pollen and nectar.
- They tend to be camouflaged to blend in to plants or petals, where they wait for prey.
- Weak fliers, but attracted to lights at night.
- Have excellent eyesight and can turn head to look over "shoulder."
- Single ear (tympanum) of mantids is found between the second and third pair of legs on the underside of the thorax. Mantids that fly at night can hear bats and evade them by diving to the ground.

Life Cycle
- Hemimetabolous.
- Eggs laid as ootheca (up to 200 eggs) on vegetation and structures.
- Overwinter as eggs and hatch in the spring.
- Nymphs look like small adults, but lack wings. Some resemble ants or tendrils on plants.

Importance to Humans
- They are predators of plant-eating insects and have been raised and released in areas to control pest insects.

Collecting and Preserving
Immatures
- Collect same way as adults.
- Preserve in 70% ethanol.

Adults
- Collect by hand, at lights, and with beat sheet.
- Pin through prothorax. Winged individuals can be spread.
- Use a support of card stock or pins for abdomen until it dries.

Resources
- *How to Know the Grasshoppers, Crickets, Cockroaches and Their Allies* by Helfer, 1987
- *Praying Mantises of the United States and Canada* by Anderson, 2018
- Mantodea Species File Online—mantodea.speciesfile.org

Male Carolina Mantis flying (*Stagmomantis carolina*) Mantidae

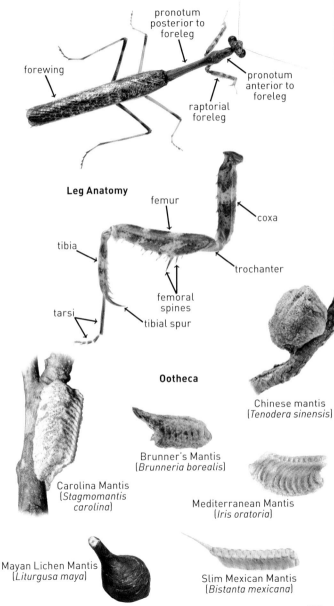

Mantis Anatomy

pronotum posterior to foreleg

forewing

pronotum anterior to foreleg

raptorial foreleg

Leg Anatomy

femur

coxa

tibia

trochanter

femoral spines

tarsi

tibial spur

Ootheca

Carolina Mantis
(*Stagmomantis carolina*)

Brunner's Mantis
(*Brunneria borealis*)

Chinese mantis
(*Tenodera sinensis*)

Mediterranean Mantis
(*Iris oratoria*)

Mayan Lichen Mantis
(*Liturgusa maya*)

Slim Mexican Mantis
(*Bistanta mexicana*)

LITURGUSID MANTISES Family Liturgusidae*

One introduced species in FL, *Liturgusa maya*. Antennae extend back to insertion of middle legs. **Pronotum tapers towards head.** Body slender and elongated. Usually found on tree trunks and can be common in urban areas, where it is often seen on wooden fence posts. Species is parthenogenetic in FL, but both sexes are known in its native range. TL 19–33 mm; World 35, NA 1.

EPAPHRODITID MANTISES Family Epaphroditidae*

Restricted to southeast and includes 1 species, *Gonatista grisea*. Mottled brown, green or gray, **pronotum longer than wide and not narrowed near head**. **Antennae not reaching backward to insertion point of middle legs.** Body broad and dilated. Front femur with deep groove between first and second spines, where tibial claw rests. Female has lobed abdomen on sides, short wings, and blends in with bark; found predominately on trees, sometimes fence posts. Males with wings that cover abdomen. TL 36–45 mm; World 10, NA 1.

COPTOPTERYGID MANTISES Family Coptopterygidae

One pecies, *Brunneria borealis*, in NA that occurs in the southeast. Has reduced wings that could be confused with walkingsticks. **Antennae thick at base.** Fine serrations on sides of thorax. Reproduces parthenogentically with no males known. TL 65–90 mm; World 24, NA 1.

SLIM MANTISES Family Thespidae

Small thin mantids, usually tan to gray. **Front tibia with few, 1–4, ventral spines** and at least 1 dorsal spine, and less than half as long as front femur. Top of head with prominent bulge between middle and each eye. Most diverse in the Neotropics. TL 25–56 mm; World 203, NA 3.

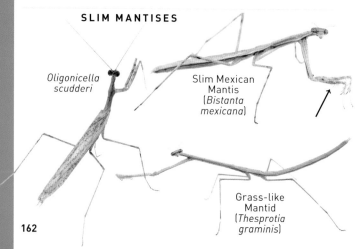

SLIM MANTISES

Oligonicella scudderi

Slim Mexican Mantis (*Bistanta mexicana*)

Grass-like Mantid (*Thesprotia graminis*)

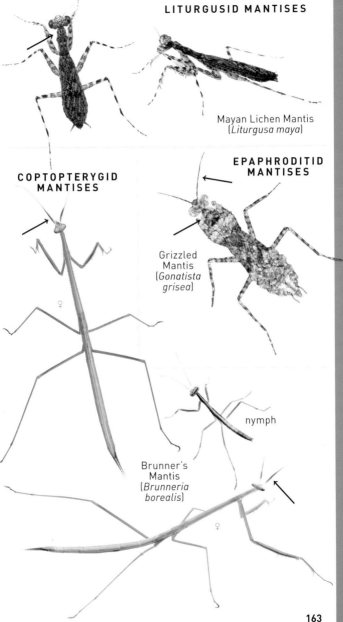

LITURGUSID MANTISES

Mayan Lichen Mantis
(*Liturgusa maya*)

EPAPHRODITID MANTISES

Grizzled
Mantis
(*Gonatista
grisea*)

COPTOPTERYGID MANTISES

♀

nymph

Brunner's
Mantis
(*Brunneria
borealis*)

♀

163

MANTOID MANTISES Family Mantoididae*

Small black to reddish brown mantid that resembles a cockroach. Eyes rounded with head capsule nearly twice as wide as pronotum. Antennae longer than body. **Pronotum convex, square shaped, about as long as wide.** Single species in NA, *Mantoida maya*, found running on the ground under brush in FL. Nymphs resemble ants or wasps. TL 10–25 mm; World 12, NA 1.

MANTIDS Family Mantidae

Most common and diverse mantid family in NA, with the largest individuals. Can be green, gray or brown. Large species with rounded eyes. FW edge shiny. **Front tibia with several teeth**, no dorsal tooth and more than half as long as femur. Most diverse in tropical regions. Historically, this represented the only family in the order; now several subfamilies have been elevated to family level. TL 40–160 mm; World 1,261, NA 19.

EREMIAPHILID MANTISES Family Eremiaphilidae*

Single, non-native species, *Iris oratoria*, found in NA from TX to CA in the south. Green or brown with **5 teeth on front femur.** Pronotum 3 times as long as wide. Female FW shorter than abdomen. **HW in both sexes dull brick-red with dark spot and lines.** TL 50–65 mm; World 259, NA 1.

EREMIAPHILID MANTISES

Mediterranean Mantis
(*Iris oratoria*)

♂

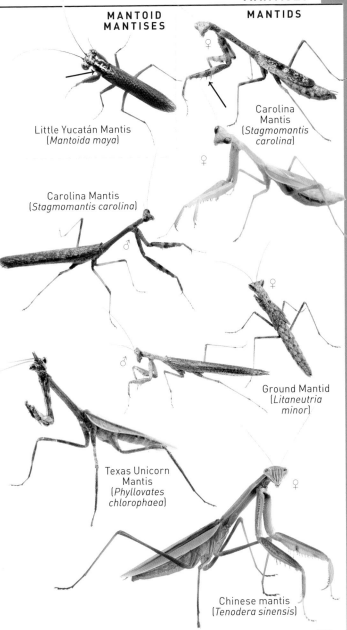

MANTOID MANTISES

MANTIDS

Little Yucatán Mantis
(*Mantoida maya*)

Carolina
Mantis
(*Stagmomantis
carolina*)

♀

Carolina Mantis
(*Stagmomantis carolina*)

♂

♀

Ground Mantid
(*Litaneutria
minor*)

♂

Texas Unicorn
Mantis
(*Phyllovates
chlorophaea*)

♀

Chinese mantis
(*Tenodera sinensis*)

MANTISES

165

Small Yellow Texas Cockroach
(*Chorisoheura texensis*)
Ectobiidae

Subterranean Termite
worker & alate nymph
(*Reticulitermes flavipes*)
Rhinotermitidae

COCKROACHES & TERMITES
ORDER BLATTODEA

Derivation of Name
- Blattodea: *blatt*, cockroach.

Dictyoptera —— mantids
—— cockroaches
—— termites

Classification
- Cockroaches and termites have been treated as separate orders historically: Blattodea/Blattaria (cockroaches) and Isoptera (termites).
- Some consider cockroaches as a suborder under the order Dictyoptera along with Mantodea.
- Recent molecular evidence places termites within the superfamily Blattoidea along with the families Blattidae and Cryptocercidae.
- Currently 3 superfamilies are recognized: Blaberoidea, Blattoidea, and Corydioidea.
- World: 7,600 species (cockroaches 4,600; termites 3,000); North America: 94 species (cockroaches 50; termites 44).

Range
- Found worldwide, mostly tropical. A number of species are cosmopolitan and have been moved around and introduced by humans.

Similar orders
- Psocodea (p. 220): tarsi 2- or 3-segmented.
- Zoraptera (p. 110): tarsi 2-segmented; antennae 9-segmented; cerci 1-segmented.
- Embioptera (p. 148): tarsi 3-segmented, basal segment of front tarsus enlarged.
- Neuroptera (p. 280): wings held roof-like, clear, heavily veined.
- Hymenoptera, ants (p. 236): abdomen constricted at base, with elbowed antennae.
- Orthoptera (p. 130): hind femur generally modified for jumping.

Pacific coast dampwood termites dispersing during nuptial flight (*Zootermopsis angusticollis*) Archotermopsidae

Cockroaches
Superfamilies Blaberoidea, Blattoidea (in part) & Corydioidea

Identification
- 2–65 mm TL.
- Body flattened and oval shaped; head usually concealed by pronotum; antennae long and slender; tarsi 5-segmented; cerci many segmented.

Food
- Scavengers, eating anything from vegetation to dead insects.

Behavior
- Cockroaches are gregarious, preferring to live together in large numbers; generally lack castes or hierarchies as in termites or ants, except for *Cryptocercus* in NA, which are subsocial.
- Some species exhibit parental care.
- Cockroaches use pheromones, for example, communicating with each other to either stay at or leave a food source.

Life Cycle
- Hemimetabolous.
- Eggs are laid in a mass called an ootheca, similar to mantids.
- Nymphs look similar to adults with wing buds.
- In some species, the ootheca is retained inside the female, where nymphs hatch, so that it appears that the female is giving live birth.

Importance to Humans
- Some species can be serious pests in homes. They have an unpleasant odor and multiply readily in cracks and crevices.
- A number of species are common in the pet trade.

Collecting and Preserving
- Found in leaf litter, under wood, and in dwellings. Some species come to lights. Much more common in the south.
- Store soft-bodied adults in 70% ethanol.
- Pin winged adults in upper right wing.

Resources
- "Catalog and Atlas of the Cockroaches (Dictyoptera) of North America North of Mexico" by Atkinson et al., 1991
- Cockroach Species File Online—cockroach.speciesfile.org
- *How to Know the Grasshoppers, Crickets, Cockroaches and Their Allies* by Helfer, 1987

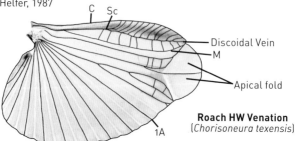

Roach HW Venation
(*Chorisoneura texensis*)

Termites
Superfamily Blattoidea (in part)

Identification
- 2–11 mm TL.
- Small, soft bodied, and elongate; tan or brown; antennae short, not much longer than head, and thread- or bead-like; wings, when present, the same size and dark brown; mouthparts chewing; tarsi 4-segmented; cerci short.

Food
- Many species eat wood and other cellulose products; others feed on grasses (*Gnathamitermes, Tenuirostritermes*).
- Have symbiotic protozoa and bacteria in gut to help digest cellulose. These symbionts are transferred to each other via anal secretions.

Behavior
- Termites are social insects with a king and queen, and the labor is divided among the different castes; grooming is common.
- Colonies include workers who care for young and soldiers for defense. Some species also have nasutes (tube on head), which defend the colony by excreting an irritating substance.
- Some live in moist rotting wood, while others prefer contact with soil or dry wood.

Life Cycle
- Hemimetabolous.
- Immatures similar in appearance to adults.
- Winged adults (alate) swarm and pair up and lay eggs in a nuptial chamber, becoming king and queen of the colony.

Importance to Humans
- A few species cause considerable damage to anything built with wood. Globally important decomposers and incorporate nutrients back into soil.

Collecting and Preserving
- Found in wood and underground.
- Collect and preserve in 70% ethanol.

Resources
- "Termites: Phylogeny, Classification, & Rise to Dominance" by Engel et al., 2009

Termite FW Venation
(*Reticulitermes flavipes*)

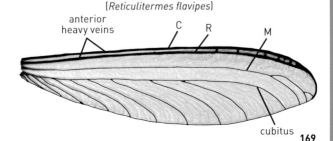

anterior heavy veins C R M

cubitus

Superfamily Blaberoidea

BLABERID COCKROACHES Family Blaberidae
Middle and hind femur variable with 1 to numerous spines or lacking spines alltogether and hairy. Front femur with 0–3 spines. Tropical group with several introduced species to NA and generally restricted to the southern states. TL 12–50 mm; World 1,200, NA 11.

WOOD COCKROACHES Family Ectobiidae
Small to medium-sized roaches. Largest family of roaches with several species invading houses. Front femur generally with 2 or 3 apical spines. Supra-anal plate and female subgenital plate not divided. *Parcoblatta* and *Chorisoneura* with **apical portion of HW that folds over when wings are at rest.** TL < 25 mm; World 2,300, NA 40.

Superfamily Blattoidea

HOUSEHOLD COCKROACHES Family Blattidae
Large roaches. All species in the US are introduced. Front femur with row of spines that either decrease in size and length distally or are the same size. **Female subgenital plate divided longitudinally.** Size 17–50 mm; World 600, NA 9.

BROWN-HOODED COCKROACHES Family Cryptocercidae*
Reddish-brown wingless cockroach with **pronotum thickened over head creating a hood.** Tip of abdomen, including cerci, covered by seventh dorsal and sixth ventral abdominal segments. Found in wet rotten wood in eastern mountains and Pacific northwest. Exhibit subsocial behavior and can live more than 12 years.
TL 22–30 mm; World 12, NA 5.

Front Femur Showing Spine Variation

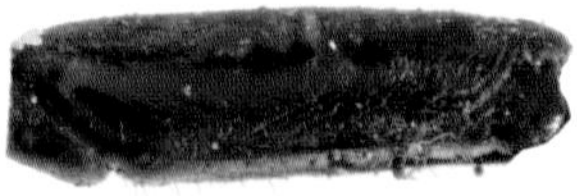

Blaberidae

Ectobiidae

Blattidae

Corydiidae

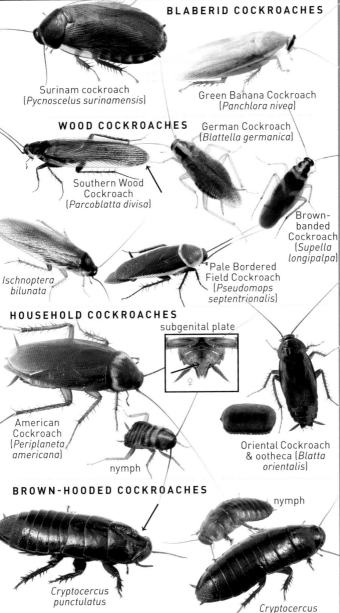

BLABERID COCKROACHES

Surinam cockroach
(*Pycnoscelus surinamensis*)

Green Banana Cockroach
(*Panchlora nivea*)

WOOD COCKROACHES

German Cockroach
(*Blattella germanica*)

Southern Wood
Cockroach
(*Parcoblatta divisa*)

Brown-
banded
Cockroach
(*Supella
longipalpa*)

*Ischnoptera
bilunata*

Pale Bordered
Field Cockroach
(*Pseudomops
septentrionalis*)

HOUSEHOLD COCKROACHES

subgenital plate

♀

American
Cockroach
(*Periplaneta
americana*)

nymph

Oriental Cockroach
& ootheca (*Blatta
orientalis*)

BROWN-HOODED COCKROACHES

nymph

*Cryptocercus
punctulatus*

*Cryptocercus
darwini*

COCKROACHES

171

Superfamily Blattoidea cont.

ROTTENWOOD TERMITES Family Archotermopsidae

Found in southwest and west coast of US inhabiting dead, rotting wood. Ocelli absent. **Wings with 3 or more heavy anterior veins;** tips of wings with branches off of R. Soldiers with more than 21-segmented antennae and tibia usually armed with spines. Cerci with 3 or more prominent segments. TL 5–40 mm; World 11, NA 3.

DRYWOOD TERMITES Family Kalotermitidae

Found in southern U.S. Leave little evidence of wood infestation in dry wood other than dust under tunnel openings. Ocelli present. **Wings with 3 or more heavy anterior veins;** tips of wings with branches off of R. Antennae with fewer than 23 segments. Short 2-segmented cerci. Soldier mandibles without marginal teeth; hind femur swollen. TL 5–10 mm; World 450, NA 18.

SUBTERRANEAN TERMITES Family Rhinotermitidae

Found all over NA and the most destructive to human structures. **Two heavy veins on anterior part of wing;** tips of wings without branches off of R. Cerci 2-segmented. **Soldier with head longer than broad** and mandibles without marginal teeth. TL 12–18 mm; World 320, NA 9.

HIGHER TERMITES Family Termitidae

Scale of front wing (remaining stub of detached wing) shorter than saddle-shaped pronotum. Two heavy veins on anterior part of wings, tips of wings without branches off of R. Cerci 1–2 segmented. **Mandibles on soldiers modified into long nose-like projection.** Worker mandibles with only 1 prominent tooth. TL 2–10 mm; World 2,000, NA 11 (1 in FL, remaining in southwest U.S.).

Superfamily Corydioidea

DESERT COCKROACHES Family Corydiidae

Found in warmer climates in NA from FL to CA. Frons is slightly thickened and bulging. **Front edge of pronotum not greatly thickened.** If FW present, anal region is flattened not folded. Many species with wingless females. TL 14–25 mm; World 250, NA 14.

DESERT COCKROACHES

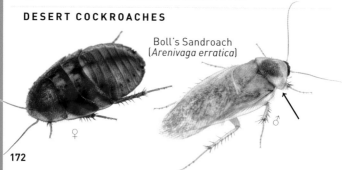

Boll's Sandroach
(*Arenivaga erratica*)

♀   ♂

ROTTENWOOD TERMITES

worker

soldier

alate

reproductive

Pacific Coast Dampwood Termite
(*Zootermopsis angusticollis*)

DRYWOOD TERMITES

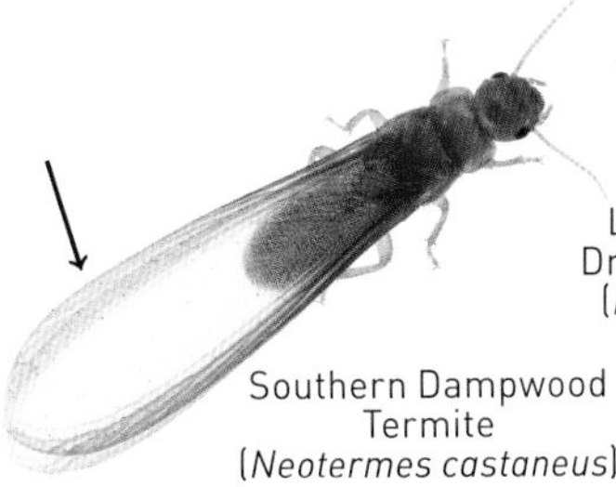

Light Western
Drywood Termite
(*Marginitermes
hubbardi*)

Southern Dampwood
Termite
(*Neotermes castaneus*)

SUBTERRANEAN TERMITES

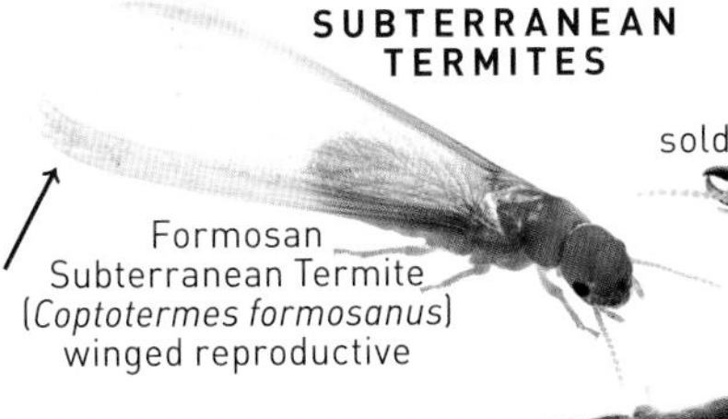

Eastern
Subterranean Termite
(*Reticulitermes flavipes*)

soldier

Formosan
Subterranean Termite
(*Coptotermes formosanus*)
winged reproductive

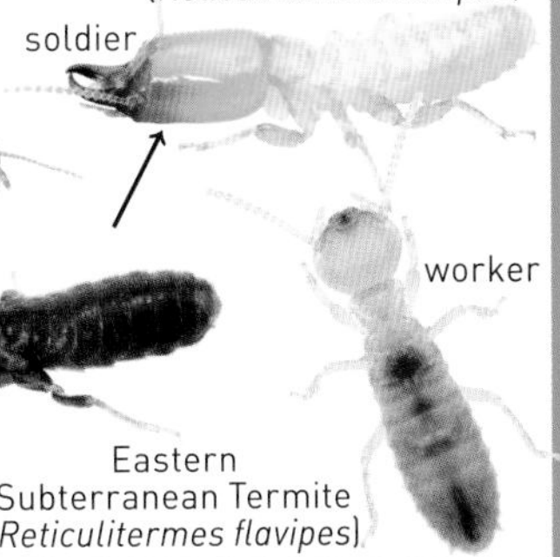

worker

reproductive
after losing
wings

Eastern
Subterranean Termite
(*Reticulitermes flavipes*)

HIGHER TERMITES

Gnathamitermes sp.
earthen tubes

Tenuirostritermes cinereus
worker and soldier

Sharpshooter
(*Cuerna costalis*)
Cicadellidae

Stink bug nymphs hatching from eggs
Pentatomidae

TRUE BUGS, CICADAS, HOPPERS, APHIDS, SCALES, & ALLIES
ORDER HEMIPTERA

Derivation of Name
- Hemiptera: *hemi*, half; *pteron*, winged. Referring to the forewings of true bugs that are leathery basally and membranous distally.

Classification
- Four suborders recognized: Heteroptera (true bugs), Auchenorrhyncha (cicadas and hoppers), Sternorrhyncha (aphids, scale insects and relatives), and Coleorrhyncha (moss bugs; not in NA).
- Historically, today's Heteroptera were placed in their own order: Hemiptera. And today's Auchenorrhyncha and Sternorrhyncha were combined and placed in their own order: Homoptera.
- World: 107,000 species (42,000 Heteroptera; 65,000 Auchenorrhyncha and Sternorrhyncha)
  North America: 10,200 species (3,850 Heteroptera; 6,350 Auchenorrhyncha and Sternorrhyncha).

Range
- Found on every continent except Antarctica.

Similar orders
- Coleoptera (p. 294): FW uniformly leathery or thickened, meeting in a straight line down middle; chewing mouthparts; antennae usually 8 or more segments.
- Orthoptera (p. 130): FW uniformly leathery; HW with many veins; antennae with many segments; chewing mouthparts.
- Psocodea (p. 220): chewing mouthparts in winged species; lack cornicles (as in aphids).

Plant-parasitic Hemipterans
Suborder Sternorrhyncha

Identification
- 1.5–3.5 mm TL.
- Jumping plant lice, whiteflies, and scale insects.
- Antennae filiform and moderately long.
- Piercing-sucking mouthparts arising from between front coxae; "Sternorrhyncha" refers to the rearward position of the mouthparts relative to the head.
- 1–2 tarsomeres (some scale insects have no legs).
- Some lack wings and or appendages (scale insects).

Food
- All families feed on plants.

Behavior
- Inactive and can be sedentary.
- Usually found close to their host plant.
- Can be found in large numbers on host plant.
- Ants can be found tending nymphs or adults.

Life Cycle
- Hemimetabolous.
- Aphids have a complex life cycle, including bisexual and parthenogenetic generations.
- Some nymphs form waxy secretions and look like blobs on host.

Importance to Humans
- Many species are serious pests on crops and other plants.

Collecting and Preserving
Immatures
- Collect by hand, with sweep net or aspirator.
- Place aphids in 70% ethanol; scale insects can be kept in 70% ethanol or mounted on a microscope slide; there are specific methods for mounting species in other groups.

Adults
- Same as immatures.

Resources
- Scale Insects—idtools.org/id/scales
- *Armored Scale Insect Pests* by Miller and Davidson, 2005
- *Scale Insects of Northeastern North America* by Kosztarab, 1996

Woolly Pine Scale
(*Pseudophilippia quaintancii*)
Coccidae

True Hoppers
Suborder Auchenorrhyncha

Identification
- 2–31 mm TL.
- Cicadas, leafhoppers, treehoppers, froghoppers, planthoppers.
- Piercing-sucking mouthparts with prominent "beak" arising from back portion of head ventrally; "Auchenorrhyncha" refers to the position of the mouthparts around the neck or throat.
- Oval ocelli, usually 2.
- Wings held roof-like over abdomen.
- Antennae 3–10 segments.
- 3-segmented tarsi.

Food
- Feed on various parts of plants.
- Some feed on fungi or mosses.

Behavior
- Feeding on plants can cause decoloration, wilting, and even death.
- Many with complex system for acoustic communication.
- Many with excellent jumping abilities.

Life Cycle
- Hemimetabolous.
- Eggs sometimes laid in twigs.
- Some groups can induce galls.

Importance to Humans
- Can cause plant damage by feeding, laying eggs, or transmitting pathogens.

Collecting and Preserving
Immatures
- Typically collected with sweep net or aspirator.
- Place immatures in 70% alcohol.

Adults
- Some groups visit lights. Collected with sweep net or aspirator.
- Pin through the scutellum to the right of center. Place on point if too small for a pin.

Resources
- Taxonomic Databases—dmitriev.speciesfile.org
- Planthoppers of NA—sites.udel.edu/planthoppers

Black Treehopper
(*Acutalis tartarea*)
Membracidae

True Bugs
Suborder Heteroptera

Identification
- 1–65 mm TL.
- Piercing-sucking mouthparts with prominent "beak" arising from anterior portion of head ventrally.
- Often dorsoventrally flattened.
- Usually 2 pairs of wings in adults, including FW (hemelytra) that is leathery basally and membranous distally; wings held flat with apical tips overlapping when at rest.
- Prominent scutellum.
- Antennae 4–5 segments.
- Many with scent glands dorsally on abdomen as nymphs and usually ventrally on thorax as adults.
- Two ocelli or none.

Food
- Plant juices, predators; some are mixed feeders.
- A few are parasites (bloodsucking).

Behavior
- Adults have a distinct smell produced by a glands on the ventral side of the thorax. This is used to repel predators.
- Eggs typically laid inconspicuously on underside of leaves or in plant tissue.
- Nymphs are frequently found in groups with adults.

Life Cycle
- Hemimetabolous.
- Egg, nymph, and adult stages; typically 5 instars.
- Nymphs closely resemble adults; some adults have reduced wings or wings are lacking alltogether.
- Mating in bed bugs is unique: the male reproductive organ punctures the abdomen of the female during insemination.

Importance to Humans
- Many are crop pests.
- Some vector plant and animal diseases.

Collecting and Preserving
Immatures
- Hand collecting, sweep net, beat sheet, UV lights, and dip netting.
- Place nymphs are placed in 70% alcohol for preservation.

Adults
- Hand collecting, sweep net, beat sheet, UV lights, and dip netting.
- Pin adults through right side of scutellum.

Resources
- *How to Know the True Bugs* by Slater and Baranowski, 1978
- *True Bugs of the World* by Schuh & Slater, 1995.
- "Identification Key to the Principal Families of Florida Heteroptera" by Choate—entnemdept.ufl.edu/choate/heteroptera_new1.pdf
- *Heteroptera of Economic Importance* by Schaefer and Panizzi, 2000

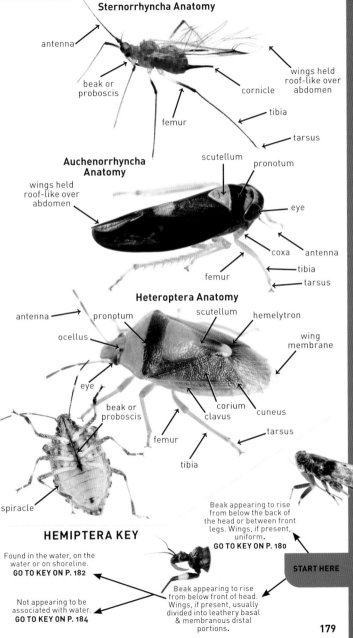

Sternorrhyncha Anatomy

antenna

beak or proboscis

femur

cornicle

wings held roof-like over abdomen

tibia

tarsus

Auchenorrhyncha Anatomy

wings held roof-like over abdomen

scutellum

pronotum

eye

coxa

antenna

tibia

femur

tarsus

Heteroptera Anatomy

antenna

pronotum

scutellum

hemelytron

ocellus

wing membrane

eye

beak or proboscis

corium

clavus

cuneus

femur

tarsus

tibia

spiracle

HEMIPTERA KEY

Found in the water, on the water or on shoreline.
GO TO KEY ON P. 182

Not appearing to be associated with water.
GO TO KEY ON P. 184

Beak appearing to rise from below the back of the head or between front legs. Wings, if present, uniform.
GO TO KEY ON P. 180

START HERE

Beak appearing to rise from below front of head. Wings, if present, usually divided into leathery basal & membranous distal portions.

179

KEY TO COMMON HOPPERS, APHIDS, SCALES, & ALLIES

Single pair of wings; 1-segmented tarsi; rarely encountered.
Scale Insects & Mealybugs
COCCOIDEA
(males), p. 186

Legs absent or with 1-segmented tarsi; under waxy covering; sessile.
Scale Insects & Mealybugs
COCCOIDEA (females), p. 186

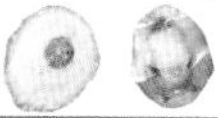

Active jumper/flier; sclerotized body; well-developed wings; antennae 5-10 segments.
PSYLLOIDEA, p. 190

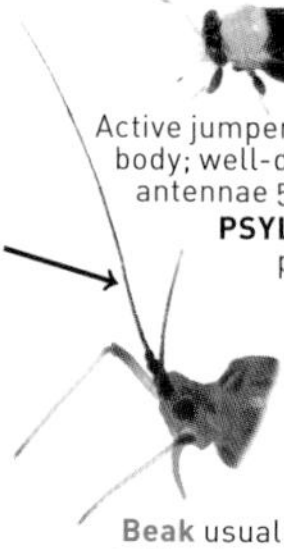

Relatively inactive and soft bodied; antennae 3-7 segments.

Wings held roof-like over body; found on conifers.
ADELGIDAE, p. 190

Wings held flat over abdomen; found in grape and hickory galls.
PHYLLOXERIDAE, p. 190

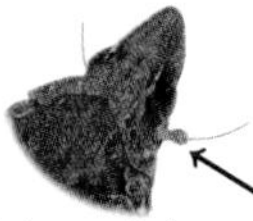

Beak usually thread-like, arising between front legs; **antennae longer than head.**
STERNORRHYNCHA

Antennae below or behind eyes; center of head often with distinct carina or ridge.
FULGOROIDEA, p. 194

Hind tibiae with 1 or 2 stout spines and **apical crown of spines**.
CERCOPOIDEA, p. 192

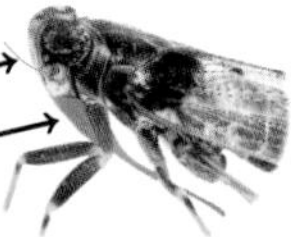

START HERE

Hind tibiae without spines; head covered above by pronotum.
AETALIONIDAE, p. 194

Beak relatively thick, arising at back of head; **antennae bristle-like,** shorter than head.
AUCHENORRHYNCHA

Hind tibiae with 0, 1 or 2 spines only.

Hind tibiae with **1 or more rows of small spines.**
CICADELLIDAE, p. 192

Antennae originate in front of eyes; lacking prominent carina or ridge.

Larger than 25 mm; **FW clear;** 3 ocelli.
CICADIDAE, p. 194

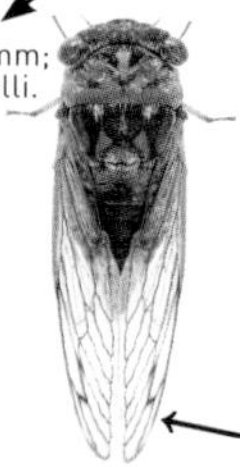

<12 mm; FW usually leathery; 2 ocelli.

Pronotum not enlarged over abdomen.

Pronotum enlarged, extending back over abdomen.
MEMBRACIDAE, p. 194

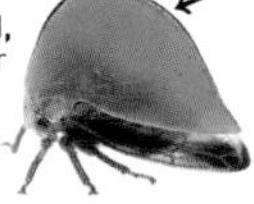

Minute (2–3 mm), white, male with 2 pairs of opaque wings; often gregarious; covered with waxy powder.

**Whiteflies
ALEYRODIDAE,**
p. 186

Two pairs of wings; generally with tubes (cornicles) arising from abdomen; often gregarious.

**Most Aphids
APHIDIDAE, p. 186**

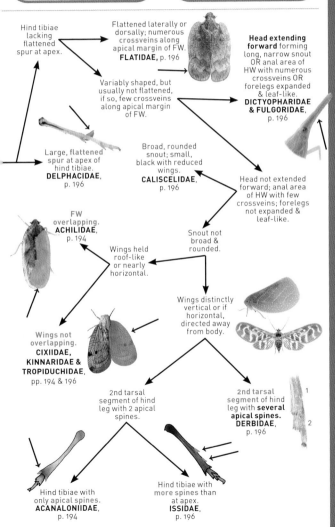

Hind tibiae lacking flattened spur at apex.

Flattened laterally or dorsally; numerous **crossveins along apical margin of FW. FLATIDAE, p. 196**

Head extending forward forming long, narrow snout OR anal area of HW with numerous crossveins OR forelegs expanded & leaf-like. **DICTYOPHARIDAE & FULGORIDAE,** p. 196

Variably shaped, but usually not flattened, if so, few crossveins along apical margin of FW.

Large, **flattened spur** at apex of hind tibiae. **DELPHACIDAE,** p. 196

Broad, rounded snout; small, black with reduced wings. **CALISCELIDAE,** p. 196

Head not extended forward; anal area of HW with few crossveins; forelegs not expanded & leaf-like.

FW overlapping. **ACHILIDAE,** p. 194

Wings held roof-like or nearly horizontal.

Snout not broad & rounded.

Wings not overlapping. **CIXIIDAE, KINNARIDAE & TROPIDUCHIDAE,** pp. 194 & 196

Wings distinctly vertical or if horizontal, directed away from body.

2nd tarsal segment of hind leg with 2 apical spines.

2nd tarsal segment of hind leg with **several apical spines. DERBIDAE,** p. 196

Hind tibiae with only apical spines. **ACANALONIIDAE,** p. 194

Hind tibiae with **more spines** than at apex. **ISSIDAE,** p. 196

KEY TO AQUATIC, SHORE, & SURFACE BUG FAMILIES

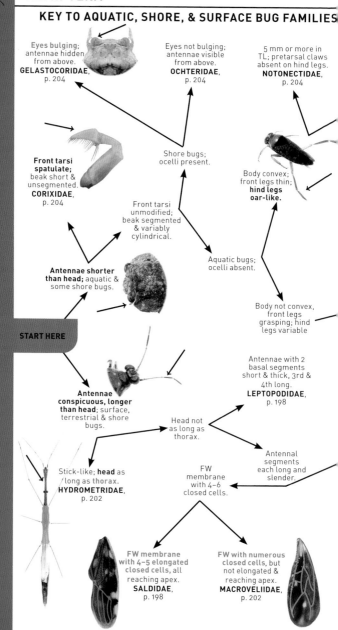

Eyes bulging; antennae hidden from above.
GELASTOCORIDAE,
p. 204

Eyes not bulging; antennae visible from above.
OCHTERIDAE,
p. 204

5 mm or more in TL; pretarsal claws absent on hind legs.
NOTONECTIDAE,
p. 204

Front tarsi spatulate; beak short & unsegmented.
CORIXIDAE,
p. 204

Shore bugs; ocelli present.

Body convex; front legs thin; **hind legs oar-like.**

Front tarsi unmodified; beak segmented & variably cylindrical.

Antennae shorter than head; aquatic & some shore bugs.

Aquatic bugs; ocelli absent.

Body not convex, front legs grasping; hind legs variable

START HERE

Antennae with 2 basal segments short & thick, 3rd & 4th long.
LEPTOPODIDAE,
p. 198

Antennae conspicuous, longer than head; surface, terrestrial & shore bugs.

Head not as long as thorax.

Antennal segments each long and slender.

Stick-like; **head** as long as thorax.
HYDROMETRIDAE,
p. 202

FW membrane with 4–6 closed cells.

FW membrane with 4–5 elongated closed cells, all reaching apex.
SALDIDAE,
p. 198

FW with numerous closed cells, but not elongated & reaching apex.
MACROVELIIDAE,
p. 202

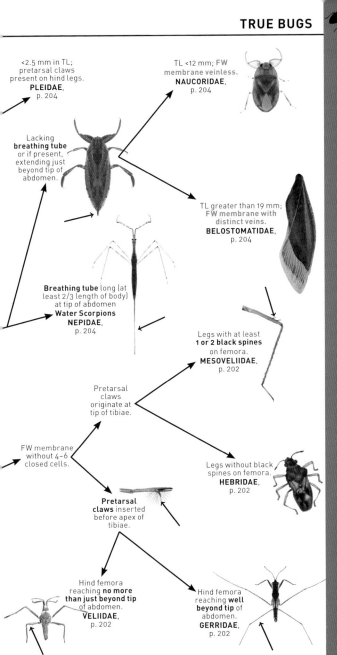

<25 mm in TL; pretarsal claws present on hind legs. **PLEIDAE**, p. 204

TL <12 mm; FW membrane veinless. **NAUCORIDAE**, p. 204

Lacking **breathing tube** or if present, extending just beyond tip of abdomen.

TL greater than 19 mm; FW membrane with distinct veins. **BELOSTOMATIDAE**, p. 204

Breathing tube long (at least 2/3 length of body) at tip of abdomen **Water Scorpions NEPIDAE**, p. 204

Legs with at least **1 or 2 black spines** on femora. **MESOVELIIDAE**, p. 202

Pretarsal claws originate at tip of tibiae.

FW membrane without 4–6 closed cells.

Legs without black spines on femora. **HEBRIDAE**, p. 202

Pretarsal claws inserted before apex of tibiae.

Hind femora reaching **no more than just beyond tip** of abdomen. **VELIIDAE**, p. 202

Hind femora reaching **well beyond tip** of abdomen. **GERRIDAE**, p. 202

HEMIPTERA

KEY TO TERRESTRIAL BUG FAMILIES

Rarely encountered families not included in the key. See text for descriptions.

Aenictopecheidae, p. 198
Ceratocombidae, p. 198
Curaliidae, p. 200
Dipsocoridae, p. 198
Enicocephalidae, p. 198
Lasochilidae, p. 199

Lyctocoridae, p. 199
Polyctenidae, p. 199
Microphysidae, p. 200
Schizopteridae, p. 198
Thaumastocoridae, p. 200

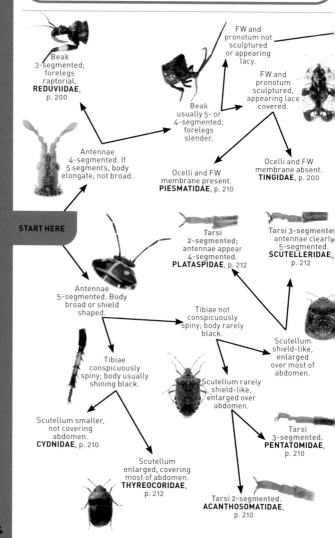

Beak 3-segmented; forelegs raptorial.
REDUVIIDAE, p. 200

FW and pronotum not sculptured or appearing lacy.

FW and pronotum sculptured, appearing lace covered.

Beak usually 5- or 4-segmented; forelegs slender.

Antennae 4-segmented. If 5 segments, body elongate, not broad.

Ocelli and FW membrane present.
PIESMATIDAE, p. 210

Ocelli and FW membrane absent.
TINGIDAE, p. 200

START HERE

Tarsi 2-segmented; antennae appear 4-segmented.
PLATASPIDAE, p. 212

Tarsi 3-segmented; antennae clearly 5-segmented.
SCUTELLERIDAE, p. 212

Antennae 5-segmented. Body broad or shield shaped.

Tibiae not conspicuously spiny; body rarely black.

Scutellum shield-like, enlarged over most of abdomen.

Tibiae conspicuously spiny; body usually shining black.

Scutellum rarely shield-like, enlarged over abdomen.

Scutellum smaller, not covering abdomen.
CYDNIDAE, p. 210

Tarsi 3-segmented.
PENTATOMIDAE, p. 210

Scutellum enlarged, covering most of abdomen.
THYREOCORIDAE, p. 212

Tarsi 2-segmented.
ACANTHOSOMATIDAE, p. 210

184

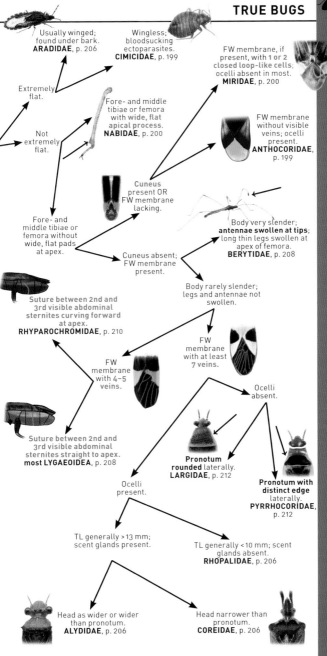

Usually winged; found under bark. **ARADIDAE**, p. 206

Wingless; bloodsucking ectoparasites. **CIMICIDAE**, p. 199

FW membrane, if present, with 1 or 2 closed loop-like cells; ocelli absent in most. **MIRIDAE**, p. 200

Extremely flat.

Fore- and middle tibiae or femora with wide, flat apical process. **NABIDAE**, p. 200

FW membrane without visible veins; ocelli present. **ANTHOCORIDAE**, p. 199

Not extremely flat.

Cuneus present OR FW membrane lacking.

Body very slender; **antennae swollen at tips**; long thin legs swollen at apex of femora. **BERYTIDAE**, p. 208

Fore- and middle tibiae or femora without wide, flat pads at apex.

Cuneus absent; FW membrane present.

Body rarely slender; legs and antennae not swollen.

Suture between 2nd and 3rd visible abdominal sternites curving forward at apex. **RHYPAROCHROMIDAE**, p. 210

FW membrane with at least 7 veins.

FW membrane with 4–5 veins.

Ocelli absent.

Suture between 2nd and 3rd visible abdominal sternites straight to apex. **most LYGAEOIDEA**, p. 208

Pronotum **rounded** laterally. **LARGIDAE**, p. 212

Pronotum with distinct edge laterally. **PYRRHOCORIDAE**, p. 212

Ocelli present.

TL generally >13 mm; scent glands present.

TL generally <10 mm; scent glands absent. **RHOPALIDAE**, p. 206

Head as wider or wider than pronotum. **ALYDIDAE**, p. 206

Head narrower than pronotum. **COREIDAE**, p. 206

STERNORRHYNCHA

Suborder Sternorrhyncha
Superfamily Aleyrodoidea

WHITEFLIES Family Aleyrodidae
Small, resembling tiny moths. Adults are winged; wings covered with waxy powder. First instar nymphs are active; remaining instars look like scale insects. Feeds on leaves of various plants.
TL 2–3 mm; World 1,600, NA 100.

Superfamily Aphidoidea

APHIDS Family Aphididae
Small, soft bodied and frequently found together in large numbers. They have a pear-like shape and a **pair of tubercles (cornicles)** arising dorsally off the end of the abdomen. Antennae generally long. May be winged or wingless (in the same species); HW much smaller than FW. TL 1–7 mm; World 4,700, NA 1,351.

Superfamily Coccoidea

COCHINEAL INSECTS Family Dactylopiidae
Resemble mealybugs. Females are broadly oval, distinctly segmented, and red; completely covered with white wax. Males winged with red body. Found in the southwest, where they live on cactus. Single genus in NA, *Dactylopius*. TL ~2 mm; World 10, NA 5.

ARMORED SCALE INSECTS Family Diaspididae
Females small, soft bodied and concealed under scale made of wax that is usually free from the body of the insect; lack eyes and legs; antennae are absent or vestigial. Males with wings and well-developed legs and antennae. Many pest species. TL 0.5–5 mm; World 2,500, NA 310.

MEALYBUGS Family Pseudococcidae
Waxy or mealy secretions covering body. Female is elongate, oval, and segmented with well-developed legs. Some species lay eggs, others give live birth. Many pest species, including in greenhouses. TL 1–8 mm; World 2,200, NA 320.

GRASS SCALES Family Aclerdidae*   Not Illus.
Females usually pink or red and heavily sclerotized at maturity; body oval with anus covered by oval or triangular plate; antennae reduced to tubercles with setae; posterior margins of body with ridges. Most species feed on grasses along stems. Generally found beneath the leaf sheath or bases of the plant. TL 1.5–15 mm; World 50, NA 16.

PIT SCALES Family Asterolecaniidae*   Not Illus.
Usually greenish or yellowish with wax filaments along margin. Most found on bark, others on leaves of host. Antennae, when present, 1-segmented. No sclerotized anal plate present. Many produce gall-like pits in bark of host. TL 0.5–5 mm; World 250, NA 17.

WHITEFLIES

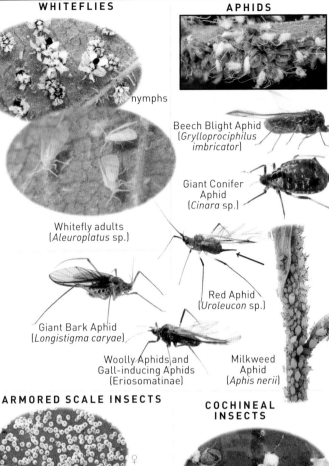

nymphs

Whitefly adults
(*Aleuroplatus* sp.)

APHIDS

Beech Blight Aphid
(*Grylloprociphilus imbricator*)

Giant Conifer
Aphid
(*Cinara* sp.)

Red Aphid
(*Uroleucon* sp.)

Giant Bark Aphid
(*Longistigma caryae*)

Woolly Aphids and
Gall-inducing Aphids
(*Eriosomatinae*)

Milkweed
Aphid
(*Aphis nerii*)

ARMORED SCALE INSECTS

♀

Diaspis echinocacti

COCHINEAL INSECTS

♀

Cochineal (*Dactylopius* sp.)

MEALYBUGS

♀

Taxus mealybug
(*Dysmicoccus wistariae*)

STERNORRHYNCHA

Superfamily Coccoidea cont.

ORNATE PIT SCALES Family Cerococcidae*   Not Illus.

Live within a sugary secretion and excretion that may be brightly colored. Antennae 1-segmented, with associated cluster of 5–7 locular pores. Anal plate present and triangular. Represented by a single genus, *Cerococcus*, in NA. Found on a variety of plants, including azaleas and cranberries. TL 2–5 mm; World 70, NA 6.

SOFT SCALES Family Coccidae

Females have elongate, oval, often convex, though sometimes flattened bodies. Scales are covered by a smooth, hard exoskeleton or a soft wax covering. Legs are usually present; antennae either absent or reduced. Males may be winged or wingless. Some species are serious pests of citrus. TL 2–8 mm; World 1,090, NA 105.

FELT SCALES Family Eriococcidae   Not Illus.

Similar to Mealybugs, but with the body bare or only lightly covered with wax. Female antennae 6 or 7 segments in most species; 10 segments in male. Usually with protruding anal lobes; anal ring usually with pores and setae. The female and her eggs generally enclosed in a felt-like sac. Some species are pests of azaleas and oak and elm trees. TL 1–5 mm; World 650, NA 55.

LAC SCALES Family Kerriidae*   Not Illus.

Species are globular, legless, and found in cells of resin. Restricted to the southwest and a single genus, *Tachardiella* in NA. Most species found on oaks, some on chinquapin, and all produce lac, some of which is highly pigmented. TL 1–4 mm; World 90, NA 7.

STEINGELIIDS Family Steingeliidae*   Not Illus.

Body elongated, yellowish, with legs and antennae; latter projecting forward in V. TL 1–3 mm; World 10, NA 3.

GIANT MEALYBUGS Family Putoidae*   Not Illus.

Large, covered with thick tufts of mealy white wax; lateral filaments broad. A single genus in NA, *Puto*. TL ~5 mm; World 50, NA 21.

GALL-LIKE SCALES Family Kermesidae*   Not Illus.

Females are small and round, heavily sclerotized and gall-like. Without protruding anal lobes. Found on twigs and leaves of oaks. TL 2–5 mm; World 90, NA 40.

KUWANIIDS Family Kuwaniidae*   Not Illus.

Small, ovoid, bright red body. Legs and antennae present. Body with obvious segmentation. Adult females produce white, waxy covering. Found in cracks and under bark. TL 1–7 mm; World 8, NA 2.

XYLOCOCCIDS Family Xylococcidae*   Not Illus.

Female body usually large, red or brown with conspicuous legs and antennae. Some found under bark. TL 5–11 mm; World 11, NA 8.

FALSE PIT SCALES Family Lecanodiaspididae*   Not Illus.

Often produce pits and swellings on twigs. Adults are yellow to reddish-brown with a waxy, papery covering with 8 transverse ridges. Common pests of azalea, holly, and other ornamentals. Single genus in NA, *Lecanodiaspis*. TL 2–5 mm; World 71, NA 5.

GROUND PEARLS Family Margarodidae*   Not Illus.

Females live on roots of plants and have a waxy pearl-like cyst. Female body usually yellowish-brown. Antenna 6–13-segmented. All species are found underground. TL 1–5 mm; World 110, NA 8.

PINE BAST SCALES Family Matsucoccidae*   Not Illus.

Females red, green, to brown with ovoid, soft bodies bearing antennae and legs. Antennae large, protruding from head in a V shape. Found wandering on pines or under bark. TL 2–8 mm; World 42, NA 18.

GIANT SCALE INSECTS Family Monophlebidae

Females large and round with segmented bodies. Legs and antennae usually conspicuous and dark. Occur on stems, branches, or foliage of plants; usually with wax covering. TL up to 13 mm; World 240, NA 8.

ENSIGN SCALES Family Ortheziidae   Not Illus.

Females with elongate, oval bodies distinctly segmented and covered by hard, waxy white plates. Found on mosses, grasses, herbaceous and woody plants, and fungi. TL 5 mm; World 210, NA 31.

PALM SCALES Family Phoenicococcidae*   Not Illus.

A single species, Red Date Scale (*Phoenicococcus marlatti*). Occurs on the date palm in the southwestern US. Reddish, ovoid individuals usually found at bases of leaf petioles or under fibrous covering of trunk. TL ~2 mm; World 1, NA 1.

PITYOCOCCID SCALES Family Pityococcidae*   Not Illus.

Body round or elongate; pink or yellow. Produces small amounts of wax around body. Found feeding on pines in the southwestern US. TL 2–5 mm; World 7, NA 5.

SCALE INSECTS

SOFT SCALES

Wax Scales
(*Ceroplastes* sp.)

GIANT SCALE INSECTS

Cottony Cushion Scale
(*Icerya purchasi*)

HEMIPTERA

Superfamily Phylloxeroidea

PINE & SPRUCE APHIDS Family Adelgidae   Not illus.
Form conical-shaped galls on spruce and white cottony tufts on
the bark and branches, twigs, needles, and cones of other conifers.
Female FW with reduced venation and 5-segmented antennae and
lacking cornicles. Wingless forms with 3-segmented antennae.
TL 2–3 mm; World 50, NA 22.

PHYLLOXERIDS Family Phylloxeridae
All forms with 3-segmented antennae. Wings held flat over
abdomen with FW stalked at base. Some produce galls and are
considered pests. Feed on non-conifer plants.
TL 2–3 mm; World 75, NA 29.

Superfamily Psylloidea

APHALARIDS Family Aphalaridae*   Not illus.
Metatibia in adults with an open crown of sclerotized apical spurs.
TL 1–11 mm; World 600, NA 77.

HOMOTOMIDS Family Homotomidae*   Not Illus.
Green or brown. Antennal flagellum thickened and densely setose.
FW costal break present. Two adventive species found in CA that
feed on *Ficus*. TL 2–5 mm; World 80, NA 2.

CALOPHYIDS Family Calophyidae
No morphological characters have been recognized to distinguish
this group. Species are associated with sumac and peppertree.
TL 1–10 mm; World 90, NA 14.

LIVIIDS Family Liviidae
Adult metatibia with an open crown of sclerotized apical spurs.
TL 1–11 mm; World 300, NA 26.

PHACOPTERONIDS Family Phacopteronidae*   Not illus.
Lanceolate setae present on abdomen margin and/or FW-pad
margin. Single genus, *Pseudophacopteron*, in NA, restricted to AZ
and FL. TL 1–11 mm; World 50, NA 3.

TRIOZIDS Family Triozidae
Variable morphology and host plant preferences. FW with veins Cu,
M and Rs intersecting at a single point; pterostigma absent. TL 1–11
mm; World 950, NA 80.

JUMPING PLANT LICE Family Psyllidae
Metacoxa with horn-shaped meracanthus, metatibia usually with
grouped apical spurs, metabasitarsus usually with 2 lateral spurs.
TL 1–11 mm; World 800, NA 134.

TRIOZIDS

*Calinda
longistylus*

CALOPHYIDS

Calophya nigripennis

PHYLLOXERIDS

Phylloxera
sp.

LIVIIDS

Livia saltatrix

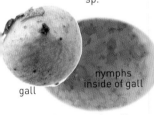

gall

nymphs
inside of gall

Phylloxera sp. in gall
formed on Hickory

Asian Citrus Liviid
(*Diaphorina citri*)
nymphs

JUMPING PLANT LICE

Gall formed by Red Bay Psyllid
(*Trioza magnoliae*)

Hackberry Psyllid
(*Pachypsylla* sp.)

Mesquite Psyllid
(*Heteropsylla texana*)

PHYLLOXERIDS • PSYLLIDS

191

Suborder Auchenorrhyncha
Superfamily Cercopoidea

APHROPHORID FROGHOPPERS Family Aphrophoridae
Hind tibiae with 1 or 2 spines laterally and a crown of short spines at tip. Nymphs, called spittlebugs, live in frothy foam (spittle) on both grasses and trees. TL 5–13 mm; World 990, NA 30.

CERCOPID FROGHOPPERS Family Cercopidae
Hind tibiae with 1 or 2 spines laterally and a crown of short spines at tip. Many reflexively bleed a distasteful substance from their tarsi. Nymphs, called spittlebugs, live in frothy foam (spittle) on grasses and other herbs. TL 7–13 mm; World 1,500, NA 4.

CLASTOPTERID FROGHOPPERS Family Clastopteridae
Hind tibiae with 1 or 2 spines laterally and a crown of short spines at tip. **Wings modified to form false heads posteriorly** as an antipredator defense. Nymphs, called spittlebugs, live in frothy foam (spittle). TL 2–6 mm; World 100, NA 30.

Superfamily Cicadoidea

LEAFHOPPERS Family Cicadellidae
Often colorful. **One or more rows of small spines extending the length of the hind tibiae.** Found on nearly every type of plant, but usually the food plant is species-specific. Many are economically important pest species. TL 2–31 mm; World 22,000, NA 3,000.

LEAFHOPPERS

Empoasca sp.

Broad-headed Sharpshooter
(*Oncometopia orbona*)

Excultanus excultus

Saddleback Leafhopper
(*Colladonus clitellarius*)

Rugosana querci

Ponana sp.

APHROPHORID FROGHOPPERS

Hill-Prairie
Spittlebug
(*Lepyronia gibbosa*)

Pine Spittlebug
(*Aphrophora cribrata*)

Spittlebug
(*Aphrophora* sp.)

Diamondback Spittlebug
(*Lepyronia quadrangularis*)

Saratoga Spittlebug
(*Aphrophora saratogensis*)

CERCOPID FROGHOPPERS

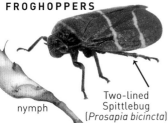

nymph

Two-lined
Spittlebug
(*Prosapia bicincta*)

CLASTOPTERID FROGHOPPERS

Clastoptera octonotata

Pecan Spittlebug
(*Clastoptera achatina*)

Sunflower Spittlebug
(*Clastoptera xanthocephala*)

LEAFHOPPERS

Glassy-winged
Sharpshooter
(*Homalodisca vitripennis*)

Gyponana octolineata

LEAFHOPPERS • FROGHOPPERS

AUCHENORRHYNCHA

Superfamily Cicadoidea cont.

AETALIONDS Family Aetalionidae*
Resemble a large cercopid, but without characteristic spines on hind tibiae and pronotum extends farther over head. **Face is vertical.** Beak extends to hind coxae. Single genus, *Aetalion*, in NA found in AZ, CA, and FL. TL 7–31 mm; World 42, NA 2.

TREEHOPPERS Family Membracidae
Enlarged pronotum that extends posteriorly over abdomen and often projecting forward beyond head. Wings are partially concealed by pronotum. Many species have horns, spikes, or keels while others appear humpbacked. TL 2–21 mm; World 3,500, NA 266.

CICADAS Family Cicadidae
Recognizable by large size, distinctive shape, and presence of 3 ocelli. Males produce a loud species-specific call using a pair of tymbals located basally and ventrolaterally on the abdomen. *Magicicada* species are periodical, emerging in huge numbers in either 13-year or 17-year cycles. TL 25–56 mm; World 3,200, NA 170.

Superfamily Fulgoroidea

ACANALONIID PLANTHOPPERS Family Acanaloniidae
Usually green. **Wings broad, prominently-veined and held vertically.** Hind tibia with only apical spines. Similar to Flatidae, but shaped differently and **lack numerous crossveins in costal area of FW.** TL 10–13 mm; World 80, NA 20.

ACHILID PLANTHOPPERS Family Achilidae
Usually brown or black with **overlapping FW.** Proboscis long, generally 5-segmented with apical segment long, extending to hind coxae. Hind tibiae with or without spine in basal half. Nymphs often found under loose bark. TL 2–16 mm; World 500, NA 55.

KINNARID PLANTHOPPERS Family Kinnaridae*   Not Illus.
Pale, similar to small Cixiidae. Row of spines on second segment of hind tarsomere. Median ocellus near frontoclypeal suture. Females have a reduced ovipositor and abdominal tergites 7–9 with chevron-shaped wax-producing plates. Single genus, *Oeclidius*, found in western NA. TL 2–5 mm; World 100, NA 6.

TROPIDUCHID PLANTHOPPERS
         Family Tropiduchidae   Not Illus.
Brown, green, or yellow. Some elongated with narrow head that usually projects forward. Others with short, stubby body and broad wings that may be ornamented and used in displaying. Anterior margin of vertex often with thick callus. Found on monocots in the understory of tropical forests. TL 7–11 mm; World 620, NA 64.

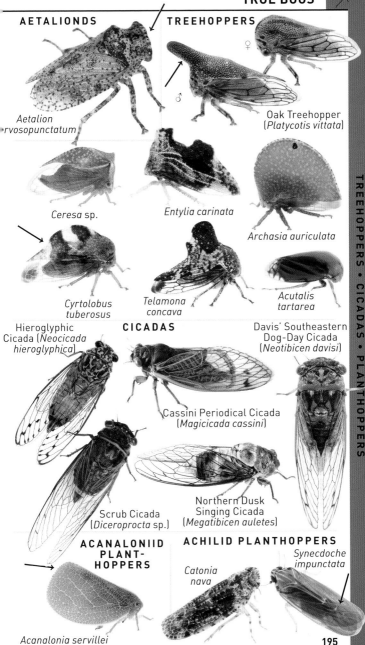

AETALIONDS

Aetalion nervosopunctatum

TREEHOPPERS

♀

♂

Oak Treehopper
(*Platycotis vittata*)

Ceresa sp.

Entylia carinata

Archasia auriculata

Cyrtolobus tuberosus

Telamona concava

Acutalis tartarea

CICADAS

Hieroglyphic Cicada (*Neocicada hieroglyphica*)

Davis' Southeastern Dog-Day Cicada (*Neotibicen davisi*)

Cassini Periodical Cicada (*Magicicada cassini*)

Scrub Cicada (*Diceroprocta* sp.)

Northern Dusk Singing Cicada (*Megatibicen auletes*)

ACANALONIID PLANT-HOPPERS

Acanalonia servillei

ACHILID PLANTHOPPERS

Catonia nava

Synecdoche impunctata

195

AUCHENORRHYNCHA

Superfamily Fulgoroidea cont.

PIGLET BUGS Family Caliscelidae
Similar to issid planthoppers. Brownish to black, small, cylindrical and most with wings reduced, shorter than abdomen. Feed on grasses. TL 5–8 mm; World 200, NA 53.

ISSID PLANTHOPPERS Family Issidae
Usually brown or black with compact bodies, often somewhat compressed. Some species with head projected into a short snout. Hind tibia with spines in addition to those at the apex; front femora and tibiae expanded into leaf-like projections in some species. Found on trees and grasses. TL 2–8 mm; World 930, NA 90.

CIXIID PLANTHOPPERS Family Cixiidae
Median ocellus present. **Hind tibiae long, with few or no lateral spines;** circle of apical spines present. Row of spines on second hind tarsomere. Both pairs of wings relatively large, held roof-like over abdomen. Females with distinct blade-like ovipositor. TL 3–10 mm; World 2,200, NA 180.

DELPHACID PLANTHOPPERS Family Delphacidae
Body generally brown, elongated and parallel sided. Large, flattened spur at apex of hind tibiae. **Basal antennal segments enlarged with single seta distally.** Many are small with reduced wings. TL 2–11 mm; World 2,200, NA 340.

DERBID PLANTHOPPERS Family Derbidae
Delicate, often moth-like in appearance, with long wings and yellow, brown, or cream-colored bodies. **Row of spines on second hind tarsal segment.** Head laterally compressed with apical segment of beak short. Antennae often flattened and short. Large compound eyes; 2 ocelli. Feed on flowering plants or on woody fungi growing on rotting logs. TL 7–11 mm; World 1,700, NA 70.

DICTYOPHARID PLANTHOPPERS Family Dictyopharidae
Green or brown, often with **head prolonged into long, slender beak-like structure. Apex of FW often narrowed; numerous small veins;** HW large, small, or absent. TL 7–13 mm; World 730, NA 85.

FLATID PLANTHOPPERS Family Flatidae
Body compressed (laterally or dorsally) with large rectangular-shaped wings. **FW usually with numerous crossveins in costal area;** submarginal vein paralleling wing margin. Hind tibiae with spines on sides as well as apical spines. TL 7–16 mm; World 1,450, NA 27.

FULGORID PLANTHOPPERS Family Fulgoridae
Many large species, some brilliantly colored. **Head often large and quadrangular.** Wings large with numerous crossveins; HW, generally with reticulated anal area. Some species excrete waxy filaments. TL 7–26 mm; World 700, NA 17.

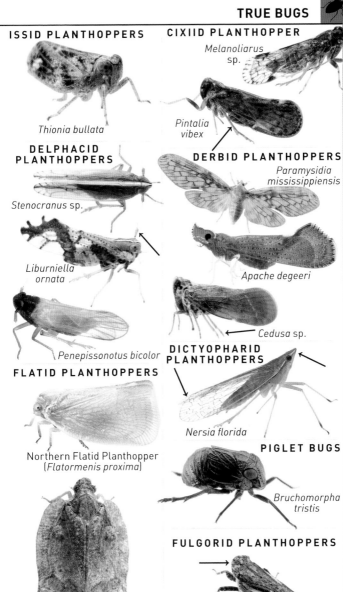

ISSID PLANTHOPPERS

Thionia bullata

CIXIID PLANTHOPPER

Melanoliarus sp.

Pintalia vibex

DELPHACID PLANTHOPPERS

Stenocranus sp.

Liburniella ornata

Penepissonotus bicolor

DERBID PLANTHOPPERS

Paramysidia mississippiensis

Apache degeeri

Cedusa sp.

FLATID PLANTHOPPERS

Northern Flatid Planthopper (*Flatormenis proxima*)

Flataloides scabrosa

DICTYOPHARID PLANTHOPPERS

Nersia florida

PIGLET BUGS

Bruchomorpha tristis

FULGORID PLANTHOPPERS

Cyrpoptus belfrage

HETEROPTERA

CERATOCOMBIDS Family Ceratocombidae*
Tiny, with **long, slender antennae**; segments 3 and 4 with hairs or bristles. Found under litter along water margins or under loose bark of fallen trees. TL 0.5–1.5 mm; World 50, NA 4.

DIPSOCORIDS Family Dipsocoridae* Not Illus.
Tiny, elongate, and flattened. Head distinctly pointed. Ventral side of abdomen with numerous short, dense hairs. Male genitalia asymmetrical. Found in moist habitats near streams under stones. Single genus, *Cryptostemma*, in NA. TL 2–3 mm; World 30, NA 2.

JUMPING SOIL BUGS Family Schizopteridae* Not Illus.
Tiny, dark, beetle-like in appearance. Head strongly deflexed with large eyes. Hind legs modified for jumping. This tropical group ranges into the south-central and eastern states.
TL 0.5–2 mm; World 360, NA 10.

GNAT BUGS Family Enicocephalidae Not Illus.
Brown, slender body with distinctly elongated head constricted behind eyes and at base. Antennae and beak each 4-segmented. FW entirely membranous. Front femora and tarsi thickened; middle and hind tarsi 2-segmented. Spines on front tibia in 1 group. Usually found under stones, bark, or debris, feeding on small insects. May aggregate in groups of thousands and swarm like midges. TL 2.5–5 mm; World 400, NA 10.

AENICTOPHECHEIDS Family Aenictopecheidae* Not Illus.
Small, presumably predaceous bugs. Spines on front tibia stout, arranged linearly in 2 groups of 3, separated by deep central notch. Fully winged or brachypterous. Found in the west and south under large, flat stones in sandy substrate.
TL ~5 mm; World 20, NA 7.

SHORE BUGS Family Saldidae
Brown or black, small, oval and flattened bugs found along shorelines of various water bodies. **Four or 5 long, closed cells in membrane of hemelytra.** Some burrow; often fly for short distances when disturbed. Most predaceous. TL 2–8 mm; World 340, NA 72.

SPINY-LEGGED BUGS Family Leptopodidae* Not Illus.
Similar to saldids, yellowish-brown to dark brown with dark transverse bands on hemelytra. Single adventive species, *Patapius spinosus*, found in the western US. Generally found along streams, but may occur far from water. TL ~2 mm; World 40, NA 1.

Infraorder Cimicomorpha

BED BUGS Family Cimicidae
Brown with flat, oval bodies. Wingless ectoparasites of birds and mammals. *Cimex lectularius* is associated with humans and may become a serious pest. Exhibit traumatic insemination wherein the males punctures the female. TL 5–13 mm; World 110, NA 15.

MINUTE PIRATE BUGS Family Anthocoridae — Not Illus.
Small, elongated, flattened bugs; often dark with white markings. Wings with faint or no veins in membranous area of hemelytra. Male with asymmetrical genitalia. Found on flowers, under bark, or on ground; most are predaceous. TL 2–5 mm; World 500, NA 90.

LASIOCHILIDS Family Lasiochilidae* — Not Illus.
Similar to anthocorids, but with dorsolateral tergites on abdominal segments 1 and 2. Male with symmetrical genitalia. Found under bark and on bracket fungi. TL 2.5–5 mm; World 60, NA 7.

LYCTOCORIDS Family Lyctocoridae* — Not Illus.
Similar to anthocorids, but differ in appearance of asymmetrical male genitalia and presence of apophyses on anterior margin of abdominal segment 7. Single genus, *Lyctocoris*, found in NA. TL 2–5 mm; World 20, NA 8.

BAT BUGS Family Polyctenidae* — Not Illus.
Wingless ectoparasites of bats, lacking compound eyes and ocelli. Front legs short and flattened; middle and hind legs long and slender. Body covered with bristles. Single genus, *Hesperoctenes*, with 2 species found in southern US. TL ~5 mm; World 32, NA 2.

SOIL • GNAT • SHORE • BED BUGS

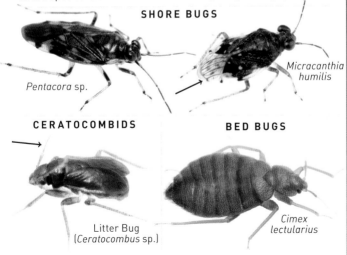

SHORE BUGS

Pentacora sp.

Micracanthia humilis

CERATOCOMBIDS

BED BUGS

Litter Bug
(*Ceratocombus* sp.)

Cimex lectularius

Infraorder Cimicomorpha

MINUTE BLADDER BUGS Family Microphysidae*   Not Illus.
Minute, dark-colored and somewhat variable in appearance, from
elongate to broadly oval and flattened. Has ocelli, symmetrical male
genitalia, and 2-segmented tarsi. TL 1.5 mm; World 30, NA 5.

DAMSEL BUGS Family Nabidae
Slender, often brown or black, with 4-segmented beak. **Front
femora slightly enlarged. FW with small cells around margin in
membranous area.** Predaceous on other small insects.
TL 2–13 mm; World 400, NA 40.

PLANT BUGS Family Miridae
Large group with variably colored slender, delicate bodies. Lack
ocelli. **Cuneus present, FW often deflected downward;** 2 closed
cells in FW membrane. Antennae long and thin. Specialized setae on
middle and hind femora. Many are herbivorous, others predaceous;
some considered pests. TL 2–16 mm; World 10,000, NA 2,000.

THAUMASTOCORIDS Family Thaumastocoridae*   Not Illus.
Small, flattened, either elongated or oval. Two introduced species,
one in CA, *Thaumastocoris peregrinus*, found on Eucalyptus, the other
in FL, *Xylastodoris luteolus*, found on Royal Palm.
TL ~2.5 mm; World 20, NA 2.

LACE BUGS Family Tingidae
Small with dorsal surface of body ornately sculptured, along with
numerous cells in wings, giving adults a lace-like appearance. Lack
ocelli. Beak and antennae with 4 segments. Scutellum absent or
greatly reduced; pronotum with posterior margins angulate.
TL 2–5 mm; World 2,350, NA 165.

ASSASSIN BUGS Family Reduviidae
Often brightly colored with **3-segmented beak** folded into striated
groove in prosternum. Head often constricted behind eyes. Antennae
long and thin. Abdomen expanded at middle in many species. Species
are either predatory or bloodsucking; some species associated with
mammal nests and burrows. TL 7–41 mm; World 7,000, NA 195.

CURALIIDS Family Curaliidae*   Not Illus.
Small, male with ruby-red head, thorax, cuneus and abdomen.
Antennae white and red banded; legs white. Single species, *Curalium
cronini*, found in LA and FL. TL ~1.5 mm; World 1, NA 1.

LACE BUGS

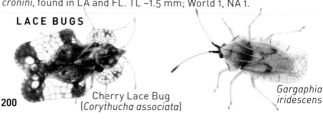

Cherry Lace Bug
(*Corythucha associata*)

*Gargaphia
iridescens*

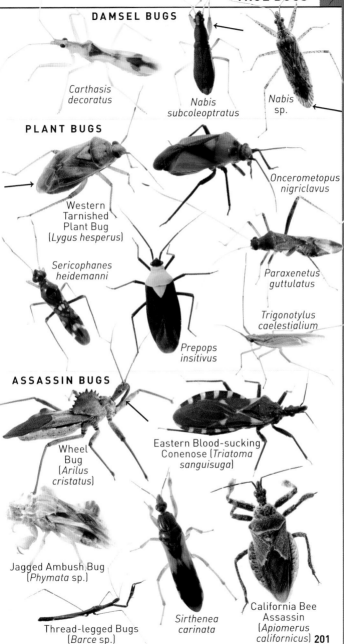

DAMSEL BUGS

Carthasis decoratus

Nabis subcoleoptratus

Nabis sp.

PLANT BUGS

Oncerometopus nigriclavus

Western Tarnished Plant Bug (*Lygus hesperus*)

Sericophanes heidemanni

Paraxenetus guttulatus

Trigonotylus caelestialium

Prepops insitivus

ASSASSIN BUGS

Wheel Bug (*Arilus cristatus*)

Eastern Blood-sucking Conenose (*Triatoma sanguisuga*)

Jagged Ambush Bug (*Phymata* sp.)

Thread-legged Bugs (*Barce* sp.)

Sirthenea carinata

California Bee Assassin (*Apiomerus californicus*) **201**

Infraorder Gerromorpha

WATER STRIDERS Family Gerridae
Dark, elongated bodies with long legs. Front legs noticeably shorter than others; **hind femur longer than abdomen.** Middle legs arising closer to hind legs than front legs. Ocelli present, but small. Tarsi 2-segmented; claws preapical. Found on surface of ponds and slow streams. TL 2–16 mm; World 750, NA 46.

SMALL WATER STRIDERS Family Veliidae
Dark, elongated bodies. Middle legs generally arising about halfway between front and hind legs. If arising closer to hind legs, tarsi 1-segmented; claws preapical. Hind femora not extending beyond apex of abdomen. Ocelli absent. Pronotum broader than abdomen. Found on surface of both ponds and streams, including areas with rapids. TL 1.5–6.4 mm; World 1,000, NA 34.

VELVET WATER BUGS Family Hebridae*
Small, oblong bodies entirely covered with velvety hairs. **Pronotum broad anteriorly.** Found on surface of shallow pools with abundance of aquatic vegetation. TL 1–3 mm; World 220, NA 15.

MACROVELIID SHORE BUGS Family Macroveliidae*
Two species found in western states. Similar to mesoveliids, but with 6 closed cells in hemelytra. **Pronotum with posteriorly projecting lobe covering scutellum.** Found along shores of springs and streams, often on moss. TL ~5 mm; World 3, NA 2.

WATER MEASURERS Family Hydrometridae
Gray or brown, slender body with **bulging eyes at midpoint of elongated head**. Legs long, thread-like. Antennae 4-segmented, inserted near front of head and slightly longer than length of head. Usually wingless. Found on emergent and floating vegetation along edges of marshes, ponds, and streams.
TL 7–16 mm; World 130, NA 9.

WATER TREADERS Family Mesoveliidae*
Small, with elongated greenish or yellowish bodies, either winged with ocelli or wingless lacking ocelli. Only anterior basal part of FW thickened; clavus membranous and **membrane area lacking veins.** Found on water surface feeding on small insects. Single genus, *Mesovelia*, in NA. TL 2–5 mm; World 50, NA 3.

WATER TREADERS

Mesovelia mulsanti

HETEROPTERA

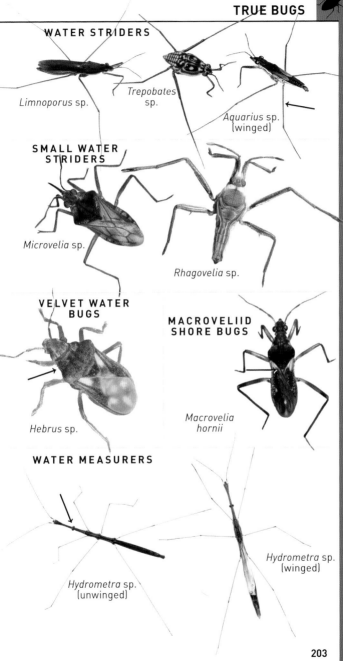

WATER STRIDERS

Limnoporus sp.

Trepobates sp.

Aquarius sp. (winged)

SMALL WATER STRIDERS

Microvelia sp.

Rhagovelia sp.

VELVET WATER BUGS

Hebrus sp.

MACROVELIID SHORE BUGS

Macrovelia hornii

WATER MEASURERS

Hydrometra sp. (unwinged)

Hydrometra sp. (winged)

HEMIPTERA

Infraorder Nepomorpha

WATER BOATMEN Family Corixidae
Body elongate and dorsoventrally flattened. Narrow dark crosslines on pronotum and hemelytra. Front legs short with **1-segmented scoop-shaped tarsi**; hind legs elongated and oar-like with numerous hairs. Most common in freshwater ponds and lakes, predaceous or feeding on algae. TL 2–10 mm; World 600, NA 125.

SAUCER BUGS Family Naucoridae
Brown to green, usually oval and somewhat flattened. Front legs with **enlarged femora**. Lack ocelli. Tarsi 2-segmented with long claws on middle and hind legs. Found in ponds and streams on submerged vegetation. TL 6–15 mm; World 400, NA 22.

GIANT WATER BUGS Family Belostomatidae
Family contains the largest species in the order. Body generally brown, elongate-oval, and flattened. Front legs raptorial. **Pair of respiratory straps** may be visible extending from underneath wings distally. Found in ponds, lakes, and streams feeding on insects, tadpoles, and fish. TL 12–77 mm; World 160, NA 19.

WATERSCORPIONS Family Nepidae
Brownish, most narrowly elongate with a **long caudal breathing tube. Forelegs raptorial.** Well-developed wings, but seldom fly. Do not swim, but crawl along substrate or on submerged plants in ponds and vegetated stream margins. TL 15–46 mm (excluding breathing tube); World 270, NA 13.

BACKSWIMMERS Family Notonectidae
Dorsal surface of body is strongly convex. Front legs unmodified (not spatulate). Hind legs modified for swimming with long hairs. Wings lightly colored; venter darkly colored. Swim upside down and body is typically angled downward when at rest. Predaceous, feeding on insects, tadpoles, and small fish. TL 5–16 mm; World 400, NA 32.

PYGMY BACKSWIMMERS Family Pleidae*   Not Illus.
Similar to notonectids, but much smaller. Dorsal surface highly convex; wings forming hard shell. TL ~2 mm; World 40, NA 5.

TOAD BUGS Family Gelastocoridae
Oval and flattened body, often cryptically colored, with **bulging eyes**. Antennae inserted beneath eyes. **Forelegs short.** Generally rest with anterior portion of body raised; capable of jumping. Found along shores of ponds and streams. TL 5–11 mm; World 110, NA 8.

VELVETY SHORE BUGS Family Ochteridae*   Not Illus.
Dark, mottled, oval body with velvety pubescent covering. Beak long, reaching to hind coxae. Antennae short, inserted beneath eyes. Two ocelli. Found along shorelines of slow streams, ponds, and seeps. Single genus, *Ochterus*, in NA. TL 5 mm; World 70, NA 6.

SAUCER BUGS

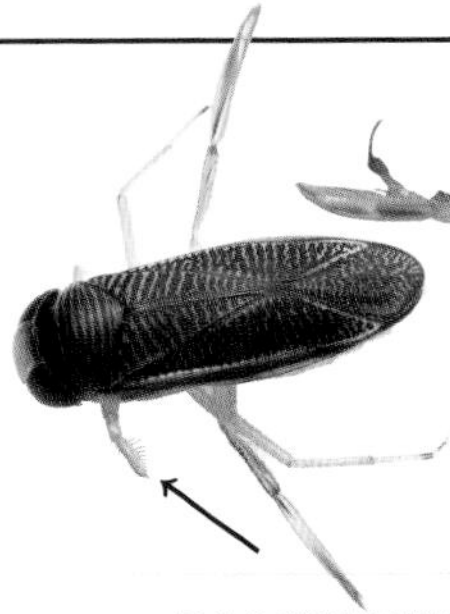

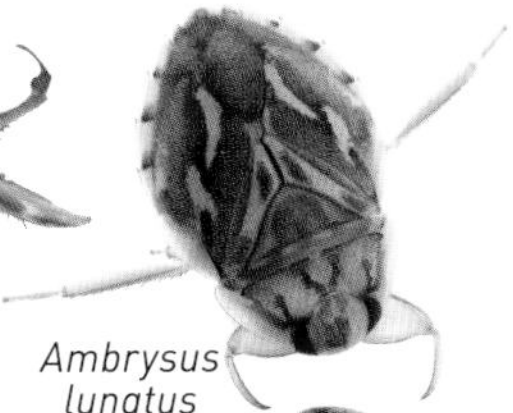

WATER BOATMAN

BACKSWIMMERS

Ambrysus lunatus

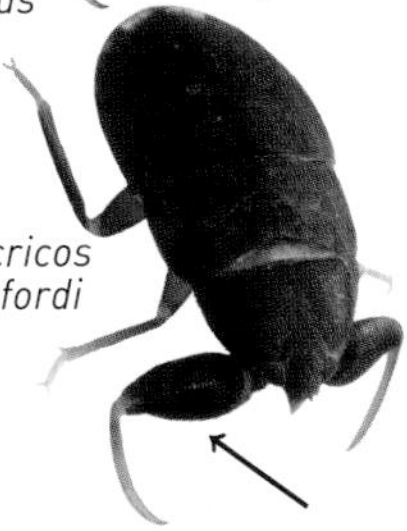

Cryphocricos hungerfordi

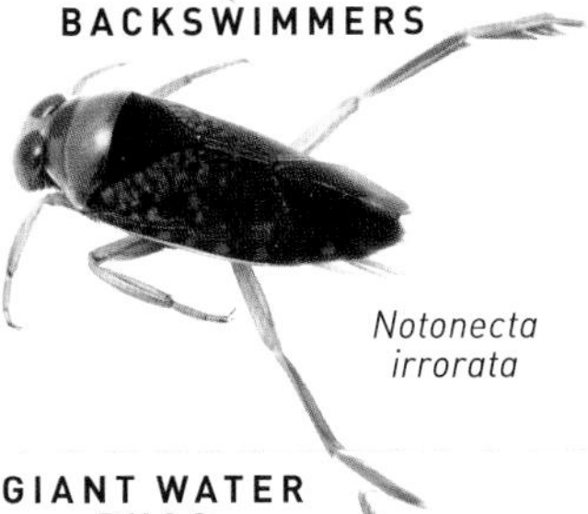

Notonecta irrorata

GIANT WATER BUGS

Belostoma sp.

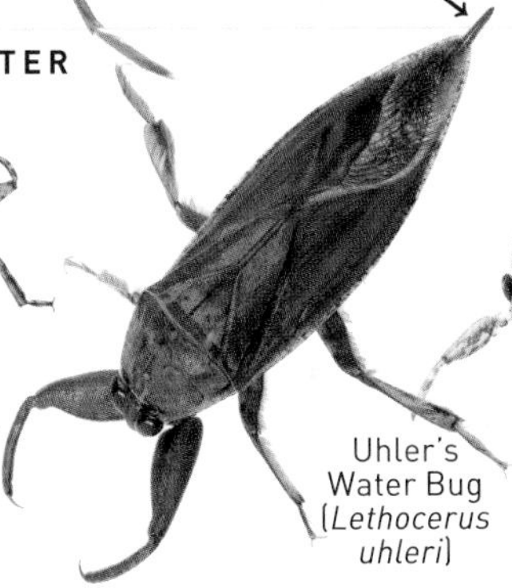

Uhler's Water Bug (*Lethocerus uhleri*)

Abedus sp.

WATERSCORPIONS

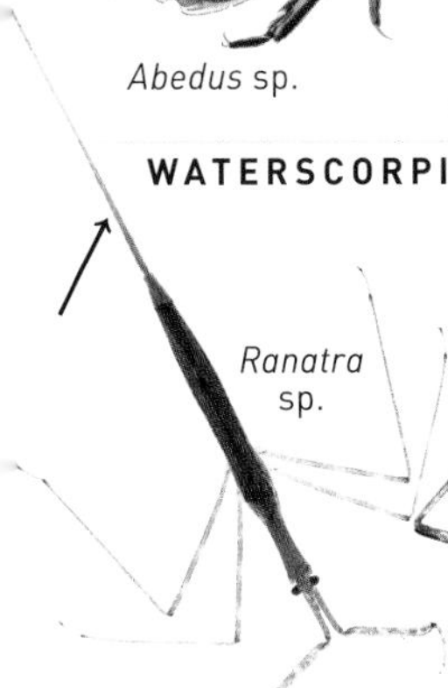

TOAD BUGS

Curicta scorpio

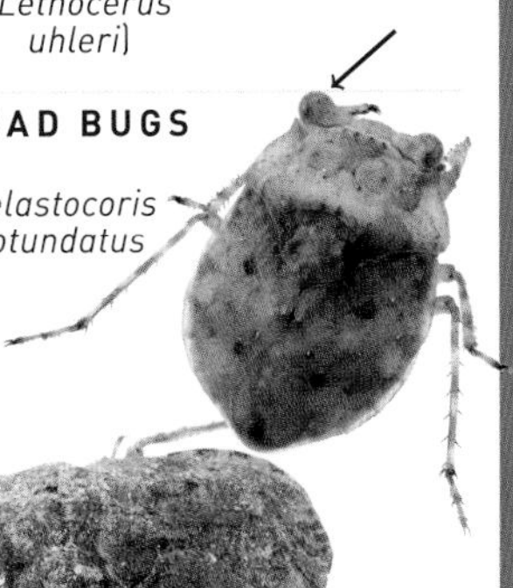

Gelastocoris rotundatus

Ranatra sp.

Nerthra sp.

Infraorder Pentatomomorpha

FLAT BUGS Family Aradidae

Brownish, flattened with **abdomen extending beyond short wings** and somewhat granular in appearance. Antennae and beak each 4-segmented. Ocelli lacking. Found under loose bark and between crevices of decaying trees. TL 2–11 mm; World 2,000, NA 127.

BROAD-HEADED BUGS Family Alydidae

Usually yellowish or brown. **Head generally as broad or broader than pronotum;** body elongate. Some with **enlarged hind femora.** Many have pungent scent glands. Found on vegetation along roadsides and in wooded areas. Some can become serious pests of crops. TL 10–18 mm; World 250, NA 30.

LEAF-FOOTED BUGS Family Coreidae

Usually dark colored; numerous parallel veins in FW membrane. Hind tibiae in many with **leaf-like expansions.** Abdominal margin raised and expanded beyond wings in many species; often revealing a light-dark pattern. Nearly all are herbivorous, and some can be serious pests. TL 7–41 mm; World 1,000, NA 88.

SCENTLESS PLANT BUGS Family Rhopalidae

Variable in color and shape. Lack scent glands. Numerous veins in FW membrane. All are herbivorous, many found on weeds. Some, like the Eastern Boxelder Bug, will enter houses in large numbers during the fall. TL 5–16 mm; World 200, NA 40.

ARTHENEIDS Family Artheneidae*   Not Illus.

Similar to lygaeids, yellowish or straw-colored and somewhat flattened. Margins of pronotum broadly expanded. Antennae 4-segmented, somewhat expanded distally. Two adventive species found in NA. Found on cattail seed heads. TL ~5 mm; World 20, NA 2.

OXYCARENIDS Family Oxycarenidae   Not Illus.

Similar to artheneidae, but lack expanded margins of pronotum. Spiracles on dorsum of abdominal segment 2. Feed on the seeds of numerous plant families. Some species are pests on cotton. TL 2–5 mm; World 150, NA 12.

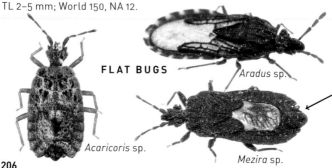

FLAT BUGS

Aradus sp.

Acaricoris sp.

Mezira sp.

HETEROPTERA

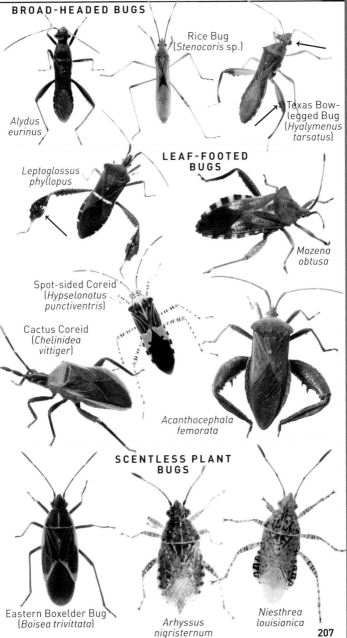

BROAD-HEADED BUGS

Rice Bug
(*Stenocoris* sp.)

*Alydus
eurinus*

Texas Bow-
legged Bug
(*Hyalymenus
tarsatus*)

**LEAF-FOOTED
BUGS**

*Leptoglossus
phyllopus*

*Mozena
obtusa*

Spot-sided Coreid
(*Hypselonotus
punctiventris*)

Cactus Coreid
(*Chelinidea
vittiger*)

*Acanthocephala
femorata*

**SCENTLESS PLANT
BUGS**

Eastern Boxelder Bug
(*Boisea trivittata*)

*Arhyssus
nigristernum*

*Niesthrea
louisianica*

HETEROPTERA

Infraorder Pentatomomorpha cont.

BLISSIDS Family Blissidae
Dark, elongated, and similar to lygaeids, but lack conspicuous punctures on hemelytra. Spiracles on abdominal segment 7 located ventrally. Some with **reduced wings**. Found on monocots, where they suck sap. TL 2–16 mm; World 435, NA 30.

CYMIDS Family Cymidae   Not Illus.
Small, elongate or oval, pale yellow or straw-colored, coarsely punctate bugs that look superficially like seeds. Bucculae short. Ocelli present. Spiracle on abdominal segment 7 located ventrally. Found on sedges and rushes. TL 2–5 mm; World 54, NA 10.

BIG-EYED BUGS Family Geocoridae
Small, short-bodied with large, **kidney-shaped eyes protruding** and often curving back over outer margins of pronotum. Head, thorax, and hemelytra conspicuously punctate. Some predatory, others herbivorous; generally found on or near the ground. TL 2–5 mm; World 300, NA 30.

STILT BUGS Family Berytidae
Slender, elongate with **long, thin legs and clubbed antennae.** Antenna 4-segmented with terminal segment enlarged. Generally herbivorous and most are host-specific. TL 5–10 mm; World 170, NA 12.

HETEROGASTRIDS Family Heterogastridae*   Not Illus.
Gray or brownish, somewhat oval and elongate. Spiracles on underside of abdomen. Large, closed cell at base of FW membrane. Three western species in NA, belonging to the genus *Heterogaster*. Feed on urticaceous plants. TL ~6 mm; World 24, NA 3.

SEED BUGS Family Lygaeidae
Many species brightly colored orange or red and black. **FW membrane with 5 distinct veins** and usually a single closed cell. All abdominal spiracles are located dorsally. Many species feed chiefly on seeds. TL 2–20 mm; World 1,000, NA 80.

NINIDS Family Ninidae*   Not Illus.
Tan or reddish with apex of scutellum bifid. Clavus partially hyaline. Beak is distinctly 2-stepped. Eat seeds, primarily of Cyperaceae and Juncaceae. Single species, *Cymoninus notabilis*, found along the Gulf from TX to FL. TL 7–11 mm; World 14, NA 1.

PACHYGRONTHIDS Family Pachygronthidae *   Not Illus.
Small, brown, coarsely punctate, with front femora thickened and finely spinose. All abdominal spiracles located ventrally. Many found on grasses, sedges, and rushes. TL 5–8 mm; World 80, NA 7.

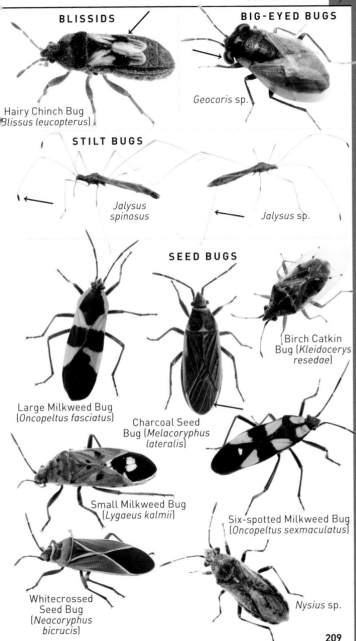

BLISSIDS

Hairy Chinch Bug
(*Blissus leucopterus*)

BIG-EYED BUGS

Geocoris sp.

STILT BUGS

Jalysus spinosus

Jalysus sp.

SEED BUGS

Birch Catkin Bug (*Kleidocerys resedae*)

Large Milkweed Bug
(*Oncopeltus fasciatus*)

Charcoal Seed Bug (*Melacoryphus lateralis*)

Small Milkweed Bug
(*Lygaeus kalmii*)

Six-spotted Milkweed Bug
(*Oncopeltus sexmaculatus*)

Whitecrossed Seed Bug
(*Neacoryphus bicrucis*)

Nysius sp.

Infraorder Pentatomomorpha cont.

ASH-GRAY LEAF BUGS Family Piesmatidae   Not Illus.

Small, gray or tan, usually with a reticulately sculptured corium and clavus and pronotum with ridges; jugum extends well beyond tylus. Feed on plants, including pigweed. TL 2–4 mm; World 44, NA 7.

DIRT-COLORED SEED BUGS Family Rhyparochromidae

Generally brown or mottled, often with **front femora enlarged and spinose.** Forward-curving abdominal suture between sterna 4 and 5 incomplete. Feed on mature seeds. TL 5–10 mm; World 2,100, NA 170.

SHIELD BUGS Family Acanthosomatidae

Similar to pentatomids, but **tarsi 2-segmented**. Females of a number of species guard eggs and young nymphs. Found feeding mostly on woody plants. TL 12–36 mm; World 180, NA 5.

BURROWING BUGS Family Cydnidae

Body usually oval, convex, black or reddish-brown, and heavily sclerotized. **Legs often spiny.** Beetle-like, but with clearly visible membranous wings. Generally found beneath stones or boards, in sand, and amongst roots of grass tufts. TL 2–8 mm; World 770, NA 41.

STINK BUGS Family Pentatomidae

Generally rounded or ovoid shape with distinctive **pronotal "shoulders."** Antennae 5-segmented. **Tarsi 3-segmented.** Sternum usually lacking median longitudinal keel. Many brightly colored. Most species phytophagous, some recognized as pests; others are predaceous. TL 5–20 mm; World 5,000, NA 220.

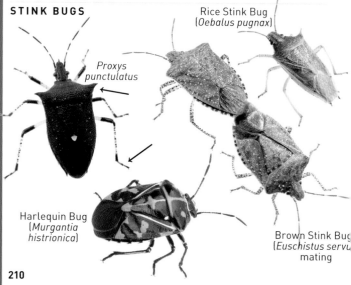

STINK BUGS

Proxys punctulatus

Rice Stink Bug
(*Oebalus pugnax*)

Harlequin Bug
(*Murgantia histrionica*)

Brown Stink Bug
(*Euschistus servu...*
mating

HETEROPTERA

210

DIRT-COLORED SEED BUGS

Eremocoris depressus

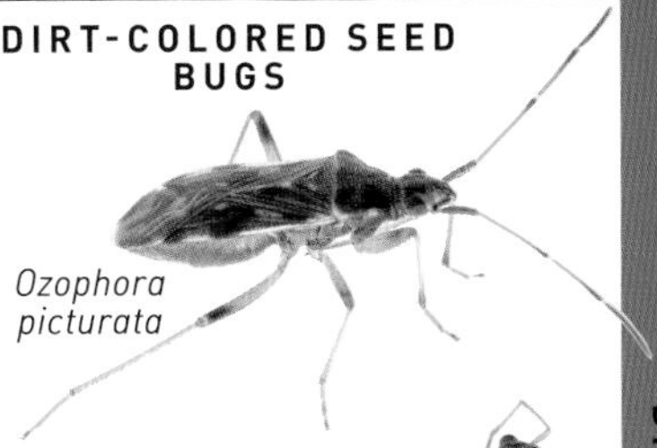

Ozophora picturata

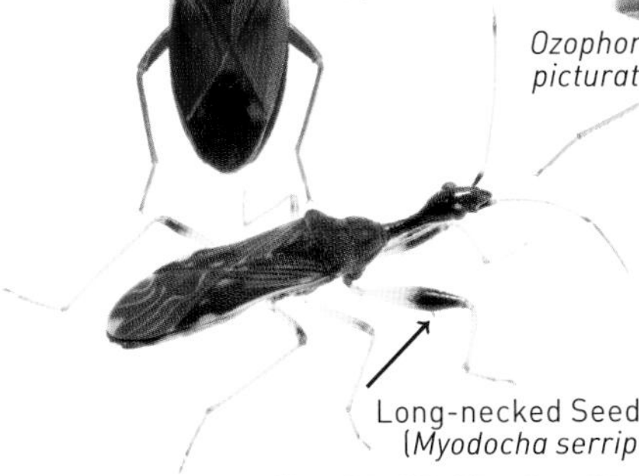

Long-necked Seed Bug (*Myodocha serripes*)

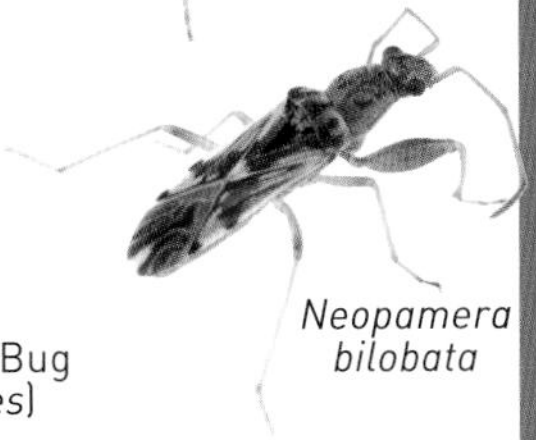

Neopamera bilobata

SHIELD BUGS

Birch Shield Bug (*Elasmostethus interstinctus*)

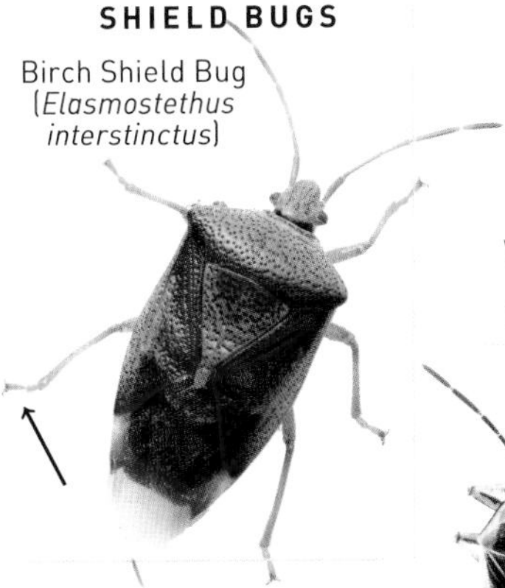

BURROWING BUGS

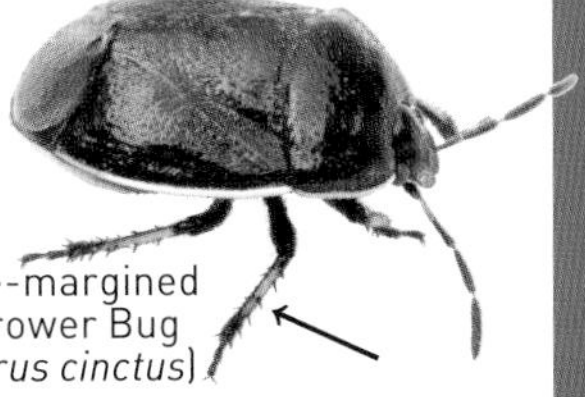

White-margined Burrower Bug (*Sehirus cinctus*)

STINK BUGS

Bagrada Bug (*Bagrada hilaris*)

Brown Marmorated Stink Bug (*Halyomorpha halys*)

Spined Soldier Bug (*Podisus maculiventris*)

Green Stink Bug (*Chinavia hilaris*)

211

Infraorder Pentatomomorpha cont.

PLATASPIDS Family Plataspidae

Strongly ovoid, convex, and beetle-like. Antennae appear 4-segmented. Scutellum covers all of abdomen. **Tarsi 2-segmented.** Wings long and folded under scutellum. Common introduced species, *Megacopta cribraria*, that feeds on kudzu. Recently a second introduced species, *Brachyplatys subaeneus*, has been discovered in FL. TL 5–8 mm; World 560, NA 2.

SHIELD-BACKED BUGS Family Scutelleridae

Strongly ovoid, convex, and beetle-like. Scutellum enlarged, covering abdomen and wings. Corium narrow, not extending to anal margin of wing. Tarsi 3-segmented. Variably colored. All are herbivorous, and some are considered pests.
TL 5–21 mm; World 500, NA 37.

EBONY BUGS Family Thyreocoridae

Small, black, shiny, convex and oval in shape. Tibiae with small spines or are spineless. Scutellum covers most of abdomen and wings. Antennae 5-segmented; beak 4-segmented. Feed on flowers and developing seeds. TL 3–6 mm; World 212, NA 40.

BORDERED PLANT BUGS Family Largidae

Usually stout-bodied with contrasting colored edges of hemelytra, but some ant-like in appearance. Ocelli lacking. Four-segmented beak. Typically ground-dwelling. TL 7–21 mm; World 190, NA 20.

RED BUGS Family Pyrrhocoridae

Elongated, brightly colored, and resembling lygaeids, but lack ocelli and with **numerous branched veins and cells in membranous area of FW.** Opening behind base of middle legs reduced. Found in the southern states. TL 7–20 mm; World 400, NA 10.

PLATASPIDS

Kudzu Bug
(*Megacopta cribraria*)

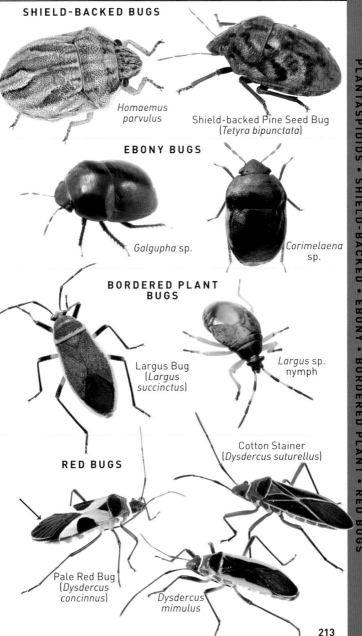

SHIELD-BACKED BUGS

Homaemus parvulus

Shield-backed Pine Seed Bug
(*Tetyra bipunctata*)

EBONY BUGS

Galgupha sp.

Corimelaena sp.

BORDERED PLANT BUGS

Largus Bug
(*Largus succinctus*)

Largus sp. nymph

RED BUGS

Cotton Stainer
(*Dysdercus suturellus*)

Pale Red Bug
(*Dysdercus concinnus*)

Dysdercus mimulus

213

Hoplothrips sp.
Phlaeothripidae

Subadults and immature
Tube-tailed Thrips
(Phlaeothripidae)

THRIPS
ORDER THYSANOPTERA

Derivation of Name
- Thysanoptera: *thysan*, a fringe; *optera*, wing. Refers to fringed edge of wings with numerous long hairs. Thrips (both singular and plural) is derived from a Greek word meaning *wood louse*.

Identification
- 0.5–5 mm TL.
- Small, slender, long and cylindrical.
- Usually with 4 very narrow wings with limited or no venation; wings with fringe of long hairs.
- Have only 1 mandible; the right mandible is vestigial and thought to have no function. Other mouthparts are modified for piercing and scraping plant material and blood of other insects.
- Antennae are short, as long as head or slightly longer and usually 7–8 segments, but ranging between 4 and 9. Segments 3 and 4 generally with sensory organ.
- Legs are short and unmodified except for the tarsi, which are bladder-like at the end.

Classification
- Classification has not changed much, and there are 2 suborders Terebrantia and Tubulifera.
- Oldest fossils to date are from the Triassic; found in VA and Kazakhstan: *Triassothrips virginicus* and *Kazacholthrips triassicus*.
- Currently 5 families in NA.
- World: 6,300 species; North America: 1,000 species.

Range
- Found throughout the world except Antarctica and extreme Arctic.

Similar orders
- Embioptera (p. 148): tarsi 3-segmented, basal segment enlarged; cerci 1–2-segmented; larger, 4–10 mm.
- Coleoptera (p. 294): with elytra usually extending over length of abdomen, but reduced in Staphylinidae.
- Diptera (p. 436): 1 pair of wings; antennae 3-segmented; mouthparts sucking; softer body.
- Hymenoptera (p. 236): antennae long; tarsi 5-segmented; HW smaller than FW; wings with more veins.

Food
- Most species feed on plant tissue, but a few feed on pollen and fungus spores, and some are predacious.

Behavior
- May be solitary, subsocial, or eusocial.
- In Panama a species that feeds on lichen lives in groups of up to 200 individuals. Adults guard eggs and larvae, and lay chemical trails to coordinate foraging.
- Some species in Phlaeothripidae are morphologically and behaviorally specialized into castes. Offspring are tended for extended periods of time. Immatures care for adults in some eusocial species. Adults are haplodiploid and have genetic sex determination (females hatch from diploid eggs and males hatch from haploid eggs) occurs in some Hymenoptera.
- Many, including wingless species, are aerial dispersers.

Life Cycle
- Considered between holometabolous and hemimetabolous.
- First 2 instars look like smaller adults but can have wingpads start to form like in hemimetabolous insects. The third and fourth instars, called proupa and pupa, respectively, have external wingpads, are not mobile, and do not feed.
- Parthenogenesis occurs in many species (females are capable of self-fertilization).
- Eggs are laid in vegetation when adults have an ovipositor, otherwise they are laid in crevices or under bark.
- Nymphs generally feed where the eggs were laid.

Importance to Humans
- Many of the plant-eating species are harmful to crops, fruits, and garden flowers by scarring the leaves, causing corky tissue development and leaving fecal droplets on leaves. Large populations can induce premature flower loss.
- Some species are vectors of tospoviruses.
- Some carnivorous species useful for biocontrol of other thrips or mites

Collecting and Preserving
Immatures & Adults
- Found in flowers, on vegetation, fungi, leaf litter, debris, grasses, roots, and soils.
- Handpick individuals using a small brush; beat vegetation over a small white tray or sheet; use Berlese funnel on leaf litter, fungus, and decaying wood.
- Best collected in AGA (10 parts of 60% ethanol, 1 part glycerine, 1 part acetic acid).
- Preserve in 60–70% ethanol, out of dark and in a freezer to maintain colors. Mount on slides for critical examination.
- Collect in 70% ethanol; permanently mount on slides.

Resources
- ThripsWiki—thrips.info, General Information
- *Thysanoptera: An Identification Guide* by Mound and Kibby, 1998

Fringed Wings

Antennal Sensory Organs

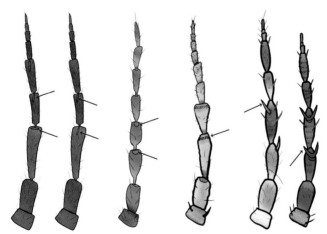

Aeolothripidae   Merothripidae   Heterothripidae   Thripidae

Suborder Terebrantia

PREDATORY THRIPS Family Aeolothripidae
Usually dark with wings broad and rounded at the tip with banded coloring. Antennae 9-segmented; segment 3 with flat sensorium either elongated longitudinally or transverse. Ovipositor curved upward. Most feed on small insects and mites and can be found on flowers. TL 1–5 mm; World 216, NA 44.

MELANTHRIPID THRIPS Family Melanthripidae*   Not illus.
Similar to Aeolothripidae, but adult females bear a pair of lobes on the posterior margin of the seventh sternite; each with a pair of setae. Thrips in this family are all phytophagous, breeding in flowers. A single genus, *Ankothrips*, is found in the western U.S. TL 1–5 mm; World 67, NA 7.

TREE THRIPS Family Heterothripidae*   Not illus.
Dark body with lots of small hairs, FW has 2 complete rows of hairs on veins. Antennae 9-segmented with small circular sensory organ on segments 3 and 4 resembling blisters. Can often be found on jack-in-the-pulpit. TL 0.5–2 mm; World 89, NA 20.

JUMPING THRIPS Family Merothripidae*   Not illus.
Antennae 8-segmented; 3 and 4 with apical tympanum-like sensorium. Pronotum with a dorsal long suture on each side. FW is narrow and pointed, front and hind legs thickened for jumping. S10 tergite usually with a pair of large trichobothria. Ovipositor curved downward and reduced. Species are fungus feeding, found on mushrooms and under bark of dead branches.
TL 0.5–1 mm; World 17, NA 4.

COMMON THRIPS Family Thripidae
FW is more narrow than predatory thrips, and pointed at the tip. Dark, slightly flattened abdomen generally with small hairs throughout. Antennae 6–9-segmented; segments 3 and 4 with a protruding forked or simple sensory cone. This group can be pests on plants, and almost all of them are plant eaters; a few are predators. They are often found on leaves, where they breed, not flowers. TL 0.5–2 mm; World 2,100, NA 260.

Suborder Tubulifera

TUBE-TAILED THRIPS Family Phlaeothripidae
Larger and thicker than other families, dark brown or black with light-colored mottled wings. **Last abdominal segment tube-like**. FW lack longitudinal veins and setae except at base. Females lack ovipositor. Many are associated with dead plant tissues, where they feed on fungi or their decaying products; others are leaf-feeding. TL 0.5–15 mm; World 3,500, NA 350.

PREDATORY THRIPS

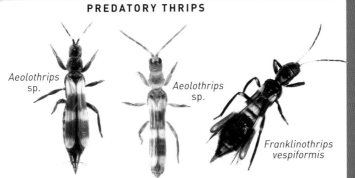

Aeolothrips sp.

Aeolothrips sp.

Franklinothrips vespiformis

COMMON THRIPS

Caliothrips sp.

Frankliniella sp.

TUBE-TAILED THRIPS

Haplothrips sp.

Clover Thrip (*Haplothrips leucanthemi*)

Adraneothrips sp.

Trachythrips watsoni

Poultry Body Louse
(*Menacanthus stramineus*)
Menoponidae

Blaste posticata nymph
Psocidae

BOOKLICE, BARKLICE, & PARASITIC LICE
ORDER PSOCODEA

Derivation of Name
- Psocodea: *psokos*, rubbed small. Refers to the gnawing habit of this group when feeding.

Classification
- At various times the true (parasitic) lice have been divided into 2 separate orders, the chewing lice (Mallophaga) and sucking lice (Anoplura), or combined into the single order Phthiraptera.
- The free-living book- and barklice were placed in Psocoptera.
- Recent molecular evidence has resulted in the free-living and parasitic groups being combined into 1 order, the Psocodea.
- Currently 3 suborders are recognized: Psocomorpha (barklice), Troctomorpha (true lice and booklice), and Trogiomorpha (barklice). The order contains 28 families of bark- and booklice and 16 families of true lice in North America.
- World: 11,000 species (parasitic 5,000; bark-/ booklice 5,922); North America: 1,220 species (parasitic 920; bark-/booklice 318).

Range
- Found worldwide. A number of species are cosmopolitan and have been introduced by humans.

Similar orders
- Blattodea, termites (p. 166): tarsi 4-segmented; antennae usually short and moniliform; FW and HW of winged forms similar in size.
- Zoraptera (p. 110): antennae 9-segmented; cerci present, 1-segmented.
- Embioptera (p. 148): tarsi 3-segmented, basal segment of front tarsi enlarged.
- Neuroptera (p. 280): tarsi 5-segmented.

Forewing Venation

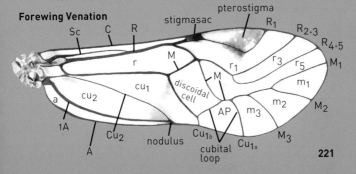

Chewing and Sucking Lice
Suborder Troctomorpha (in part)

Identification
- Usually less than 5 mm TL.
- Small, soft-bodied; compound eyes small or absent and lack ocelli; antennae short, 3–5-segmented and thread-like or variously modified, often concealed in grooves on head in chewing lice; legs short (1–2 segmented); wings absent; cerci lacking. Chewing lice with: chewing mouthparts; head as wide or wider than thorax; 1–2-tarsal claws. Sucking lice with: sucking mouthparts; head narrower than thorax; 1 tarsal claw.

Food
- Ectoparasites of birds, mammals, and humans.

Behavior
- Chewing lice spend their entire life on their host; transmission from one host to another usually occurs when hosts come in contact.
- Chewing lice feed on feathers of birds or hair and skin of mammals.
- Sucking lice feed on blood of mammals, and bites are often irritating.
- Individuals are unable to survive long off a host.
- Species are generally specific to a species or group of species and often to a particular area of the host's body.

Life Cycle
- Hemimetabolous. Immatures similar in appearance to adults.
- Eggs are laid on host's body; usually attached to hair or feathers. The Human Body Louse lays eggs on clothing.
- Three instars.

Importance to Humans
- Can be irritating pests of their host. Often heavily infested host can become emaciated.
- One species, Body Louse (*Pediculus humanus*), is an important disease vector of epidemic typhus and relapsing fever.
- Sucking lice are often found on domestic animals, but not birds.

Collecting and Preserving
- Look for host; collect in 70% ethanol; permanently mount on slides.

Resources
- General Information—phthiraptera.info
- *The Sucking Lice of North America* by Kim et al., 1986
- *Chewing and Sucking Lice as Parasites of Mammals and Birds* by Price and Graham, 1997

Dog Biting Louse
(*Trichodectes canis*)
Trichodectidae

Booklice and Barklice
Suborder Trogiomorpha, Troctomorpha (in part), Psocomorpha

Identification
- 2–13 mm TL; most less than 5 mm.
- Small, soft-bodied; compound eyes present and variously pronounced; ocelli present or absent; antennae usually long and slender, filiform; mouthparts chewing; front of head (postclypeus) somewhat bulging in appearance; legs with 2–3-segmented tarsi; most winged, but some reduced or absent; wings, when present, numbering 4 and usually held roof-like over abdomen, venation reduced; cerci lacking.

Food
- Feed chiefly on decaying organic material, molds, fungi, and lichen.
- Some species feed on paper products.

Behavior
- Typically encountered in soil litter or on and around dead tree branches, but also found on stones, logs, and fences.
- Some species are gregarious, living in silken webs surrounding tree trunks and branches.
- Males may perform a mating dance.

Life Cycle
- Hemimetabolous.
- Immatures are similar in appearance to adults with wings smaller or absent.
- Generally 6 instars.
- After fertilization, females lay eggs under bark, leaves, or a silk mat.

Importance to Humans
- Some species may be considered a nuisance when locally abundant.
- Some booklice occurring indoors will feed on stored grains. Others are library pests feeding on glue of old books.

Collecting and Preserving
- Collect in 70% ethanol; permanently mount on slides.

Resources
- *North American Psocoptera* by Mockford, 1993
- Psocodea Species File—psocodea.speciesfile.org
- Psoco Net—www.psocodea.org
- Psocoptera of Texas—sam-diane.com/psocopteraoftexas.html

Ectopsocus californicus
Ectopsocidae

Suborder Trogiomorpha
Antennae with more than 18 segments. FW lacking distinct
pterostigma. Veins Cu$_2$ and 1A in FW ending
separately on wing margin.

SCALY WINGED BARKLICE Family Lepidopsocidae
These can be called "hairy" psocids. Body and wings covered with
scales, occasionally with dense hairs. **Wings usually pointed at the
tip** and held in shallow tent-like fashion, but rounded when they are
reduced. Most found in leaf litter.
TL 1–5 mm; World 216, NA 14.

GRANARY BOOKLICE Family Trogiidae
Wings usually lacking or reduced to small pads; if well developed
then FW broadly rounded at the tip and lacking visible veins. Body
and wings without scales but with **abundant setae.** Hind femur
slender. Can be found in leaf litter, bark, and foliage.
TL 1–2.5 mm; World 59, NA 8.

BIRD NEST BARKLICE Family Psoquillidae
**Wings held relatively flat over abdomen, wings rounded at tip and
never with scales.** Veins visible even if wings reduced. Frequently
found in bird nests, also occur on dead palm leaves and leaf litter.
TL 0.5–1 mm; World 37, NA 7.

Suborder Troctomorpha
Book- and Barklice with 13-segmented antennae; labial palps
2-segmented; tarsi 3-segmented; FW lacking distinct pterostigma.

TROPICAL BARKLICE Family Amphientomidae
Large, brown with mottled pattern of scales on wings and body.
Wings not held flat over abdomen. FW M$_3$ branched and 2A usually
present. Found on limestone outcrops. Large tropical family with
few representatives in North America.
TL 3–5 mm; World 156, NA 3.

BOOKLICE Family Liposcelididae
Body elongate and flattened. Legs short, **hind femora thickened.**
Usually wingless. If present, wings lay flat over abdomen with only
4 veins. Very common family most often found in buildings or in
grain. TL 0.2–2 mm; World 193, NA 37.

THICK BARKLICE Family Pachytroctidae
Body short and arched. Wings when present folded flat over
abdomen, venation complete. Compound eyes located near back of
head. **Mesothorax and metathorax distinctly separated.** Found in
leaf litter and dead palm fronds. TL 1-2 mm; World 92, NA 5.

SCALY WINGED BARKLICE

Echmepteryx hageni

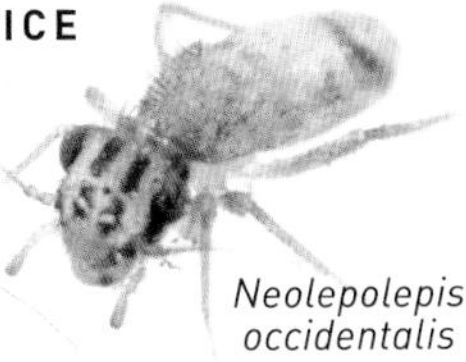

Neolepolepis occidentalis

GRANARY BOOKLICE

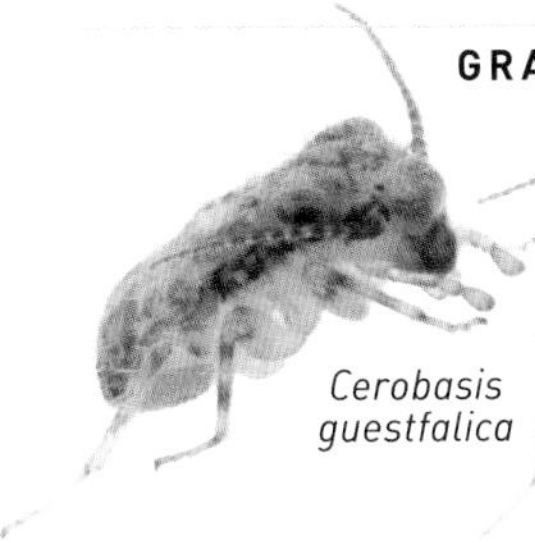

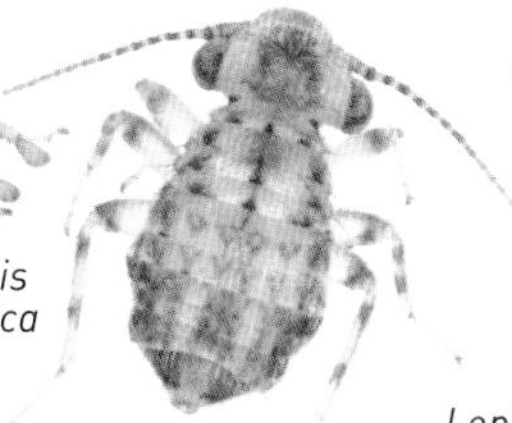

Cerobasis guestfalica

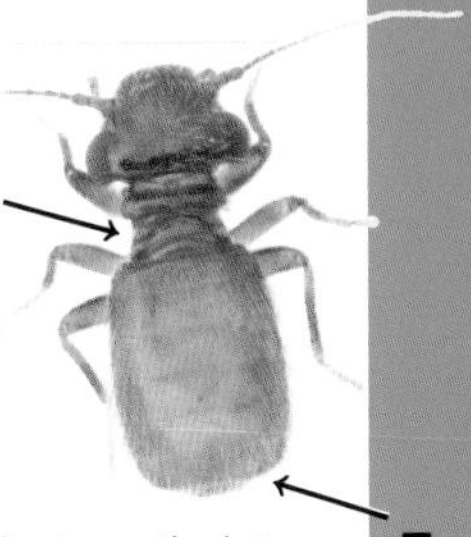

Lepinotus reticulatus

BIRD NEST BARKLICE

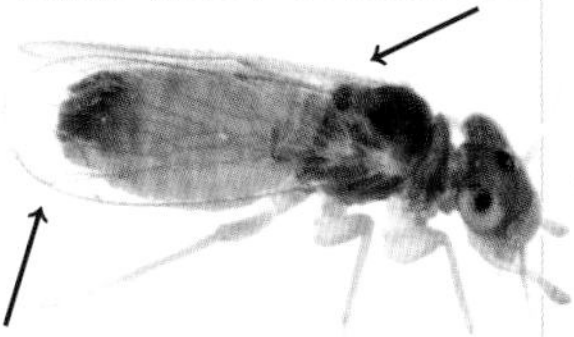

Rhyopsocus eclipticus

TROPICAL BARKLICE

Lithoseopsis hellmani

BOOKLICE

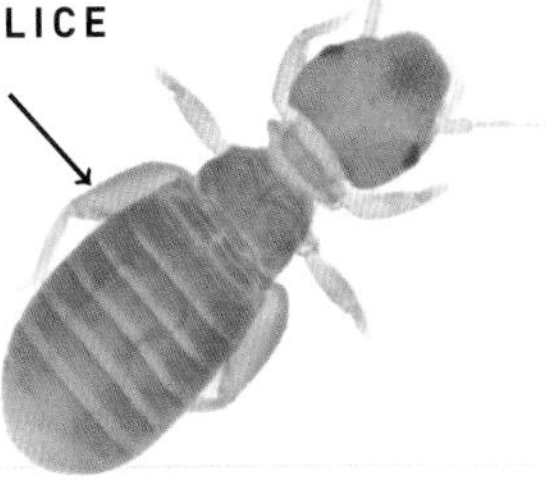

Liposcelis bostrychophila

THICK BARKLICE

Nanopsocus oceanicus

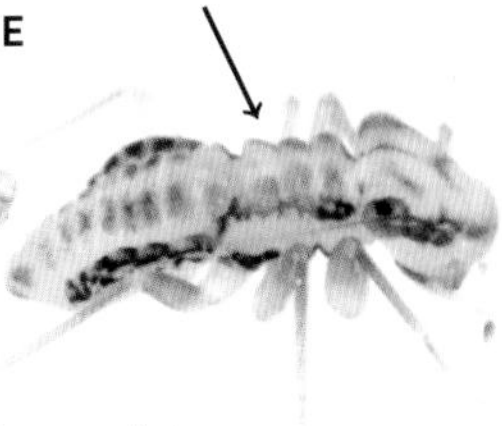

Tapinella maculata

TROCTOMORPHA

Suborder Troctomorpha cont.
Ischnocera Chewing Lice

BIRD LICE Family Philopteridae
Large family; several can be found on poultry and can be pests. Two
tarsal claws on all legs. TL 1.5–5 mm; World 2,945, NA 500.

MAMMAL CHEWING LICE Family Trichodectidae
Tarsi with 1 claw. Antennae usually 3-segmented. This group
parasitizes mammals instead of birds and is of little medical
importance. TL 1–3 mm; World 413, NA 85.

Amblycera Chewing Lice

HAWK LICE Family Laemobothriidae
Similar to Poultry Lice but **head less triangular, with a swelling
on each side at base of antennae;** antennal grooves open ventrally.
Parasites of birds. TL 7–10 mm; World 22, NA 8.

POULTRY LICE Family Menoponidae
Tarsi with 2 claws. Head broadly triangular, expanded behind eyes,
antennae in grooves. Parasite of birds.
TL 2–3 mm; World 1,140, NA 250.

Sucking Lice

UNGULATE LICE Family Haematopinidae
Head has prominent ocular points but no eyes. Several species are
pests of domestic animals such as cows, deer, pigs, horses, and
zebras. TL ~5 mm; World 21, NA 4.

SPINY RAT LICE Family Polyplacidae
**Middle and hind legs are similar in size and shape compared to
the front legs.** This is one of the largest families and is found on
rodents, moles, shrews, hares, rabbits, and monkeys.
TL 1–2 mm; World 192, NS 26.

BODY LICE Family Pediculidae
Narrower and longer than the other group of lice found on humans.
Found on head and body of humans. Live in and lay eggs on clothing,
not human host. TL ~2.5 mm; World 6, NA 1.

PUBIC LICE Family Pthiridae
Claws on middle and hind pair of legs and abdomen are shorter
and oval-shaped compared to the body lice. Found on humans in the
pubic area, but is not known to transmit any disease.
TL ~1.5 mm; World 2, NA 1.

BIRD LICE

Common Peafowl Louse
(*Goniodes pavonis*)

MAMMAL CHEWING LICE

Deer Louse
(*Tricholipeurus parallelus*)

HAWK LICE

Hawk Louse
(*Laemobothrion* sp.)

POULTRY LICE

Poultry Body Louse
(*Menacanthus stramineus*)

UNGULATE LICE

Hog Louse
(*Haematopinus
suis*)

SPINY RAT LICE

Spiny Rat Louse
(*Polyplax spinulosa*)

BODY LICE

Head Louse
(*Pediculus humanus capitis*)

PUBIC LICE

Crab Louse
(*Pthirus pubis*)

PARASITIC LICE

227

TROCTOMORPHA

Parasitic Lice not illustrated
Suborder Troctomorpha
Amblycera Chewing Lice

MARSUPIAL CHEWING LICE Family Boopiidae
Abdominal tergum 1 fused with metanotum. *Heterodoxus spiniger* has transferred to domestic dogs and become widely distributed. TL 0.5–3 mm; World 57, NA 1.

GUINEA PIG LICE Family Gyropidae*
Most species have either no tarsal claw on the first pair of legs or the claw is modified for clasping hair. Found on cavimorph rodents such as the guinea pig. TL ~7 mm; World 90, NA 4.

HUMMINGBIRD LICE Family Ricinidae*
Similar to Laemobothriidae, but without swellings on head at base of antennae. Found on hummingbirds and many passerine birds. TL 2–5 mm; World 100, NA 32.

MARSUPIAL LICE Family Trimenoponidae*
Tarsi with 2 claws. Occurs on marsupials and rodents in South America. Single species, *Trimenopon hispidum*, in NA found on guinea pigs. TL 0.5-1 mm; World 18, NA 1.

Sucking Lice

PALE LICE Family Linognathidae*
Found on a variety of domesticated animals including cows, deer, giraffes, dogs, and hyraxes. Small tubercles posterior to each spiracle. TL 1–3 mm; World 69, NA 3.

SQUIRREL LICE Family Enderleinellidae
Front and middle legs similar in size and shape, in contrast to the hind legs. Found on rodents and squirrels. TL 0.5–3 mm; World 49, NA 10.

SEAL LICE Family Echinophthiriidae*
Body thickly covered with short stout spines or with spines and scales. Found on sea lions, seals, walrus, and river otters. TL 0.5–3 mm; World 12, NA 3.

ARMORED LICE Family Hoplopleuridae*
Hind legs larger than fore- and mid-legs. Found on rodents, picas, hares, moles, and shrews. TL 0.5–2 mm; World 189, NA 16.

PECCARY LICE Family Pecaroecidae*
Head long and narrow, longer than thorax. One species, *Pecaroecus javalii*, found on peccary. TL ~4 mm; World 1, NA 1.

Barklice and Booklice not illustrated
Suborder Trogiomorpha

CAVE BARKLICE Family Psyllipsocidae*
Moderately small pale psocids with wings hairy and reduced in size, when present. Occur in dark, damp places such as caves and basements. TL ~2 mm; World 50, NA 6.

Large-winged Psocids Family Prionoglaridae*
Adults with large wings and sparse hairs. FW with Sc forming large arc and fusing with R1. Many species wingless. Found in caves in the southwest. TL ~2 mm; World 23, NA 3.

Suborder Troctomorpha

HIDDEN BARKLICE Family Sphaeropsocidae*
Body not flattened, spherical in shape. Females with elytraform FW, no hindwings. This family is rare and non-native to NA. TL ~0.5 mm; World 15, NA 6 (+1 intercepted at an airport and likely not established).

Suborder Psocomorpha

ELLIPTICAL BARKLICE Family Epipsocidae*
Tarsi 2-segmented. Pterostigma without crossveins. Found in forest ground litter or shaded outcroppings. TL ~10 mm; World 129, NA 3.

PERPLEXING BARKLICE Family Asiopsocidae*
Pterostigma without crossveins. Found in Arizona and Florida. Found in twigs of small trees. TL 1–3 mm; World 16, NA 3.

FALSE LIZARD BARKLICE Family Pseudocaeciliidae*
Forewing with large hairs. Margins of body brown or yellow. Solitary, living under sparse webbing. Found on citrus trees in FL. TL 2–3 mm; World 300, NA 2.

MIDDLE BARKLICE Family Mesopsocidae*
Relatively large species. Tarsi 3-segmented. Short hairs on body, but not on wings. Found early in the spring from the Appalachians to the Pacific coast on conifers and broad-leafed trees. TL 3–5 mm; World 74, NA 3.

DAMP BARKLICE Family Elipsocidae*
If wings present, Cu1a usually present in FW and usually free from M. Tarsi 3-segmented. Ovipositor valvulae usually complete. TL 1–5 mm; World 120, NA 12.

HAIRY-VEINED BARKLICE Family Ptiloneuridae*
FW with hairs on ventral side. Body color generally brown or reddish-brown. Found in forests and on rocks. TL 4–6 mm; World 60, NA 2.

PARASITIC LICE

PSOCOMORPHA

Suborder Psocomorpha
Antennae with 11 segments. Labial palps 1-segmented either rounded or triangular. Tarsi 2- or 3-segmented. FW generally with distinct pterostigma.

ANCIENT BARKLICE Family Archipsocidae
Found commonly in colonies under webbing. Individuals are relatively small and shades of red-brown to orange-brown. **Rows of short hairs pointing backward on top of thorax.** Hairs usually covering entire body. TL 1–3 mm; World 81, NA 7.

LEAF LITTER BARKLICE Family Hemipsocidae
This is a tropical family with few species in NA. Wings with vein Cu_{1a} in FW joined to M by a crossvein; **vein M_2 branched** and wing margin with at least 1 setae. Found in leaf litter.
TL ~3 mm; World 38, NA 2.

COMMON BARKLICE Family Psocidae
Wings present with variable colors or lacking pigment. Pterostigma and cubital loop in FW; vein Cu_{1a} fused to M forming discoidal cell. Largest family in NA. TL 3–8 mm; World 1274, NA 68.

OUTER BARKLICE Family Ectopsocidae
Pterostigma is rectangular; no cubital loop. Cu_{1a} absent in FW. Dark spots distally on veins at wing margin. Body brown and wings clear or with few markings. These can be found among dead leaves. Can also do well in agricultural situations.
TL 1–3 mm; World 224, NA 14.

FATEFUL BARKLICE Family Lachesillidae
Hairs on wings sparse or absent on veins. **Cu_{1a} vein present, not attached to M.** Relatively small compared to other species that live on plants and grasses. TL ~3 mm; World 390, NA 56.

LASH-FACED PSOCIDS Family Trichopsocidae
Body yellow with **brown band across thorax.** Cu_{1a} present. Hairs on wings sparse except on wing margins. **Dark spots distally on veins at wing margin.** Pterostigma not distinctly rectangular. Can be found on leaves. Few species in the US.
TL ~3 mm; World 11, NA 2.

MOUSE-LIKE BARKLICE Family Myopsocidae
Wings are mottled with veins alternating light and dark on this relatively large species. **Tarsi are 3-segmented.** Labial palps have only one segment. Can be found on shaded stone outcroppings. TL 3–5 mm; World 182, NA 11.

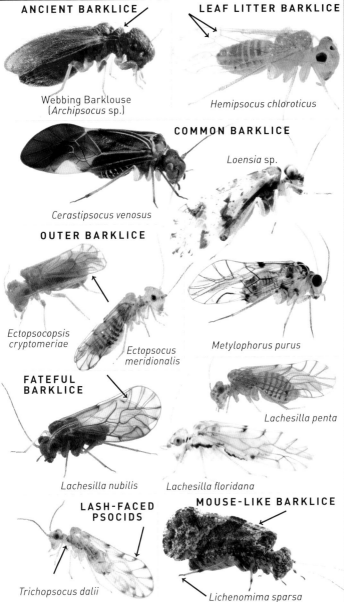

ANCIENT BARKLICE

LEAF LITTER BARKLICE

Webbing Barklouse
(*Archipsocus* sp.)

Hemipsocus chloroticus

COMMON BARKLICE

Loensia sp.

Cerastipsocus venosus

OUTER BARKLICE

Ectopsocopsis cryptomeriae

Ectopsocus meridionalis

Metylophorus purus

FATEFUL BARKLICE

Lachesilla penta

Lachesilla nubilis

Lachesilla floridana

LASH-FACED PSOCIDS

MOUSE-LIKE BARKLICE

Trichopsocus dalii

Lichenomima sparsa

PSOCOMORPHA

Suborder Psocomorpha cont.

STOUT BARKLICE Family Peripsocidae
Body usually brown. Wings usually clear with diffused color. **No cubital loop in FW** and lacks hairs. Can be found on all parts of conifers and broad-leafed trees. TL 2–5 mm; World 335, NA 13.

LOVING BARKLICE Family Philotarsidae
Body shades of brown and gray. **Hold wings flat over back** as opposed to tent-like in other families. FW with hairs on margins and on crossveins. **Cu_{1a} is present and not attached to M. Tarsi are 3-segmented.** Can be abundant in summer on trees located on rock outcroppings. TL ~3 mm; World 133, NA 5.

HAIRY-WINGED BARKLICE Family Amphipsocidae
Hairs present on distal end of HW on top and bottom. Hairs on entire margin of FW. M in FW has 2 branches. Relatively large species. TL 2–5 mm; World 270, NA 1.

SHAGGY PSOCID Family Dasydemellidae
Hairs on HW restricted to cell r_3 margin. M in FW is 3-branched. Hairs found on FW only on top side of wing. Single wide-ranging species in NA, *Teliapsocus conterminus*. Found on foliage. TL ~5 mm; World 30, NA 1.

NARROW BARKLICE Family Stenopsocidae
Short hairs on HW in r_3 cell. FW Pterostigma-r_{2+3} crossvein and m-cu_{1a} present. Found on tree leaves. Few species in the US restricted to the Gulf and Pacific coasts. TL 2–5 mm; World 189, NA 1.

LIZARD BARKLICE Family Caeciliusidae
Most have long wings, but some individuals found in ground litter have short wings. Hairs on FW veins are short and slanted distally. Relatively large family distributed throughout NA. TL ~2 mm; World 779, NA 35.

Metylophorus novaescotiae
Psocidae

BOOKLICE, BARKLICE, & PARASITIC LICE

STOUT BARKLICE

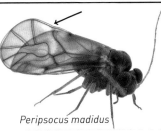 

Peripsocus madidus

Peripsocus sp.

LOVING BARKLICE

Aaroniella badonneli

HAIRY-WINGED BARKLICE

♀

♂ *Polypsocus corruptus*

SHAGGY PSOCID

Teliapsocus conterminus

NARROW BARKLICE

Graphopsocus cruciatus

LIZARD BARKLICE

Stenocaecilius casarum

Valenzuela subflavus

233

WING VENATION CHARACTERS

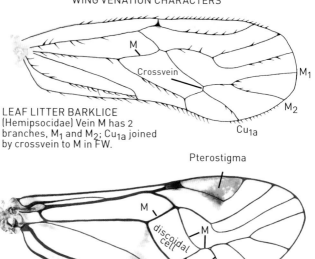

LEAF LITTER BARKLICE
(Hemipsocidae) Vein M has 2
branches, M_1 and M_2; Cu_{1a} joined
by crossvein to M in FW.

COMMON BARKLICE (Psocidae)
Pterostigma and cubital loop in
the FW, Cu_{1a} fused to M forming
discoidal cell.

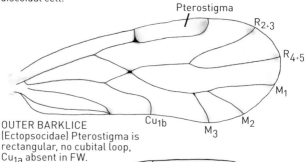

OUTER BARKLICE
(Ectopsocidae) Pterostigma is
rectangular, no cubital loop,
Cu_{1a} absent in FW.

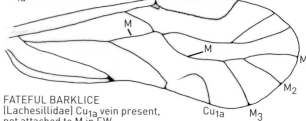

FATEFUL BARKLICE
(Lachesillidae) Cu_{1a} vein present,
not attached to M in FW.

WING VENATION CHARACTERS

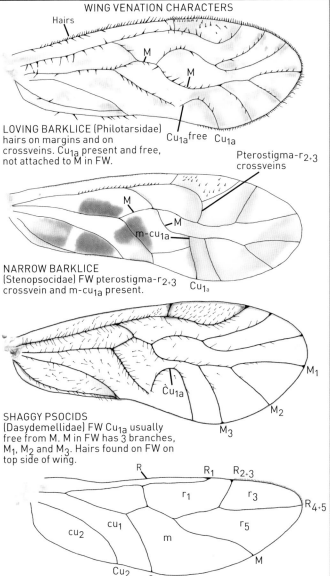

Hairs

M

M

Cu_{1a} free Cu_{1a}

LOVING BARKLICE (Philotarsidae) hairs on margins and on crossveins. Cu_{1a} present and free, not attached to M in FW.

Pterostigma-r_{2+3} crossveins

M

M

m-cu_{1a}

Cu_{1a}

NARROW BARKLICE (Stenopsocidae) FW pterostigma-r_{2+3} crossvein and m-cu_{1a} present.

Cu_{1a}

M_1

M_2

M_3

SHAGGY PSOCIDS (Dasydemellidae) FW Cu_{1a} usually free from M. M in FW has 3 branches, M_1, M_2 and M_3. Hairs found on FW on top side of wing.

R

R_1

R_{2+3}

r_1

r_3

R_{4+5}

cu_2

cu_1

r_5

m

Cu_2

Cu_{1a}

M

SHAGGY PSOCIDS (Dasydemellidae) HW sometimes has hairs on margin of r_3 and sometimes none.

Paper Wasp guarding nest
(*Polistes exclamans*)
Vespidae

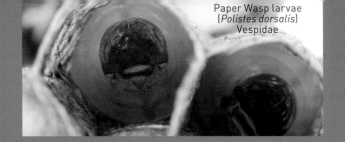

Paper Wasp larvae
(*Polistes dorsalis*)
Vespidae

BEES, WASPS, ANTS, & SAWFLIES
ORDER HYMENOPTERA

Derivation of Name
- Hymenoptera: *hymeno*, god of marriage, referring to the marriage of the front and hindwings working as 1; *hymen*, membrane; *ptera,* wings.

Identification
- TL 0.2–50 mm.
- Antennae usually long, with more than 10 segments; often 13 in males and 12 in females, but range from 3 to 60.
- Mouthparts chewing.
- Four membranous wings when present; FW larger than HW. Usually wings joined by a row of tiny hooks called *hamuli*.
- Female often with prominent ovipositor; in some groups, modified into a stinger.

Classification
- Historically 2 suborders, Symphyta and Aculeata, have been recognized. General greement that Aculeata are a monophyletic group derived from the paraphyletic Symphyta.
- World: 153,000 species; North America: more than 18,000 species. Numbers represent described species; actual numbers are much higher.

Range
- Found on every continent except Antarctica.

Similar orders
- Diptera (p. 436): 2 membranous wings and 2 halteres; mouthparts piercing-sucking or lapping-sucking; antennae often not filiform.
- Lepidoptera (p. 378): wings covered with scales at least in part; usually with a coiled proboscis; antennae generally long and somewhat clubbed.
- Numerous species in different groups mimic ants and wasps.

Ants carrying pupae
Formicidae

Food
- Adults usually feed on nectar and pollen; others are predators, with food usually intended for larvae.
- Larvae are parasitic, fed prey by adults, or eat plants.

Behavior
- Some species in this group have a highly developed social system, like the honey bee; others are solitary.
- Parasitic wasps have elaborate adaptations and behaviors to find hosts.
- Many wasps, solitary and social, create elaborate burrows or nests to hold prey for young.
- Females typically have a specialized ovipositor for inserting eggs into hosts; modified into a stinger in many groups.

Life Cycle
- Holometabolous.
- Many sawfly larvae have legs and prolegs (caterpillar-like).
- Pupation may take place in an isolated cell, a silk cocoon, or a cell constructed by an adult.
- Males come from unfertilized eggs and are usually short-lived.

Importance to Humans
- One of the most beneficial orders of insects, e.g., pollination of crops.
- Parasitoids or predators of many insect and spider pests.

Collecting and Preserving
Immatures
- Collect larvae from nests.
- Permanently store in 80% ethanol, though not ideal as a fixative.

Adults
- Collect by sweeping, especially flowers.
- Raise parasites from hosts, especially caterpillars.
- Malaise traps are essential to collect parasitic species.
- Generally pinned or point mounted. Microhymenoptera can be pointed, stored in alcohol, or stored dry in gelatin capsules.

Resources
- *Hymenoptera of the world: An identification guide to families* ed. by Goulet and Huber, 1993
- Sawfly GenUS—idtools.org/id/sawfly
- Wasp Web—WaspWeb.org; Ant Web—Antweb.org
- *Annotated Keys to the Genera of Nearctic Chalcidoidea (Hymenoptera)* ed by Gibson et al., 1997
- *The Bees in Your Backyard* by Wilson and Messinger Carril, 2016

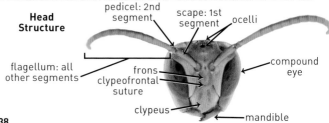

Head Structure

pedicel: 2nd segment
scape: 1st segment
ocelli
flagellum: all other segments
frons
clypeofrontal suture
clypeus
compound eye
mandible

Wing Venation

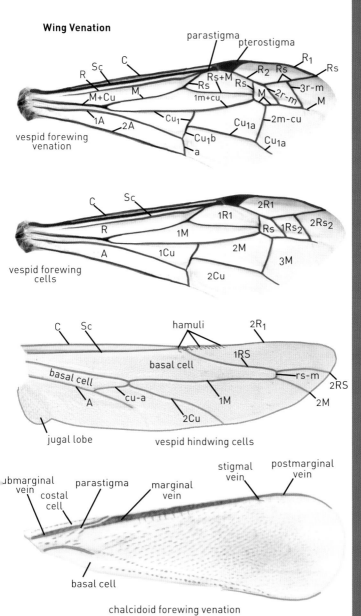

vespid forewing venation

parastigma
pterostigma
C
Sc
R
R
R_2
Rs
R_1
Rs
Rs+M
M+Cu
M
Rs
Rs
1m+cu
M
2r-m
3r-m
r-m
M
1A
2A
Cu_1
Cu_1b
Cu_1a
2m-cu
Cu_1a
a

vespid forewing cells

C
Sc
R
$2R_1$
$1R_1$
1M
Rs
$1Rs_2$
$2Rs_2$
A
1Cu
2M
2Cu
3M

vespid hindwing cells

C
Sc
hamuli
$2R_1$
basal cell
1RS
basal cell
A
cu-a
1M
rs-m
2RS
2M
2Cu
jugal lobe

chalcidoid forewing venation

submarginal vein
costal cell
parastigma
marginal vein
stigmal vein
postmarginal vein
basal cell

239

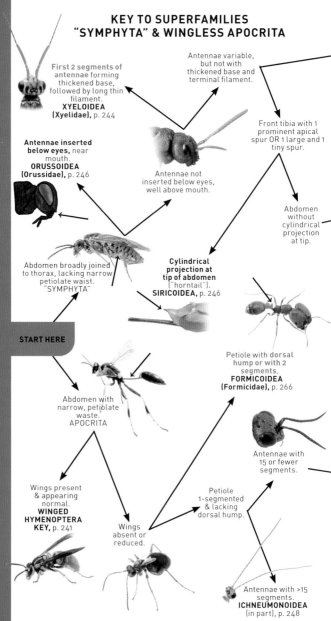

KEY TO SUPERFAMILIES
"SYMPHYTA" & WINGLESS APOCRITA

Antennae variable, but not with thickened base and terminal filament.

First 2 segments of antennae forming thickened base, followed by long thin filament.
XYELOIDEA (Xyelidae), p. 244

Front tibia with 1 prominent apical spur OR 1 large and 1 tiny spur.

Antennae inserted below eyes, near mouth.
ORUSSOIDEA (Orussidae), p. 246

Antennae not inserted below eyes, well above mouth.

Abdomen without cylindrical projection at tip.

Abdomen broadly joined to thorax, lacking narrow petiolate waist.
"SYMPHYTA"

Cylindrical projection at tip of abdomen ("horntail").
SIRICOIDEA, p. 246

START HERE

Petiole with dorsal hump or with 2 segments.
FORMICOIDEA (Formicidae), p. 266

Abdomen with narrow, petiolate waste.
APOCRITA

Antennae with 15 or fewer segments.

Wings present & appearing normal.
WINGED HYMENOPTERA KEY, p. 241

Wings absent or reduced.

Petiole 1-segmented & lacking dorsal hump.

Antennae with >15 segments.
ICHNEUMONOIDEA (in part), p. 248

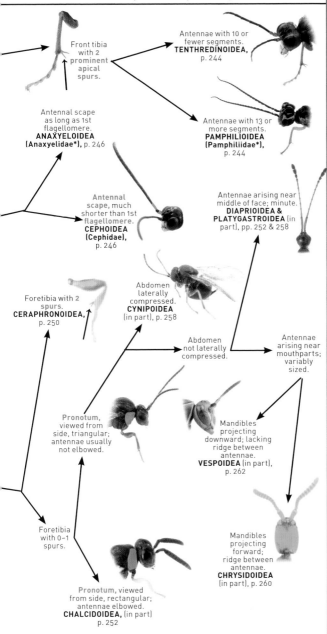

Front tibia with 2 prominent apical spurs.

Antennae with 10 or fewer segments. **TENTHREDINOIDEA,** p. 244

Antennal scape as long as 1st flagellomere. **ANAXYELOIDEA (Anaxyelidae*),** p. 246

Antennae with 13 or more segments. **PAMPHILIOIDEA (Pamphiliidae*),** p. 244

Antennal scape, much shorter than 1st flagellomere. **CEPHOIDEA (Cephidae),** p. 246

Antennae arising near middle of face; minute. **DIAPRIOIDEA & PLATYGASTROIDEA** (in part), pp. 252 & 258

Foretibia with 2 spurs. **CERAPHRONOIDEA,** p. 250

Abdomen laterally compressed. **CYNIPOIDEA** (in part), p. 258

Abdomen not laterally compressed.

Antennae arising near mouthparts; variably sized.

Pronotum, viewed from side, triangular; antennae usually not elbowed.

Mandibles projecting downward; lacking ridge between antennae. **VESPOIDEA** (in part), p. 262

Foretibia with 0–1 spurs.

Mandibles projecting forward; ridge between antennae. **CHRYSIDOIDEA** (in part), p. 260

Pronotum, viewed from side, rectangular; antennae elbowed. **CHALCIDOIDEA,** (in part) p. 252

KEY TO SUPERFAMILIES
WINGED APOCRITA

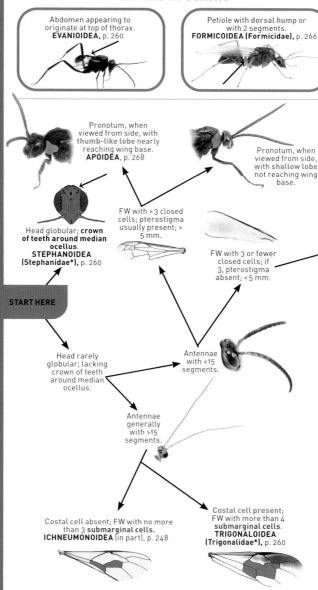

Abdomen appearing to originate at top of thorax. **EVANIOIDEA**, p. 260.

Petiole with **dorsal hump** or with 2 segments. **FORMICOIDEA (Formicidae)**, p. 266.

Pronotum, when viewed from side, with **thumb-like lobe** nearly reaching wing base. **APOIDEA**, p. 268

Pronotum, when viewed from side, with **shallow lobe** not reaching wing base.

Head globular; **crown of teeth around median ocellus**. **STEPHANOIDEA (Stephanidae*)**, p. 260

FW with >3 closed cells; pterostigma usually present; > 5 mm.

FW with 3 or fewer closed cells; if 3, pterostigma absent; < 5 mm.

START HERE

Head rarely globular; lacking crown of teeth around median ocellus.

Antennae with <15 segments.

Antennae generally with >15 segments.

Costal cell absent; FW with no more than 3 **submarginal cells**. **ICHNEUMONOIDEA** (in part), p. 248

Costal cell present; FW with more than 4 **submarginal cells**. **TRIGONALOIDEA (Trigonalidae*)**, p. 260

BEES, WASPS, ANTS, & SAWFLIES

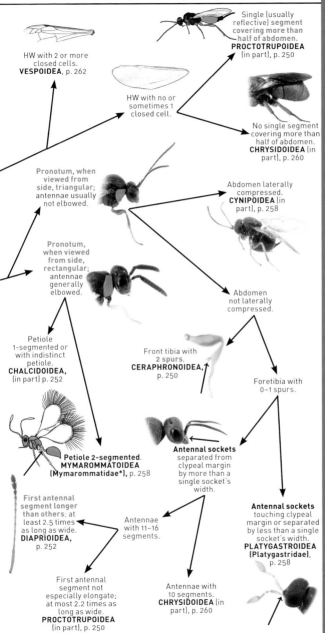

HW with 2 or more closed cells.
VESPOIDEA, p. 262

HW with no or sometimes 1 closed cell.

Single (usually reflective) segment covering more than half of abdomen.
PROCTOTRUPOIDEA (in part), p. 250

No single segment covering more than half of abdomen.
CHRYSIDOIDEA (in part), p. 260

Pronotum, when viewed from side, triangular; antennae usually not elbowed.

Abdomen laterally compressed.
CYNIPOIDEA (in part), p. 258

Pronotum, when viewed from side, rectangular; antennae generally elbowed.

Abdomen not laterally compressed.

Petiole 1-segmented or with indistinct petiole.
CHALCIDOIDEA, (in part) p. 252

Front tibia with 2 spurs.
CERAPHRONOIDEA, p. 250

Foretibia with 0–1 spurs.

Petiole 2-segmented.
MYMAROMMATOIDEA (Mymarommatidae*), p. 258

Antennal sockets separated from clypeal margin by more than a single socket's width.

Antennal sockets touching clypeal margin or separated by less than a single socket's width.
PLATYGASTROIDEA (Platygastridae), p. 258

First antennal segment longer than others; at least 2.5 times as long as wide.
DIAPRIOIDEA, p. 252

Antennae with 11–16 segments.

First antennal segment not especially elongate; at most 2.2 times as long as wide.
PROCTOTRUPOIDEA (in part), p. 250

Antennae with 10 segments.
CHRYSIDOIDEA (in part), p. 260

243

"Symphyta"
Thorax and abdomen broadly attached; larvae caterpillar-like.

Superfamily Xyeloidea

XYELID SAWFLIES Family Xyelidae*

Antennae with **segment 3 long, slender with multi-segmented (11–20 segments) terminal filament.** FW with 3 marginal cells; costal cell divided by a longitudinal vein forming subcosta. **Female with protruding ovipositor.** Found on various trees, including pines, and occasionally on flowers. Larvae caterpillar-like with obvious prolegs; antennae 6–7 segments. TL 5–10 mm; World 80, NA 30.

Superfamily Pamphilioidea

WEB-SPINNING & LEAFROLLING SAWFLIES

Family Pamphiliidae*

Antennae long and thread-like, 18–24 segments. Costal cell divided by a longitudinal vein forming subcosta. Pronotum relatively long with a nearly straight posterior margin. **Mid- and hind tibiae with preapical spurs.** Abdomen strongly compressed dorsoventrally with lateral carina. Larvae lack distinct prolegs; 3-segmented subanal appendages present; 7-segmented antennae. Larvae spin webs on various woody plants. TL 7–15 mm; World 200, NA 72.

Superfamily Tenthredinoidea

Pronotum narrowed medially and curved posteriorly; 2 spurs apically on front tibia; transscutal groove absent on mesonotum.

ARGID SAWFLIES Family Argidae

Stout bodied. **Antennae 3-segmented with third segment longer than previous 2;** males with last segment sometimes split as a Y or U. Generally dark colored, often with red or orange pronotum. Middle tibia with or without preapical spurs. Ovipositor short. Larvae with 6–8 pairs of prolegs and 1-segmented antennae; found mainly on various types of foliage. TL 8–15 mm; World 800, NA 60.

CIMBICID SAWFLIES Family Cimbicidae

Robust, often resembling a bumble bee. **Antennae 6–7 segments and clavate.** Middle tibia lacking preapical spurs. Abdomen with lateral carina. Ovipositor short, barely extending beyond abdomen. Larva with 8 pairs of prolegs and 2-segmented antennae.
TL 15–25 mm; World 200, NA 12.

COMMON SAWFLIES Family Tenthredinidae

Wasp-like with 7–10, usually 9, segmented simple antennae. Middle tibia lacking preapical spurs. Sc and R usually fused with narrow costal cell. Ovipositor barely extends beyond abdomen. Larvae with 6–8 pairs of prolegs; 3–5-segmented antennae; found mainly on various types of foliage; some are leaf miners. TL 5–20 mm; World 7,500, NA 900.

BEES, WASPS, ANTS, & SAWFLIES

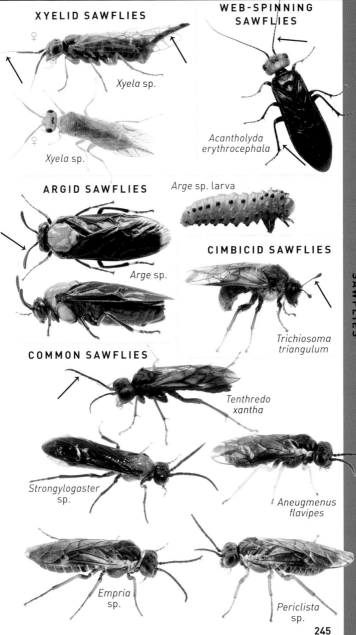

XYELID SAWFLIES

♀ *Xyela* sp.

♀ *Xyela* sp.

WEB-SPINNING SAWFLIES

Acantholyda erythrocephala

ARGID SAWFLIES

Arge sp. larva

Arge sp.

CIMBICID SAWFLIES

Trichiosoma triangulum

COMMON SAWFLIES

Tenthredo xantha

Strongylogaster sp.

Aneugmenus flavipes

Empria sp.

Periclista sp.

SAWFLIES

245

Superfamily Tenthredinoidea cont.

CONIFER SAWFLIES Family Diprionidae
Antennae 14–32 segments; pectinate or bipectinate in males, serrate in females. Mid-tibia without preapical spurs. Ovipositor not projecting beyond abdomen. Larvae caterpillar-like, with 8 pairs of prolegs and 3-segmented antennae. TL 6–12 mm; World 140, NA 40.

PERGID SAWFLIES Family Pergidae*
Antennae 6-segmented (2 basal segments and 4 flagellomeres); may be somewhat serrate. FW lacking vein 2R and usually anal cell. Larvae feed on a wide range of plants. TL 6–10 mm; World 440, NA 10.

Superfamily Cephoidea

STEM SAWFLIES Family Cephidae
Antennae filiform, 16–30 segments. **Posterior margin of pronotum nearly straight.** Front tibia with single apical spur; middle tibia usually with preapical spurs. **Abdomen cylindrical or laterally compressed; ovipositor visible from above.** Larvae with vestigial true legs; lack prolegs; 4–5-segmented antennae. TL 5–18 mm; World 160, NA 16.

Superfamily Siricoidea
Tergite on S1 medially divided; mid-tibia without preapical spurs. Larvae wood borers; vestigial true legs; prolegs absent.

HORNTAILS Family Siricidae
Large, elongated with 14–30-segmented antennae. Both sexes with **short, dorsal abdominal projection** (horn). **Females with ovipositor in sheath, located ventrally.** Front tibia with single spur. Larvae with 1-segmented antennae. TL 10–50 mm; World 120, NA 28.

XIPHYDRIID WOOD WASPS Family Xiphydriidae*
Similar to horntails, but lack dorsal abdominal projection. Antennae 13–19 segments. Front tibia with single or pair of unequal spurs. Larvae with 3-segmented antennae. TL 12–20 mm; World 80, NA 9.

Superfamily Anaxyeloidea

INCENSE CEDAR WOOD WASP Family Anaxyelidae* Not Illus.
Large, dark with brown-orange legs and margin of abdominal tergites pale. Female with projecting, ventral ovipositor. Single species, *Syntexis libocedrii*, found in western NA. Oviposit in burnt *Calocedrus*, *Thuja*, and *Juniperus*. TL ~15 mm; World 1, NA 1.

Superfamily Orussoidea

PARASITIC WOOD WASPS Family Orussidae*
Dark, elongated with 10–11-segmented **antennae inserted below eyes.** FW with single submarginal cell. Larvae lack legs and prolegs, have 1-segmented antennae, and are ectoparasitoids of wood-boring insect larvae (primarily buprestids). TL 8–14.5 mm; World 85, NA 7.

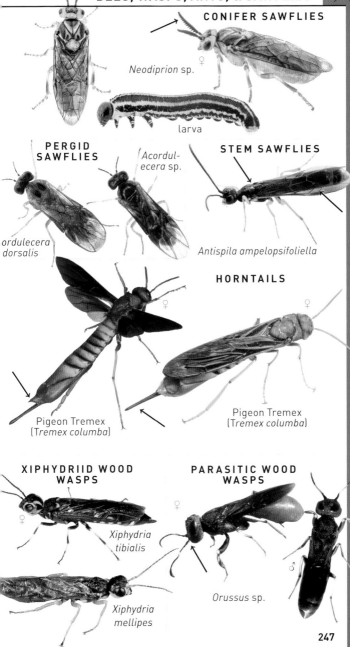

CONIFER SAWFLIES

Neodiprion sp. ♀

larva

PERGID SAWFLIES

Acordul-ecera sp.

ordulecera dorsalis

STEM SAWFLIES

Antispila ampelopsifoliella

HORNTAILS

♀

♀

Pigeon Tremex
(*Tremex columba*)

Pigeon Tremex
(*Tremex columba*)

XIPHYDRIID WOOD WASPS

♀

Xiphydria tibialis

Xiphydria mellipes

PARASITIC WOOD WASPS

Orussus sp.

♂

247

Parasitic Apocrita—Have narrow waist or constriction.
Superfamily Ichneumonoidea
Pronotum often triangular in profile. Antennae filiform, usually 16 segments. Costal cell absent. Hind trochanter 2-segmented. Sternite 1 of gaster divided. Ovipositor arises in front of apex of abdomen, often as long or longer than body.

BRACONID WASPS Family Braconidae

FW with no more than 1 m-cu crossvein. First 2 abdominal tergites beyond petiole usually fused. **Abdomen usually only slightly longer than thorax.** A large diverse group containing both endo- and ectoparasites; many solitary, others gregarious. They attack all stages of the host. TL 1–15 mm; World 17,000, NA 1,700.

ICHNEUMON WASPS Family Ichneumonidae

FW usually with 2 m-cu crossveins; Rs+M absent. First 2 abdominal tergites beyond petiole usually independent and overlapping. **Abdomen usually notably longer than thorax.** Most species are parasitoids of Lepidoptera, Hymenoptera, Diptera, Coleoptera, Neuroptera, and Mecoptera larvae as well as spiders and their eggs. They are tremendously diverse, some being species-specific, others attacking a wide range of hosts; most are solitary, others are gregarious as larvae.
TL 2.5–76 mm; World 25,000, NA 5,000.

ICHNEUMON WASPS

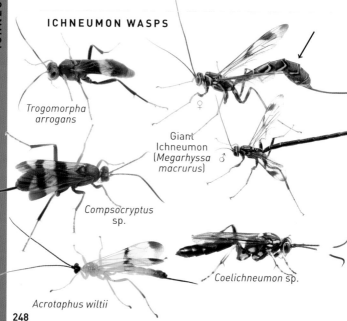

Trogomorpha arrogans

Giant Ichneumon (*Megarhyssa macrurus*) ♂

Compsocryptus sp.

Coelichneumon sp.

Acrotaphus wiltii

BRACONID WASPS

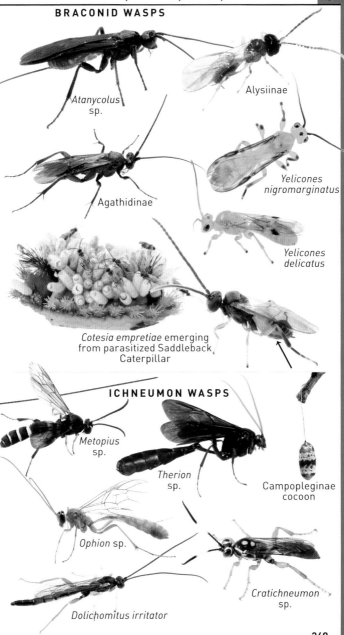

Atanycolus sp.

Alysiinae

Agathidinae

Yelicones nigromarginatus

Yelicones delicatus

Cotesia empretiae emerging from parasitized Saddleback Caterpillar

ICHNEUMON WASPS

Metopius sp.

Therion sp.

Campopleginae cocoon

Ophion sp.

Cratichneumon sp.

Dolichomitus irritator

249

Superfamily Ceraphronoidea
Geniculate antennae with long scape; foretibia with 2 spurs apically; C+R fused into solid bar; second metasomal segment very large.

CERAPHRONID WASPS Family Ceraphronidae
Male antennae with 10–11 segments; female with 9–10. **Middle tibia with single spur.** Highly reduced venation, with long marginal vein, small linear stigma, and curved stigmal vein; many wingless. Median groove in mesoscutum. Found in leaf litter. TL 1–5 mm; World 360, NA 52.

MEGASPILID WASPS Family Megaspilidae
Antennae 11 segments in both sexes. Middle tibia with 2 apical spurs; **large apical spur of front tibia forked apically.** Reduced venation, with stigma usually large and semicircular; rarely absent or linear. Can be wingless. TL 1–5 mm; World 450, NA 61.

Superfamily Proctotrupoidea
Diverse, strongly sclerotized, non-metallic; metasomal segment 2 often largest; ovipositor external or housed in internal sheath.

HELORID WASPS Family Heloridae*
Robust, black with long, sickle-shaped, scissored mandibles. Labrum long and narrow, finger-like. Tarsal claws pectinate. Metasomal segment 1 distinctly elongate. Two species of *Helorus* found in NA. TL ~7.5 mm; World 7, NA 2.

PELECINID WASPS Family Pelecinidae
Female with long, slender abdomen; male (rarely seen) with shorter, clubbed abdomen. Generally black, including legs and antennae. Single, distinctive species in NA, *Pelecinus polyturator.* Parasitoids of soil-inhabiting scarab beetle larvae. TL 25–50 mm; World 3, NA 1.

PROCTOTRUPID WASPS Family Proctotrupidae
Robust, generally black with smooth sculpture except propodeum. Antennae filiform with 11 flagellomeres in both sexes. FW with large, distinct pterostigma followed by small submarginal cell. Terminal abdominal segment narrow, elongated, and curved downward in most species. TL 3–15 mm; World 400, NA 45.

ROPRONIID WASPS Family Roproniidae*   Not Illus.
Robust, coarsely sculptured body. FW with first discoidal cell 6-sided. Abdomen laterally compressed and strongly petiolate. Single genus, *Ropronia*, in NA. At least some parasitize sawfly cocoons. TL 5–10 mm; World 18, NA 3.

VANHORNIID WASPS Family Vanhorniidae*   Not Illus.
Black, antennae inserted between eyes, right above clypeus. Metasoma fused to form large carapace. Single species, *Vanhornia eucnemidarum*, in NA that parasitizes eucnemid larvae in old, dying maple trees. TL ~6 mm; World 5, NA 1.

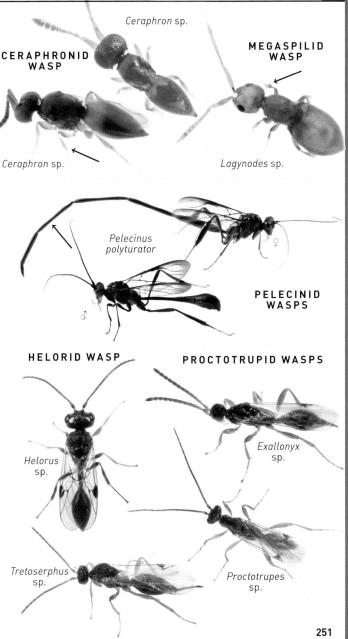

CERAPHRON sp.

CERAPHRONID WASP

Ceraphron sp.

MEGASPILID WASP

Lagynodes sp.

Pelecinus polyturator

♀

PELECINID WASPS

♂

HELORID WASP

PROCTOTRUPID WASPS

Helorus sp.

Exallonyx sp.

Tretoserphus sp.

Proctotrupes sp.

Superfamily Diaprioidea

DIAPRIID WASPS Family Diapriidae

Small, black, and shiny. Antennae 11–15 segments; **first segment longer than second and third combined; arising from protuberance on front of head.** Trochanters 2-segmented. Wing venation reduced or absent. Larvae are parasitoids of fungus gnats and other Diptera. TL 1–76 mm; World 4,000, NA 300.

Superfamily Chalcidoidea

Antennae generally elbowed with elongated scape and fewer than 13 segments; pronotum, when viewed laterally, squarish; most less than 4 mm; often dark, green or blue, some with yellow.

FIG WASPS Family Agaonidae* Not Illus.

Species include pollinators and non-pollinators of fig trees. Larval stage found inside fruits of fig trees. Pollinators mutualistic with fig trees; non-pollinators are parasitic. TL 1–3 mm; World 750, NA 13.

APHELINID WASPS Family Aphelinidae Not Illus.

Dark, yellowish or brown, with thorax and abdomen broadly joined. Large eyes. Antennae with 8 or fewer segments. Tarsi 5-segmented. Wings with setal tracts; FW marginal vein relatively long. TL 0.6–1.4 mm; World 1,200, NA 210.

AZOTID WASPS Family Azotidae* Not Illus.

Similar to aphelinids, only recently recognized as a family. Antennal club 1-segmented. Parastigma with single campaniform sensilla. Foretibia with horizontal spur at apex. Single genus, *Ablerus*, with 6 species in NA. TL <1.5 mm; World 100, NA 6.

CHALCIDID WASPS Family Chalcididae

Black, brown, yellow, or red without metallic luster. **Head and mesosoma heavily sclerotized, often coarsely punctate.** Antennal flagellum shorter than length of eye (5–7 segments) or conspicuously longer (9–11 segments). Tarsi 5-segmented. Hosts primarily include Lepidoptera pupae and Diptera larvae. TL 2.5–9 mm; World 1,500, NA 90.

ENCYRTID WASPS Family Encyrtidae

Pronotum visible and transverse dorsally. Mesopleuron large and convex. Middle coxae inserted at or anterior to middle of mesopleuron (vs. inserted near hind coxae). FW marginal vein short. Hosts include a wide range of insects and stages, including being hyperparasitoids. TL 2–3 mm; World 3,700, NA 480.

ERIAPORID WASPS Family Eriaporidae* Not Illus.

Similar to aphelinids, only recently recognized as a family. Three-segmented maxillary palp. Clypeus transverse. Single species, *Myiocnema comperei*, in NA. TL <1.5 mm; World 22, NA 1.

DIAPRIID WASPS

Aclista sp.

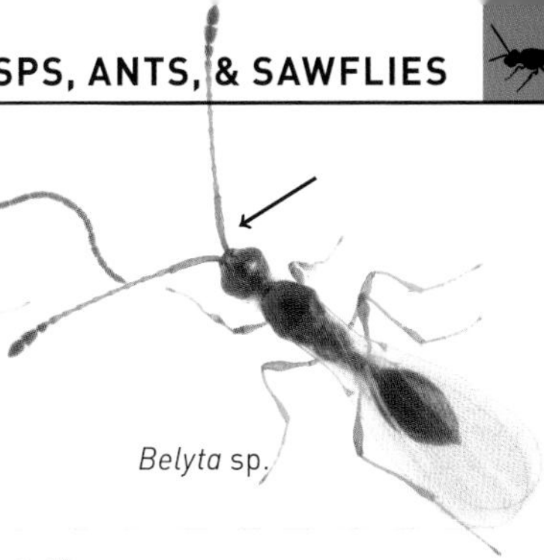

Belyta sp.

CHALCID WASPS

Brachymeria sp.

Conura sp.

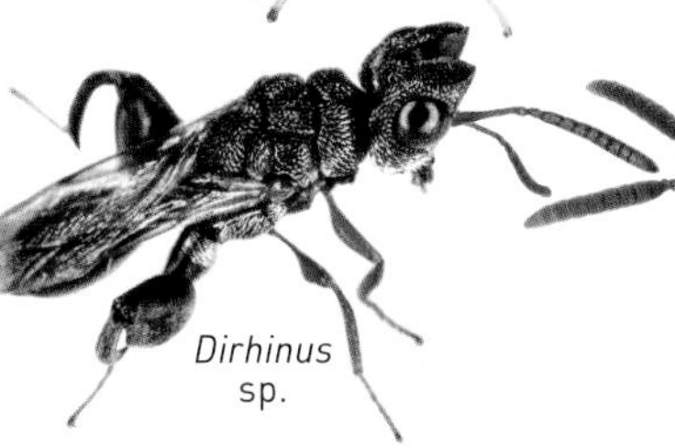

Dirhinus sp.

Brachymeria tegularis

Encyrtid Wasp

ENCYRTID WASPS

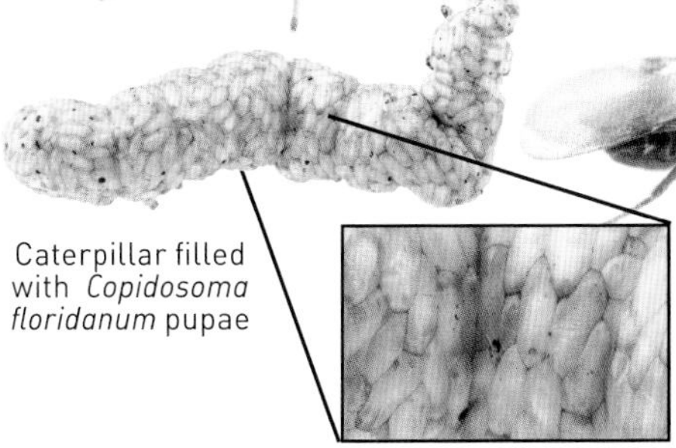

Caterpillar filled with *Copidosoma floridanum* pupae

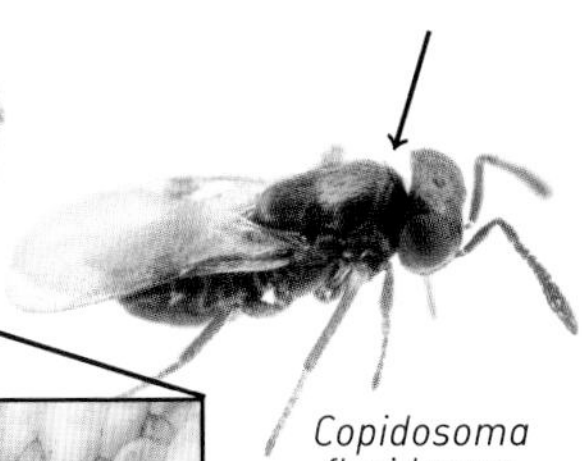

Copidosoma floridanum

DIAPRIID • CHALCID • ENCYRTID

253

Superfamily Chalcidoidea cont.

EUCHARITID WASPS Family Eucharitidae*   Not Illus.
Body punctate with small head, large mandibles, bulging humpbacked thorax, and stalked abdomen attached low on thorax. Pronotum usually concealed by head dorsally. Gaster small with long petiole. FW marginal vein moderately long. External parasitoids of ant larvae and pupae. TL 2–5.4 mm; World 420, NA 35.

EULOPHID WASPS Family Eulophidae
Antennae with 10 or fewer segments, including 2–4-segmented apical club. Dark with metallic luster. Tarsi 4-segmented with protibial spur short and straight. Primarily parasitoids of leaf-mining Lepidoptera and Hymenoptera larvae. TL 1–2.5 mm; World 4,300, NA 830.

EUPELMID WASPS Family Eupelmidae
Elongated, somewhat ant-like, metallic-colored with flat mesonotum. Wings with long marginal vein. Front and hind coxae widely separated. Middle tibial spur long and stout. Some wingless or brachypterous. Parasitoids of numerous orders. TL 2–5 mm; World 900, NA 120.

EURYTOMID WASPS Family Eurytomidae
Usually black, some yellowish with hairy antennae, head, and thorax. Hind coxae not enlarged; hind femur lacking teeth. Similar to perilampids (but more slender); sometimes treated as subfamily of Pteromalidae. Many are phytophagous, others parasitic on species confined to galls, stems and seeds. TL 1–5 mm; World 1,400, NA 280.

LEUCOSPID WASPS Family Leucospidae
Stout-bodied with wings folded lengthwise at rest; generally black and yellow. Hind femur swollen and toothed ventrally. Ovipositor long, curving forward over abdomen. Single genus, *Leucospis*, in NA. Parasitoids of aculeate Hymenoptera. TL 4–17 mm; World 130, NA 6.

MEGASTIGMID WASPS Family Megastigmidae   Not Illus.
Yellowish, brown, and/or black with scattered bristle-like setae; sometimes with metallic luster. Antennae with 11 flagellomeres; male antennae may have whorls of setae. Some parasitoids of gall-inducing insects, others phytophagous. TL 1–2.5 mm; World 202, NA 28.

FAIRYFLIES Family Mymaridae   Not Illus.
Minute. HW narrowly elongate and stalked; both wings with hairs around margin. Antennae long and clubbed in females, filiform in males. Venation reduced. Tarsi 4–5 segments. Parasitoids of insect eggs. TL 0.2–1 mm; World 1,400, NA 200.

ORMYRID WASPS Family Ormyridae
Body generally heavily sclerotized with metallic luster; distinctively convex in profile. Antennae with 11 flagellomeres. FW with relatively long marginal vein and short post-marginal vein. Parasitoids of gall-forming insects. Only *Ormyrus* in NA. TL ~2.5 mm; World 140, NA 20.

CHALCIDOIDEA

EULOPHID WASPS

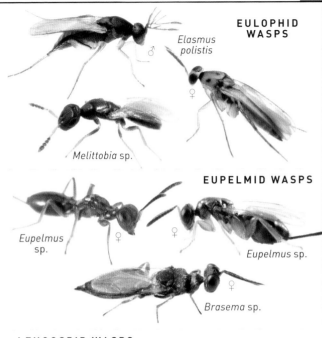

Elasmus polistis ♂

♀

Melittobia sp.

EUPELMID WASPS

Eupelmus sp. ♀

♀ *Eupelmus* sp.

♀ *Brasema* sp.

LEUCOSPID WASPS

Leucospis affinis

ORMYRID WASPS

Ormyrus sp.

EURYTOMID WASPS

Sycophila sp.

Sycophila sp.

255

HYMENOPTERA

Superfamily Chalcidoidea cont.

PERILAMPID WASPS Family Perilampidae
Body robust, usually metallic blue or green and strongly convex when viewed laterally. Thorax coarsely punctate. Abdomen relatively short and somewhat triangular. Wings with long marginal and submarginal veins; short pterostigma. Generally attack pupae of tachinids, braconids, and ichneumonids. TL 1.3–5.5 mm; World 300, NA 36.

PTEROMALID WASPS Family Pteromalidae
A diverse family that will likely be split. Body generally black, blue, or green with metallic reflections. **Pronotum quadrate when viewed dorsally;** some species with coarse punctures. Hind coxae not enlarged. Abdomen small. Parasitoids on a wide range of hosts; some are hyperparasitoids. TL 1–7.5 mm; World 3,500, NA 600.

SIGNIPHORID WASPS Family Signiphoridae*   Not illus.
Antennal club large and undivided. Body distinctly dorsoventrally flattened. Scutellum and metanotum transverse. Propodeum with triangular median area that is easily mistaken for a scutellum. Previously considered in the family Encyrtidae.
TL 1–2.5 mm; World 80, NA 20.

TANAOSTIGMATID WASPS Family Tanaostigmatidae*
Body squat and robust with protruding prepectus. Mesonotum strongly arched, making pronotum appear nearly vertical. Nearly all phytophagous, forming galls in plant stems, leaves, and seeds.
TL 1–5 mm; World 100, NA 2.

TETRACAMPID WASPS Family Tetracampidae*   Not illus.
Distinctly elongated. Antennae with fewer than 13 segments, including at least 5 making a funicle. Protibial spur moderately strong and cleft. Posterior margin of pronotum thin, closely appressed to mesoscutum. Primarily parasitoids of the larvae and eggs of Diptera and Hymenoptera plant miners. TL 1.2–2.5 mm; World 50, NA 10.

TORYMID WASPS Family Torymidae
Metallic green or red with **sculptured thorax.** FW with long marginal vein and short stigmal cell; long setae along margins. Female with swollen metafemur. Abdomen laterally compressed. **Ovipositor length of body or longer.** Many are parasitoids on gall-forming insects, others are phytophagous. TL 1–2.5 mm; World 960, NA 160.

TRICHOGRAMMATID WASPS Family Trichogrammatidae
Minute, often yellowish or brown. Antennae with 1–2 **elongated segments apically.** Tarsi 3-segmented. Primarily endoparasitoids of insect eggs. TL 0.3–1.2 mm; World 840, NA 135.

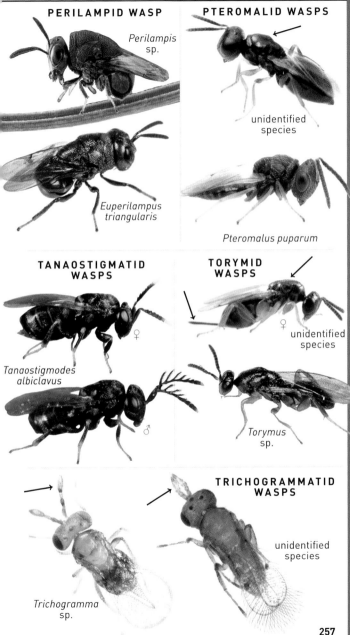

PERILAMPID WASP

Perilampis sp.

Euperilampus triangularis

PTEROMALID WASPS

unidentified species

Pteromalus puparum

TANAOSTIGMATID WASPS

Tanaostigmodes albiclavus ♀

♂

TORYMID WASPS

♀ unidentified species

Torymus sp.

TRICHOGRAMMATID WASPS

unidentified species

Trichogramma sp.

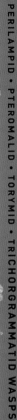

PERILAMPID • PTEROMALID • TORYMID • TRICHOGRAMMATID WASPS

HYMENOPTERA

Superfamily Mymarommatoidea

FALSE FAIRY WASPS Family Mymarommatidae*   Not Illus.
Minute, yellow to brown lacking any metallic luster. Antennae with
2-segmented petiole; prominent club in males. FW with non-
smooth surface and marginal fringe of long setae. Face appearing
triangular in frontal view. Likely parasitoids of insect eggs;
Psocodea (barklice) recently confirmed as a host.
TL .4–0.7 mm; World 10, NA 6.

Superfamily Platygastroidea

PLATYGASTRID WASPS Family Platygastridae
Black, often mixed with yellow; shiny, with **elbowed antennae and
8-segmented flagellum.** Wings without venation; sometimes with
fringe of hairs. Second metasomal segment longest and widest.
Female almost always with 6 apparent tergites. Often parasitize
insect larvae. TL 0.45–4 mm; World 4,000, NA 500.

Superfamily Cynipoidea
Generally black; antennae filiform; female with 11 flagellomeres,
male with 12–13; pronotum extends back to tegulae; shiny,
compressed abdomen; ovipositor arises from apex of abdomen.

GALL WASPS Family Cynipidae
Hind tarsi with first segment longer than following 2 or 3. Most
with **distinct humpbacked appearance.** Abdomen with 2 segments
visible from above. Species either form galls or live as inquilines in
galls; larvae always in closed cavities.
TL 2–8 mm; World 1,400, NA 750.

FIGITID WASPS Family Figitidae
Diverse family with largely undescribed species. Head and
mesosoma usually strongly sculptured. Metasomal tergum 3
largest. Scutellum often modified with spines, pits, or ridges.
Primarily parasitoids of Diptera and Neuroptera.
TL <5 mm; World 1,400, NA 60.

IBALIID WASPS Family Ibaliidae*
Body generally red or orange; abdomen strongly compressed
laterally. Wings occasionally banded; FW with long, thin radial cell.
Parasitoids of larval siricids. Single genus, *Ibalia*, in NA.
TL 8–16 mm; World 19, NA 6.

LIOPTERID WASPS Family Liopteridae*   Not Illus.
Head and thorax coarsely sculptured. Gaster and often entire body
reddish. Antennae 13 segments in female, 14 in male. Wings often
banded. Parasitoids of wood-boring beetles; FW with submarginal
cell usually complete. TL 5–12.5 mm; World 150, NA 3.

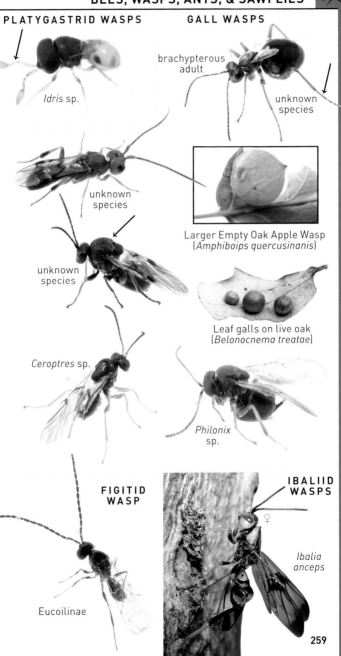

BEES, WASPS, ANTS, & SAWFLIES

PLATYGASTRID WASPS

Idris sp.

unknown species

unknown species

Ceroptres sp.

GALL WASPS

brachypterous adult

unknown species

Larger Empty Oak Apple Wasp (*Amphiboips quercusinanis*)

Leaf galls on live oak (*Belonocnema treatae*)

Philonix sp.

FIGITID WASP

Eucoilinae

IBALIID WASPS

♀

Ibalia anceps

PLATYGASTRID • GALL • FIGITID • IBALIID WASPS

Superfamily Evanioidea
Likely not a monophyletic grouping. Metasoma attaches high on the propodeum.

AULACID WASPS Family Aulacidae
Generally black and slender with **reddish metasoma** and **long antennae**; 12 flagellomeres in female, 11 in male. FW with 2 m-cu crossveins. Ovipositor long. Adults often found around logs; larvae endoparasitoids of wood-boring beetle larvae. TL 5–18 mm; World 170, NA 30.

ENSIGN WASPS Family Evaniidae
Usually black with **small, oval, laterally compressed metasoma** attached by slender propodeum above base of hind coxae. Larval parasitoids of cockroach oothecae. TL 3–7 mm; World 400, NA 11.

CARROT WASPS Family Gasteruptiidae
Similar to ichneumonids, but with short antennae and a costal cell in FW. Body slender with very long **metasoma with brown or orange markings.** Head markedly separated from pronotum. **Hind tibia strongly clavate** in both sexes. Ovipositor generally very long. Adults can be common on flowers, especially in the carrot family. Kleptoparasites of solitary wasps and bees. TL 13–40 mm; World 500, NA 15.

Superfamily Stephanoidea

CROWN WASPS Family Stephanidae*
Slender, resembling ichneumonids. Head globular and set apart distinctly on a neck-like pronotum; crown of teeth around median ocellus. Antennae long; usually more than 20 flagellomeres. Hind coxae long; hind femora swollen and toothed below. Parasitoids of wood-boring beetle larvae. TL 5–19 mm; World 340, NA 6.

Superfamily Trigonaloidea

TRIGONALID WASPS Family Trigonalidae*
Often brightly colored with stout bodies. **Antennae long and many-segmented** (more than 16); inserted on frons under small lobe. FW with 10 closed cells. Hyperparasitoids of tachinid flies or ichneumonid wasps. TL 10–13 mm; World 90, NA 5.

Stinging Apocrita (Aculeata)—Ants, Bees, & Stinging Wasps
Superfamily Chrysidoidea
Antennae often with same number of flagellomeres in both sexes; usually 8 or 11; pronotum with posterolateral apex generally reaching tegula, sometimes separated by distinct cuticular gap.

BETHYLID WASPS Family Bethylidae
Antennae usually with 11 flagellomeres (rarely 10). Metasoma with 6 or 7 exposed terga. Females in many species wingless and antlike. Parasitoids of Coleoptera and Lepidoptera larvae.
TL 2–5 mm (rarely 10 mm); World 2,200, NA 200.

AULACID WASPS

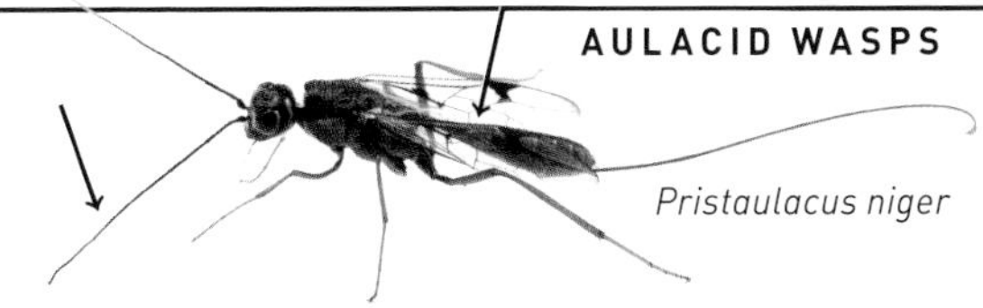

Pristaulacus niger

ENSIGN WASPS

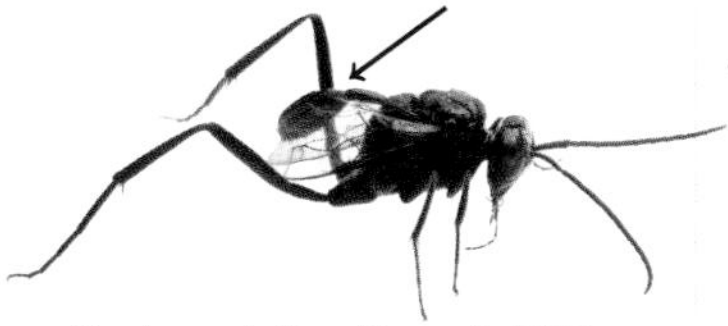

Cockroach Egg Parasitoid Wasp
(*Evania appendigaster*)

CARROT WASPS

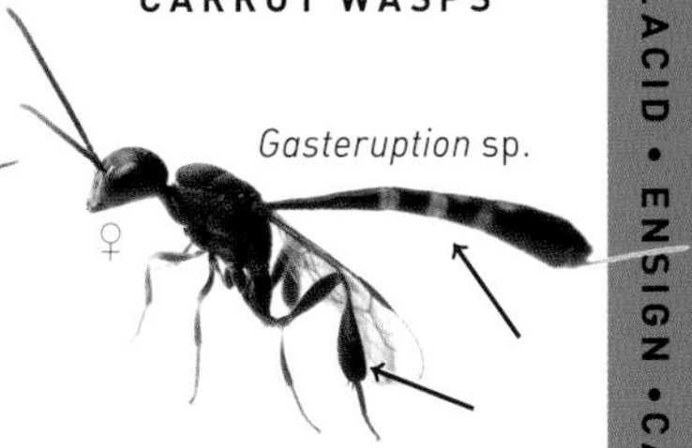

Gasteruption sp.

♀

CROWN WASPS

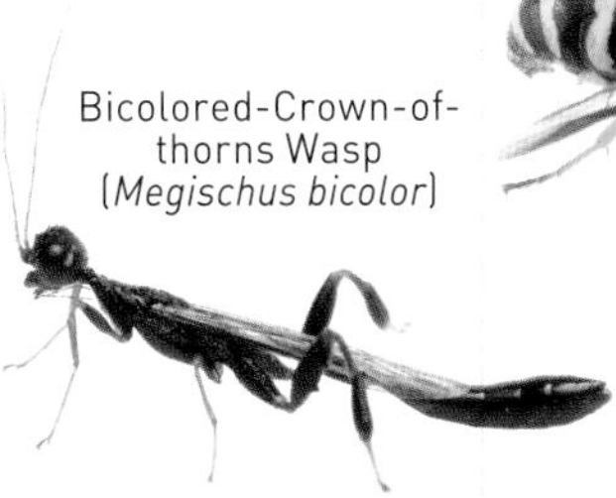

Bicolored-Crown-of-
thorns Wasp
(*Megischus bicolor*)

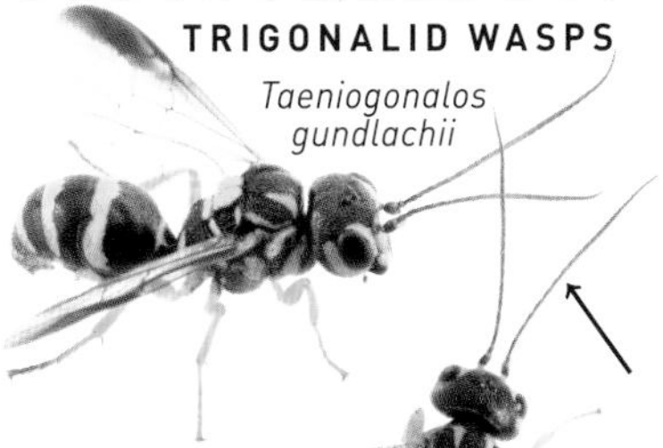

TRIGONALID WASPS

*Taeniogonalos
gundlachii*

Taeniogonalos gundlachii

BETHYLID WASPS

♀

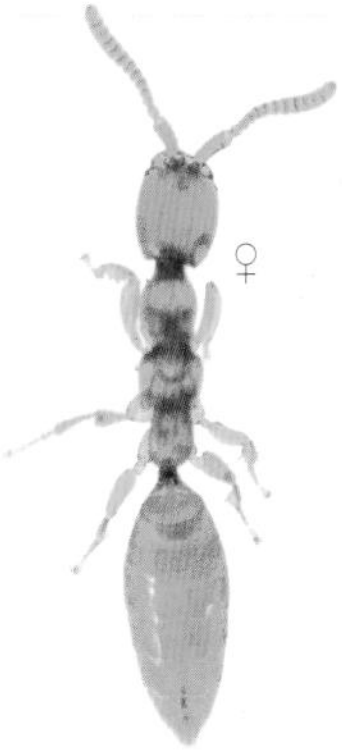

Dissomphalus sp.

♀

261

Superfamily Chrysidoidea cont.

CUCKOO WASPS Family Chrysididae
Robust, blue or green with **coarse sculpturing.** Antennae with 11 flagellomeres in males; 10 in females. Apex of abdomen often with tooth-like projections. Abdomen concave beneath, capable of rolling into defensive ball. Most are external parasitoids of Hymenoptera larvae. TL 6–12 mm; World 3,000, NA 277.

DRYINIDS Family Dryinidae
Females often wingless, mimicking ants. **Foretarsi of female generally modified for grasping.** Antennae with 8 flagellomeres; inserted near clypeus on relatively flat frons. Parasitoids of Auchenorrhyncha. TL 2.5–5 mm; World 1,900, NA 150.

EMBOLEMID WASPS Family Embolemidae*   Not Illus.
Antennae 8 flagellomeres; arising on frontal shelf similar to Dryinidae. Poorly known, but some parasitoids of Achilidae; females collected in ant nests and mammal burrows. TL 5–8 mm; World 20, NA 2.

SCEROGIBBID WASPS Family Sclerogibbidae*   Not Illus.
Females wingless and ant-like; males winged. Head distinctly projecting forward. Antennae has more than 13 flagellomeres. Female generally with enlarged profemur and lacking deep ventral constriction between meso- and metathorax. Larval ectoparasites of Embioptera nymphs. TL 2.5–5 mm; World 20, NA 5.

Superfamily Vespoidea

Antennae with 10 flagellomeres in female, 11 in male. Pronotum with posterolateral apex reaching or exceeding tegula. FW usually with 9–10 closed cells; HW with 2. Metasomal sterna 1 and 2 generally separated by constriction. Females have a painful sting.

YELLOWJACKETS, HORNETS, PAPER WASPS, POTTER WASPS, MASON WASPS, POLLEN WASPS
Family Vespidae   Most yellow or red mixed with black or brown and hold wings longitudinally when at rest. **Inner margin of eye usually notched.** Pronotum extends rearward to tegulae, appearing U-shaped from above and triangular in lateral view. Many are eusocial, others solitary. Many feed on caterpillars. TL 7–25 mm; World 5,000, NA 300.

CUCKOO WASPS

DRYINIDS

Hedychrum sp.

Anteoninae

larva in hemiptera nymph

YELLOWJACKETS, HORNETS, PAPER WASPS, POTTER WASPS, MASON WASPS, POLLEN WASPS

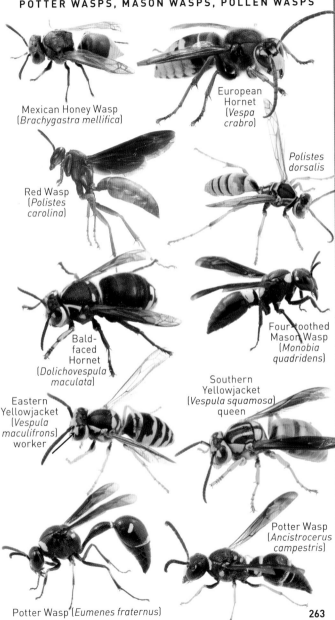

Mexican Honey Wasp
(*Brachygastra mellifica*)

European Hornet
(*Vespa crabro*)

Polistes dorsalis

Red Wasp
(*Polistes carolina*)

Bald-faced Hornet
(*Dolichovespula maculata*)

Four-toothed Mason Wasp
(*Monobia quadridens*)

Eastern Yellowjacket
(*Vespula maculifrons*)
worker

Southern Yellowjacket
(*Vespula squamosa*)
queen

Potter Wasp
(*Ancistrocerus campestris*)

Potter Wasp (*Eumenes fraternus*)

263

VESPOIDEA

Superfamily Vespoidea cont.

RHOPALOSOMATID WASPS Family Rhopalosomatidae*

Not Illus.

Three rare, variable species in NA; one brachypterous (resembling an ant), one resembles *Polistes* and one resembles ichneumon wasps. Apices of 2 or more flagellomeres usually with spines. Ectoparasitoids of cricket nymphs. TL 4.2–17 mm; World 68, NA 3.

SIEROLOMORPHID WASPS Family Sierolomorphidae*

Not Illus.

Generally dark brown or black. Coxa of mid- and hind legs arise near one another. HW lacks claval or jugal lobe. Females sometimes wingless (more robust than Bethylidae). Solitary; larvae thought to be ectoparasitoids of other insects. TL 4–7 mm; World 10, NA 6.

TIPHIID WASPS Family Tiphiidae

Males winged with an **upcurved hook at tip of abdomen.** Female usually with **mesotibia and metatibia stout and heavily spined.** HW with distinct clavate and jugal lobes. Primarily parasitoids of subterranean beetle larvae. Adults visit flowers and will eat pollen. TL 6–26 mm; World 1,500, NA 200.

CHYPHOTID WASPS Family Chyphotidae

Similar to nocturnal mutillids, but females with a suture separating the pro- and mesonotum. Usually brownish or black, some with red. One genus nocturnal, the other diurnal. Biology poorly known. Found in southwestern US. TL 5–15.5 mm; World 65, NA 51.

THYNNID WASPS Family Thynnidae

Similar to Tiphiidae. Usually dark black, brown, or red, often marked with yellow. Bodies elongated, slender (females sometimes more robust). **Some males with curved "pseudostinger" at end of abdomen.** Some females wingless, resembling ants. Antennae straight. Larvae are parasitoids of Coleoptera larvae, especially Scarabaeoidea and Cicindelinae (tiger beetles). TL 5–25.5 mm; World 1,000, NA 83.

VELVET ANTS Family Mutillidae

Females wingless, notably hairy, and appear like ants, but lacking petiole. Males winged, not as hairy, appearing more like wasps. Inner margin of eye often convex in female; more or less straight or emarginate in male. Tubercle often present between eyes. Both sexes with hair-lined grooves on side of metasoma called *felt lines*. Ectoparasitoids of immature ground-nesting insects, especially bees and solitary wasps. TL 6–30 mm; World 4,300, NA 480.

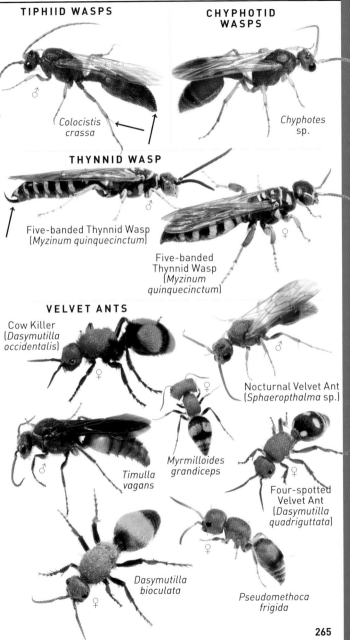

TIPHIID WASPS

Colocistis
crassa

♂

CHYPHOTID WASPS

Chyphotes
sp.

THYNNID WASP

Five-banded Thynnid Wasp
(*Myzinum quinquecinctum*)

♂

Five-banded
Thynnid Wasp
(*Myzinum
quinquecinctum*)

♀

VELVET ANTS

Cow Killer
(*Dasymutilla
occidentalis*)

♀

♂

Nocturnal Velvet Ant
(*Sphaeropthalma* sp.)

♂

*Timulla
vagans*

*Myrmilloides
grandiceps*

♀

♀

Four-spotted
Velvet Ant
(*Dasymutilla
quadriguttata*)

*Dasymutilla
bioculata*

♀

♀

*Pseudomethoca
frigida*

MYRMOSID WASPS Family Myrmosidae   Not Illus.
Similar to Mutillidae with wingless females, but less hairy. Thorax
divided into 2 distinct parts. Males usually uniformly black;
females reddish or brown. Felt lines found in Mutillidae are absent.
Ectoparasitoids of immature ground-nesting bees and wasps.
TL 5–15.5 mm; World 50, NA 16.

SPIDER WASPS Family Pompilidae
Robust, with long spiny legs. Usually dark with smoky, yellowish
or orange wings. **Hind tibiae with 2 prominent spines at apex.**
Mesopleuron with a transverse suture. Found on flowers or searching
on ground. Larvae feed on a single paralyzed spider.
TL 5–40 mm; World 5,000, NA 300.

SAPYGID WASPS Family Sapygidae*
Generally black with yellow bands or spots, with relatively short legs.
Eyes usually deeply emarginate. Hind margin of pronotum straight.
Larvae are kleptoparasites or ectoparasitoids of larval Megachilidae,
Anthophoridae and Vespidae. TL 10–15 mm; World 80, NA 19.

SCOLIID WASPS Family Scoliidae
Robust, moderately hairy, usually with dark wings. Wings with
longitudinal wrinkles toward apex. Posterior margin of pronotum
U-shaped. Hind coxae wellseparated. Meso- and metasternum
divided by transverse suture. Larvae ectoparasitoids of scarab grubs.
TL 10–50 mm; World 560, NA 20.

Superfamily Formicoidea

ANTS Family Formicidae
Generally black, brown, or reddish and live in colonies. **Second and
sometimes also third segment of metasoma forming hump(s) known
as petiole. Antennae elbowed,** 6–13 segments. Only reproductives
have wings. Social with castes. TL 12–25.5 mm; World >16,500, NA 700.

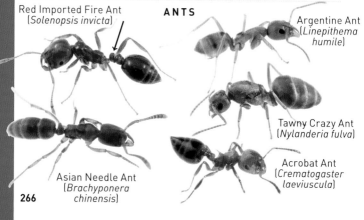

ANTS

Red Imported Fire Ant
(*Solenopsis invicta*)

Argentine Ant
(*Linepithema humile*)

Tawny Crazy Ant
(*Nylanderia fulva*)

Acrobat Ant
(*Crematogaster laeviuscula*)

Asian Needle Ant
(*Brachyponera chinensis*)

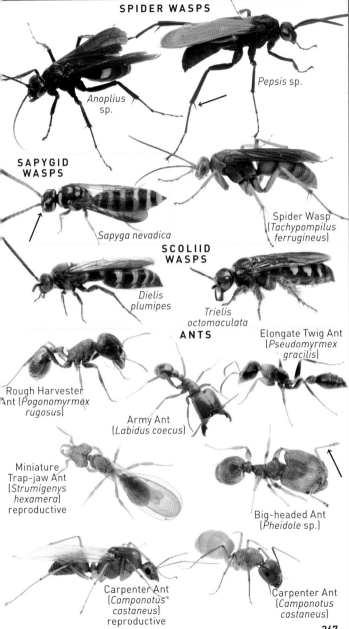

SPIDER WASPS

Anoplius sp.

Pepsis sp.

SAPYGID WASPS

Sapyga nevadica

Spider Wasp (*Tachypompilus ferrugineus*)

SCOLIID WASPS

Dielis plumipes

Trielis octomaculata

ANTS

Elongate Twig Ant (*Pseudomyrmex gracilis*)

Rough Harvester Ant (*Pogonomyrmex rugosus*)

Army Ant (*Labidus coecus*)

Miniature Trap-jaw Ant (*Strumigenys hexamera*) reproductive

Big-headed Ant (*Pheidole* sp.)

Carpenter Ant (*Camponotus castaneus*) reproductive

Carpenter Ant (*Camponotus castaneus*)

Superfamily Apoidea

Antennae with 12 segments in females; 13 in males. Posterior margin of pronotum U-shaped. FW with 9–10 closed cells; HW with 2 or fewer. Pronotum, when viewed laterally, with thumb-like lobe not extending to wing base; jugal lobe usually present on HW.

COCKROACH WASPS Family Ampulicidae   Not Illus.

Black, often with some red; round, prognathic head set off by distinct neck. Tarsal claws toothed or cleft along inner margin; middle tibiae with 2 apical spurs. Nest in twigs, under bark, and in leaf litter. Adults capture cockroach nymphs as food for larvae.
TL 5–16 mm; World 200, NA 4.

THREAD-WAISTED WASPS Family Sphecidae

Metasoma petiolate and composed of sternum only. Body usually black, sometimes with metallic blue or green, often with red, yellow, or white stripes and markings. FW with 3 (rarely 2) submarginal cells. **Middle tibiae with 2 apical spurs**. Most are solitary; some are kleptoparasites. Larvae feed on paralyzed arthropods captured by adult. TL 10–30 mm; World 800, NA 125.

CRABRONID WASPS Family Crabronidae

Large, diverse family; usually black or brown, often with yellow or pale markings. If metasoma petiolate, composed of both sternum and tergum. FW with 0–3 submarginal cells. **Middle tibiae with 1 apical spur.** Adults capture insects as prey for larvae; a few are kelptoparasitc. TL 5–41 mm; World 9,000, NA 1,225.

MINING BEES Family Andrenidae

Usually brown, reddish-brown, or yellow. Head often broader than in sweat bees; females with facial fovea. Two grooves below each antennal socket. Tongue is short; labial palps usually with all segments subequal in length. Females and most males with flat, triangular sclerite apically on upperside of metasoma. Often seen visiting flowers; many nest in the ground. TL 10–20 mm; World 4,500, NA 1,200.

PLASTERER BEES, MASKED BEES Family Colletidae

Often boldly colored bees, others all black, and generally with few hairs. Single groove below each antennal socket. Some species with yellow face. Tongue is short; glossa very reduced, usually broader than long with end split or truncate. Labial palps with all segments subequal in length. FW with 2–3 submarginal cells.
TL 2.5–13 mm; World 2,500, NA 160.

SWEAT BEES Family Halictidae

Generally with black or brown, sometimes green or blue bodies. FW in many with curved basal wing vein. Tongue is short; labial palps with all segments subequal in length. Single groove below each antennal socket. Most species are ground nesters. Visit many different flowers, and some are attracted to sweat. TL 4–11 mm; World 4,300, NA 520.

BEES, WASPS, ANTS, & SAWFLIES

THREAD-WAISTED WASPS

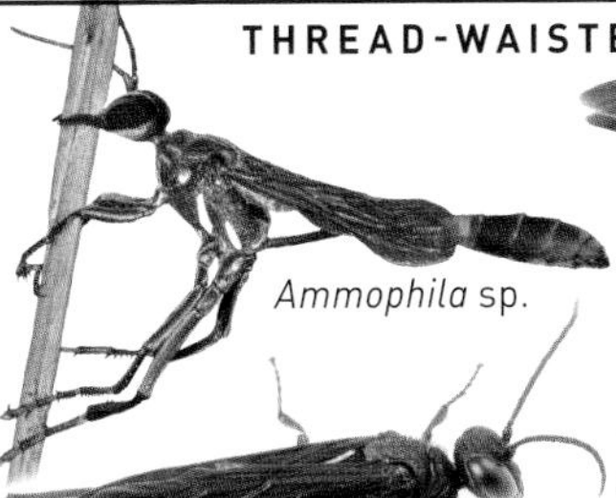

Ammophila sp.

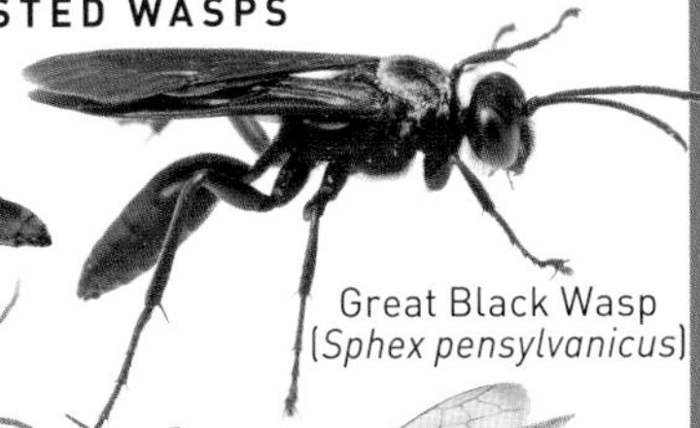

Great Black Wasp
(*Sphex pensylvanicus*)

Blue Mud
Wasp
(*Chalybion
californicum*)

Yellow-legged
Mud-dauber
(*Sceliphron
caementarium*)

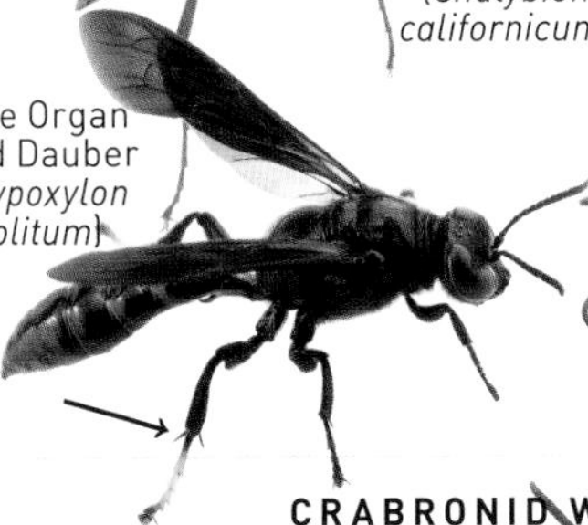

Pipe Organ
Mud Dauber
(*Trypoxylon
politum*)

*Sphex
lucae*

CRABRONID WASPS

*Tachytes
distinctus*

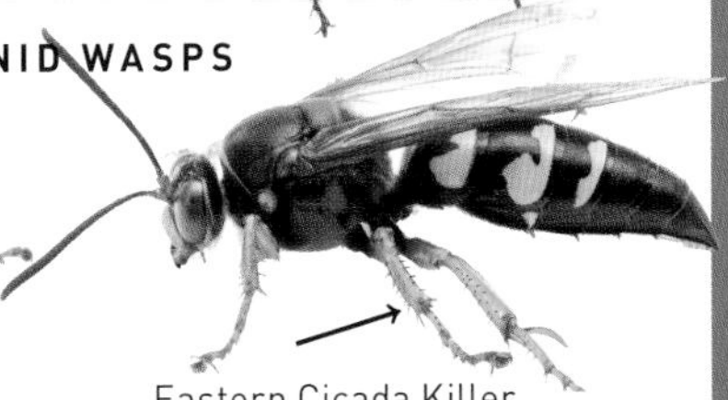

Eastern Cicada Killer
(*Sphecius speciosus*)

MINING BEES

Andrena
sp.

PLASTERER & MASKED BEES

Goldenrod
Cellophane Bee
(*Colletes solidaginis*)

Hylaeus
sp.

SWEAT BEES

Lasioglossum sp.
collecting sweat

Dark-winged
Striped-Sweat
Bee (*Agapostemon
splendens*)

Ligated Sweat
Bee (*Halictus
ligatus*)

269

Superfamily Apoidea cont.

HONEY BEES, BUMBLE BEES, & ALLIES Family Apidae

Largest and most diverse family of bees, which includes some of our most recognizable species. Many are fast-flying, large, and hairy. They are long-tongued bees with first 2 segments of labial palps longer than second 2. TL 12–25.5 mm; World 5,750, NA 1,000.

LEAFCUTTER BEES, MASON BEES, RESIN BEES

Family Megachilidae

Robust, often cigar shaped, and generally dark colored. FW with 2 submarginal cells. Labrum large, longer than wide, and hinged across its entire length. Females, except in parasitic species, with pollen-collecting scopa on underside of abdomen. Long-tongued bees with first 2 segments of labial palps longer than second 2. TL 6–18 mm; World 4,100, NA 630.

MELITTID BEES Family Melittidae*   Not Illus.

Small group of bees separated from others by characters on the tongue and genitalia. Tongue is short; labial palps with all segments subequal in length. Middle coxa sometimes slightly longer than in other bee families. Some species (*Hesperapis*) with extremely flattened abdomen, others (*Macropis*) notably rounded. TL 7.5–13 mm; World 180, NA 32.

APOIDEA

LEAFCUTTER BEES, MASON BEES, RESIN BEES

Leaf-cutter Bee
(*Megachile* sp.)

Leaf-cutter Bee
(*Megachile* sp.)

Northern Rotun
Resin Bee
(*Anthidiellum
notatum*)

Blueberry
Mason Bee
(*Osmia
ribifloris*)

HONEY BEES, BUMBLE BEES, & ALLIES

Brown-belted Bumble Bee (*Bombus griseocollis*)

Yellow-fronted Bumble Bee (*Bombus flavifrons*)

Blueberry Digger Bee (*Habropoda laboriosa*)

American Bumble Bee (*Bombus pensylvanicus*)

Red-legged Oil Digger (*Centris rhodopus*)

Arizona Carpenter Bee (*Xylocopa californica arizonensis*)

Eastern Carpenter Bee (*Xylocopa virginica*)

Horsefly-like Carpenter Bee (*Xylocopa tabaniformis*)

Western Honey Bee (*Apis mellifera*)

Long-horned Bee (*Melissodes tepaneca*)

Nomad Bee (*Nomada* sp.)

CARPENTER • DIGGER • BUMBLE • HONEY • LEAF-CUTTER • MASON BEES

271

Agulla sp. female
Raphidiidae

Agulla sp. larva
Raphidiidae

SNAKEFLIES
ORDER RAPHIDIOPTERA

Derivation of Name
- Raphidioptera: *raphio*, needle; *ptera*, wings. A winged needle, referring to the female's long ovipositor.

Identification
- 5–20 mm TL.
- Prothorax elongated with unmodified front legs arising posteriorly.
- Can raise head and prothorax up above body, giving them a cobra-like appearance.
- Females have a long needle-like ovipositor.

Classification
- Historically considered Neuroptera, but most modern revisions place them in their own order.
- Most closely related to Megaloptera and Neuroptera.
- Two NA families: Raphidiidae, 2 genera; Inocelliidae,1 genus.
- World: 260 species; North America: 22 species.

Range
- Northwestern US and western Canada, south to TX.

Similar orders
- Neuroptera (p. 280): Mantispidae have front legs attached to anterior part of prothorax; other families lack elongated prothorax.
- Megaloptera (p. 276): HW wider at base; HW anal area folded under FW at rest.
- Mecoptera (p. 430): long face forming proboscis; few costal crossveins.

Agulla bicolor female
Raphidiidae

Food

- Adults are predators and normally eat small insects like barklice, springtails, mites, and aphids. They have also been observed feeding on pollen, which is often found in gut contents.
- Larvae are predatory on soft-bodied insects; often feeding on eggs and larvae of various orders; found in leaf litter and under bark.

Behavior

- Found in all types of arboreal habitats from sea level up to the timberline.
- Females ritualistically groom after eating and tend to move ovipositor as they eat.
- Mating behavior has been documented with male hanging headfirst from females.
- Mating takes from 1.5 to 3 hours.
- Larvae can move rapidly both forward and reverse.

Life Cycle

- Holometabolous.
- Usually found in areas where temperatures get close to freezing to complete development.
- Females lay eggs under bark in the west.
- Eggs last from a few days to 3 weeks.
- Larvae last from 1 to 6 years depending on the species and complete their life cycle under rocks or bushes around roots or under the bark of trees where they pupate.
- Pupation usually occurs in the spring, and the pupal stage lasts a few days to 3 weeks.
- Some species pupate in the summer or fall, and the pupal stage lasts up to 10 months.
- Adults live 1–3 months.

Importance to Humans

- They are predators on potential pest species and have been considered for biological control in areas.

Collecting and Preserving

Immatures

- Hand collect under rocks or bark.
- Preserve in 70% ethanol.

Adults

- Collect with small aerial net or beating sheet in the spring or when adults are flying.
- Pin through mesothorax.

Resources

- "The Biology of Raphidioptera: A Review of Present Knowledge" by Aspöck, 2002

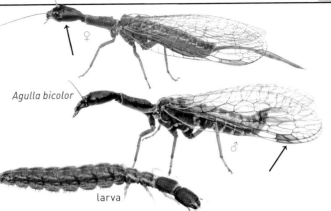

Agulla bicolor

♀

♂

larva

SNAKEFLIES Family Raphidiidae

Ocelli present. FW pterostigma with crossvein. Prothorax more elongated than in Inocelliidae and antennae relatively short. Adult head triangular. Found from CA to TX and north to BC.
TL 5–18 mm; World 230, NA 19.

♂

♂

♀

Negha sp.

SQUARE-HEADED SNAKEFLIES Family Inocelliidae

Ocelli absent. Usually larger than Raphidiidae with longer antennae that are thicker at the base. **Head square shaped. Pterostigma tends to be larger with no crossvein dividing it.** Larvae typically found under bark. Found west of the Rockies (CA, OR, WA, NV, and BC). TL 10–20 mm; World 32, NA 3.

Alderfly
(*Sialis* sp.)
Sialidae

Eastern Dobsonfly
(*Corydalus cornutus*) larva
Corydalidae

DOBSONFLIES, FISHFLIES, & ALDERFLIES
ORDER MEGALOPTERA

Derivation of Name
- Megaloptera: *megalo*, large; *ptera*, wings. Wings are larger than in other neuropteroid insects like lacewings and snakeflies.

Identification
- 7–90 mm TL.
- Head and thorax are sclerotized; abdomen long and soft bodied; lacking cerci.
- Four membranous wings with numerous veins and crossveins. HW is much wider than FW and folded under FW at rest.
- Tarsi 5-segmented.
- Chewing with pronounced, well-developed mandibles.

Classification
- Previously placed in Neuroptera, but generally recognized as its own order in recent works.
- Most closely related to Neuroptera and Raphidioptera.
- Two families, Corydalidae and Sialidae; 9 genera.
- World: 400 species; North America: 46 species.

Range
- New World, South Africa, Madagascar, Asia, and Australia. Most diverse in subtropical and tropical areas.

Similar orders
- Odonata (p. 96): prominent wings held horizontal from body or together over abdomen at rest; different wing venation; tarsi 3-segmented; antennae short.
- Plecoptera (p. 120): flattened body; most with 4 membranous wings; tarsi 3-segmented; prominent cerci present.
- Neuroptera (p. 280): FW and HW similar in size and shape; HW not folded at rest; smaller in size.
- Raphidioptera (p. 272): prothorax elongated; forelegs attached at posterior part of prothorax.

Food

- Larvae are predators; normally not taking large prey, but feed all year long. They are opportunistic feeders.

Behavior

- Found in aquatic habitats of varying type and quality. Found in streams, ponds and organic waste areas.
- Adults are found resting on trees and shrubs, and many come to lights (except *Sialis*).
- Adults do not eat solids, but may take nectar.
- The large mandibles of male dobsonflies are used for holding the female when mating.

Life Cycle

- Holometabolous.
- Female lays 300–3,000 eggs in masses overhanging water on rocks or vegetation at night. Eggs hatch in 1–2 weeks.
- Larvae are aquatic and found in all types of aquatic habitats; 10–12 molts over 1–5 years.
- Pupae are terrestrial with functioning legs and mandibles.
- Adults are short-lived and generally active only at night.

Importance to Humans

- Not a pest or nuisance, but larvae are predators that help maintain diversity in aquatic habitats. Dobsonfly larvae, hellgrammites, are used as fishing bait and sold in some areas.

Collecting and Preserving

Immatures

- Collect from aquatic habitats using dip net.
- Permanently store in 70% ethanol.

Adults

- Hand collect or use aerial net to capture, especially at lights.
- Pin or preserve in clear envelopes using same method as Odonata.

Resources

- Lacewing Digital Library—lacewing.tamu.edu
- "Species Catalog of the Neuroptera, Megaloptera, and Raphidioptera of America North of Mexico" by Penny et al., 1997

eggs

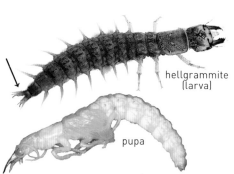

hellgrammite (larva)

pupa

underside of eggs

Eastern Dobsonfly (*Corydalus cornutus*)

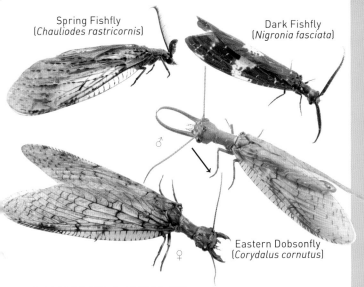

Spring Fishfly
(*Chauliodes rastricornis*)

Dark Fishfly
(*Nigronia fasciata*)

♂

Eastern Dobsonfly
(*Corydalus cornutus*)

♀

DOBSONFLIES & FISHFLIES Family Corydalidae

Found throughout NA but not as common in the west. Larvae of dobsonflies are often called hellgrammites and can be found in swift-moving currents of streams to shallow pools of water. Ocelli present. **Fourth tarsal segment simple.** Larvae have **2 anal prolegs with a claw at tip of abdomen.** TL 20–90 mm; World 80, NA 24.

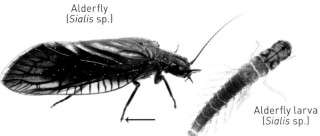

Alderfly
(*Sialis* sp.)

Alderfly larva
(*Sialis* sp.)

ALDERFLIES Family Sialidae

Adults found near water; eggs laid on the underside of vegetation. Soft, dark bodies with wings that are held roof-like over abdomen. Lack ocelli. **Fourth tarsal segment widened with lobes.** Larvae are similar to dobsonflies, but lack a claw on anal prolegs and have an elongated process at tip of abdomen. TL 7–23 mm; World 75, NA 24.

Owlfly
(*Ascaloptynx appendiculata*)
Ascalaphidae

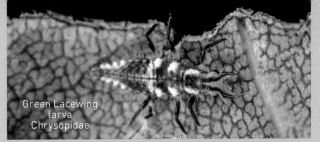
Green Lacewing
larva
Chrysopidae

LACEWINGS, ANTLIONS, & ALLIES
ORDER NEUROPTERA

Derivation of Name
- Neuroptera: *neuro*, nerve or sinew; *ptera*, wings. Describing their densely veined wings.

Identification
- 2–100 mm TL.
- Soft-bodied insects, usually with an elongated simple abdomen.
- Chewing mouthparts; variable antennae, but usually many-segmented and long; tarsi 5-segmented; FW and HW similar in shape and size; prothorax not elongated except in Mantispidae; cerci absent.

Classification
- One of the oldest orders with complete metamorphosis. Likely greater diversity in this group historically than in present day.
- The order formerly included Megaloptera and Raphidioptera, but taxonomists now treat these as separate orders, as we do here.
- Eleven families, and 24 genera in NA.
- World: 4,700 species; North America: 400 species.

Range
- Found worldwide except Antarctica.

Similar orders
- Megaloptera (p. 276): HW broader at base than FW.
- Raphidioptera (p. 272): head, but not forelegs, at end of long neck.
- Odonata (p. 96): wings held above body at rest or perpendicular to body; tarsi 3-segmented; antennae short and bristle-like; different wing venation.
- Mecoptera (p. 430): long face forming beak; few costal crossveins.
- Plecoptera (p. 120): flattened body; tarsi 3-segmented; found near streams; cerci present.
- Trichoptera (p. 364): body with hair; wings not clear.

Green Lacewing
(*Abachrysa eureka*)
in flight

Food
- Some adults are predators on other insects but will eat pollen and nectar as well.
- Larvae are predators with mandibles modified into sucking sickle-shaped tubes.

Behavior
- Found in various habitats: some antlion larvae create divots in the sand where insects like ants fall in and are eaten; some mantspid larvae feed on the larvae of wasps and bees, others on spider eggs; spongillafly larvae feed on freshwater sponges and bryozoans.
- Most are weak fliers and are found near the larval habitat. Antlions are the exception and fly similar to damselflies. Many species will come to lights but will not travel far.

Life Cycle
- Holometabolous.
- Eggs laid in various environments tend to be attached or "glued" to a surface and hatch in 5–15 days unless they are overwintering.
- Most larvae are terrestrial, but spongillaflies live in streams.
- Usually 3 larval instars lasting from a few weeks to months. Some species overwinter in the larval stage.
- Pupae are formed in a silken, double-walled cocoon.

Importance to Humans
- Many larvae are predators on small plant-eating insects like aphids and other insects. Green lacewing larvae and adults are used for biological control.

Collecting and Preserving
Immatures
- Hand collect from various habitats.
- Preserve in 70% ethanol.

Adults
- Easiest collected at lights or beat sheets.
- Pin through mesothorax; some groups may be spread like butterflies.

Resources
- Lacewing Digital Library, lacewing.tamu.edu
- "Species Catalog of the Neuroptera, Megaloptera, and Raphidioptera of America North of Mexico" by Penny et al.,1997

Beaded Lacewing
(*Lomamyia* sp.) in flight

Wing Venation

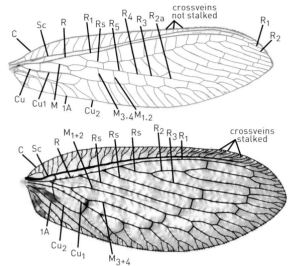

Green lacewing venation (top) showing FW costal crossveins simple and not branched. Brown lacewing venation (bottom) showing FW costal crossveins stalked.

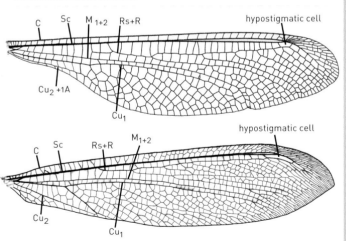

Owlfly wing venation (top) showing FW hypostigmatic cell not 2–3 times as long as wide. Antlion wing venation (bottom) showing FW hypostigmatic cell 2–3 times as long as wide.

HEMEROBIIFORMIA

Suborder Hemerobiiformia

BEADED LACEWINGS Family Berothidae*
Antennae shorter than forewing. FW tip scooped. Females of some species have scales on wings and thorax. Eggs on stalks laid on wood near termite mounds. Larvae are predaceous on termites, physically injecting them with an immobilizing venom from the tip of their mandibles. TL 7–16 mm; World 100, NA 12.

GREEN LACEWINGS Family Chrysopidae
Most are green, but can be brown, yellow, or white. Eyes sometimes golden. **FW costal crossveins simple and not branched.** Very common living on the grass and foliage of trees. The western genus, *Eremochrysa,* is often tan, while the *Abachrysa* in the southeast is largely white. Eggs are laid on stalks and are commonly encountered on door frames and windows. Larvae are predators and often carry debris on their backs for camouflage. TL 15–30 mm; World 1,200, NA 87.

DUSTYWINGS Family Coniopterygidae*
Wings and body covered in a white powder. Antennae are 16–34-segmented and filiform. Lack ocelli. Wing venation is reduced compared to other neuropterans. Smallest body size of all neuropterans. Most species are rare but can be locally common. Eggs are laid in trees near aphids and scale insects their food source. TL <5 mm; World 300, NA 55.

PLEASING LACEWINGS Family Dilaridae*
Wings rounded and hairy. Lands with wings out like moths, not over back. Antennae pectinate in males and filiform in females. Female with ovipositor as long as body. Larvae live on dead logs and likely feed on beetle larvae. One eastern species, *Nallachius americanus,* and one western, *N. pulchellus.* TL 2–8 mm; World 70, NA 2.

BROWN LACEWINGS Family Hemerobiidae
Resemble green lacewings, but brown and smaller, with hairs on wings. **FW crossveins are stalked,** not simple like in green lacewings and spongillaflies. Have 2 or more Rs veins. Larvae are found in wooded areas and do not carry debris on their backs. Eggs are not stalked, but laid on leaves. TL 5–16 mm; World 600, NA 61.

MOTH LACEWINGS Family Ithonidae*   Not Illus.
Only 1 rare species in NA, *Oliarces clara,* known from southern CA, NV, and AZ. Our species has bleached wings and a greenish-blue body. It is associated with creosote bush, where larvae likely feed on roots and vegetation. Adults swarm in large numbers for hill-topping behavior. TL 20–40 mm; World 8, NA 1.

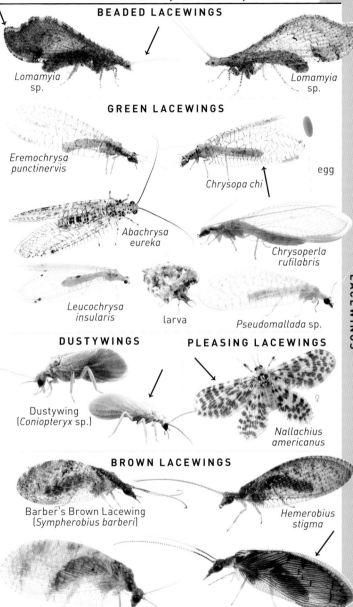

LACEWINGS, ANTLIONS, & ALLIES

BEADED LACEWINGS

Lomamyia sp.

Lomamyia sp.

GREEN LACEWINGS

Eremochrysa punctinervis

Chrysopa chi

egg

Abachrysa eureka

Chrysoperla rufilabris

Leucochrysa insularis

larva

Pseudomallada sp.

DUSTYWINGS

Dustywing (*Coniopteryx* sp.)

PLEASING LACEWINGS

♀

Nallachius americanus

BROWN LACEWINGS

Barber's Brown Lacewing (*Symperobius barberi*)

Hemerobius stigma

Megalomus fidelis

Symperobius occidentalis

HEMEROBIIFORMIA • MYRMELEONTIFORMIA

MANTIDFLIES Family Mantispidae

Prothorax is elongated. Front legs modified like praying mantids to catch prey and attached to prothorax near the head. They feed on spider eggs, scarab beetle, moth and some social wasp larvae. TL 20–35 mm; World 400, NA 15.

GIANT LACEWINGS Family Polystoechotidae*

Only 2 species in the west; larger in size than green lacewings. Rs in FW has many branches. The most common species, *Polystoechotes punctatus,* is known to be attracted to smoke and was once common in eastern NA as well. Some authorities place this group within the Ithonidae. TL 32–77 mm; World 4, NA 2.

SPONGILLAFLIES Family Sisyridae

Looks similar to a small brown lacewing. Antennae filiform in both sexes. Costal crossveins not branched. HW with free basal part of MA, Sc and R_1 in front wing fused or separate apically; Rs branched. Usually found near lakes. Larvae are aquatic and feed on freshwater sponges. When larvae are full-grown they crawl out of the water and create a netted globe-like structure in which they pupate. TL 5–11 mm; World 60, NA 6.

Suborder Myrmeleontiformia

Some authors treat these as subfamilies, while others use separate families (as we do); the matter is not yet fully resolved.

OWLFLIES Family Ascalaphidae

Large species that resemble dragonflies and damselflies, but have **clubbed antennae the length of the** abdomen. When they land they place their head and wings along the branch and **raise their abdomen perpendicular to the branch and wings**. Eggs are laid on twigs, then larvae crawl into leaf litter where they prey on other insects. TL 38–82 mm; World 430, NA 8.

ANTLIONS Family Myrmeleontidae

Similar to owlflies, but with **hooked antennae no longer than head and thorax combined.** Larvae are often called "doodlebugs," and some hide in shallow cone-shaped pits made in the sand and eat insects that fall in them. They pupate in the soil in a cocoon made of sand and silk. TL 38–82 mm; World 2,000, NA 94.

Mantidfly larvae hatching.

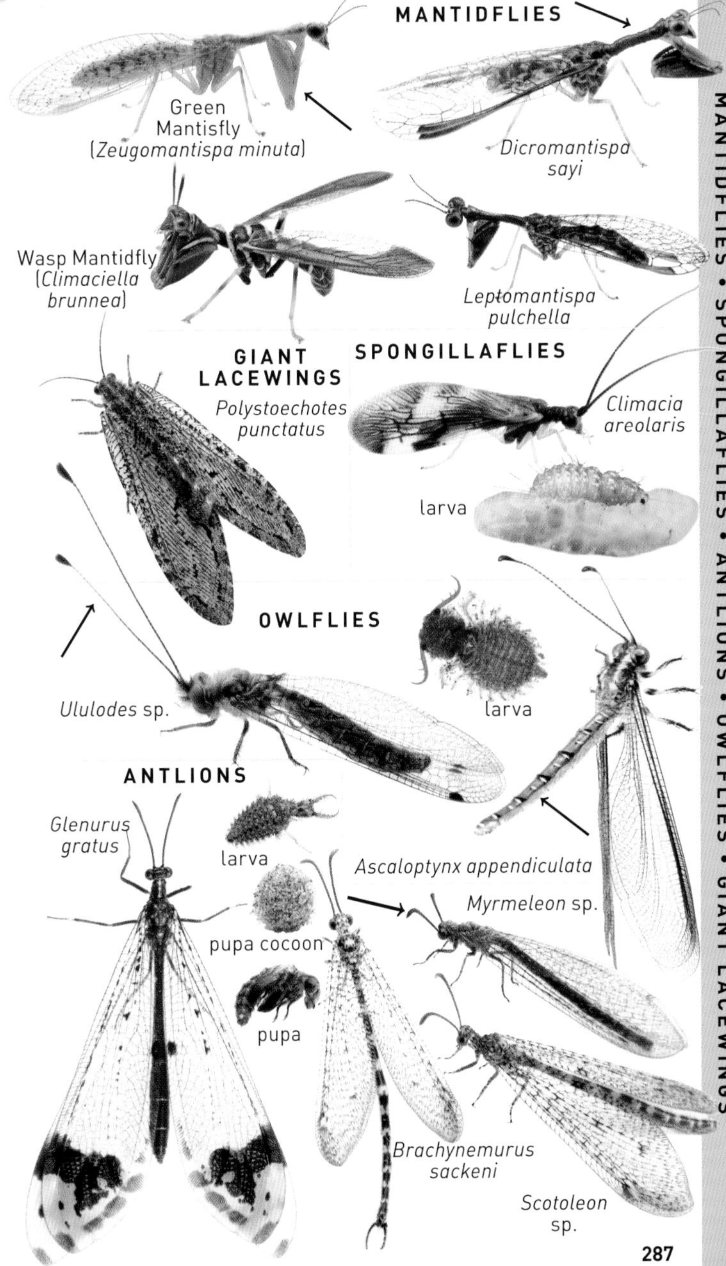

MANTIDFLIES

Green Mantisfly
(*Zeugomantispa minuta*)

Dicromantispa sayi

Wasp Mantidfly
(*Climaciella brunnea*)

Leptomantispa pulchella

GIANT LACEWINGS

Polystoechotes punctatus

SPONGILLAFLIES

Climacia areolaris

larva

OWLFLIES

Ululodes sp.

larva

ANTLIONS

Glenurus gratus

larva

pupa cocoon

pupa

Ascaloptynx appendiculata

Myrmeleon sp.

Brachynemurus sackeni

Scotoleon sp.

Male *Stylops* sp. mating
Stylopidae

Twisted-winged Insect
(*Xenos* sp.) male pupa
Stylopidae on *Polistes* wasp

TWISTED-WINGED INSECTS
ORDER STREPSIPTERA

Derivation of Name
- Strepsiptera: *strepsi*, a turning or twisting; *ptera*, wings. Refers to the twisting of the male's forewings when dried.

Identification
- 1.0–6 mm TL (males); females up to 30 mm (Myrmecolacidae).
- Males dark brown to black.
- Male FW reduced to club-like structure with irregular HW resembling a fan with reduced venation and no crossveins.
- Males and females are parasites of other insects.
- Male head is irregular with large raspberry-like eyes.
- Male antennae are 4–7-segmented (NA species) and flabellate.
- Male tarsi vary depending on family, ranging from 2–5 segmented with the same number on each leg.
- Adult female of NA species resembles a sac, lacking legs and antennae; have vestigial mouthparts, but do not use them to feed.
- Found on various insect hosts.

Classification
- Relationship of this group has been a mystery, historically placed in Lepidoptera, Diptera, Hymenoptera, and Coleoptera at various times.
- Currently placed in its own order and most closely related to Coleoptera.
- They have highly specialized larvae and adults.
- Five families in NA separated by tarsi, antennal characters, and host.
- World: 600 species; North America: 84 species.

Range
- Can be found worldwide except Antarctica.

Similar orders
- Diptera (p. 436): 1 pair of FW; HW reduced to halteres.
- Coleoptera (p. 294): FW modified into hard elytra.

STREPSIPTERA

Food
- Larvae feed as parasites in the haemocoel of host.
- Females don't feed, but live off reserves from larval stages.
- Males have reduced mouthparts and live only for a few hours using nutritional stores from larval stage.

Behavior
- Males and first instar larvae are free-living in NA species.
- Rarely seen unless attracted to female pheromones.
- Males of some families will come to lights.

Life Cycle
- Holometabolous, exhibiting hypermetamorphosis.
- Male mates with female through brood canal while she remains inside the host.
- Female gives live birth inside the host, usually inside the abdomen.
- Females give birth to 1,000 to 750,000 offspring.
- First instar larvae have legs, are very active, and leave the female to find a host; subsequent instars (usally 2) are maggot-like.
- Primary larvae use mandibles to enter the host cuticle.
- Larvae and females live inside the host without killing it.
- Male secondary larva protrudes its head and thorax from abdomen of host; secondary larva exuvium sclerotizes and pupation occurs within.

Importance to Humans
- Not known to be pests.

Collecting and Preserving
- Collect parasitized host and rear the Strepsiptera adult out.
- Males are collected in malaise traps and sometimes at lights.
- Preserve adults and larvae in 70% ethanol.

Resources
- Strepsiptera database—tinyurl.com/StrepsipteraDB

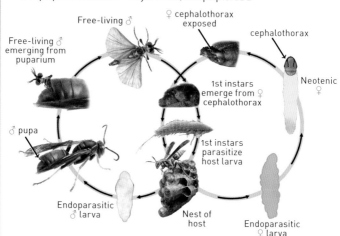

Generalized life cycle (multiple families represented). Modified from Erezylimaz et al. (2014).

Anatomy

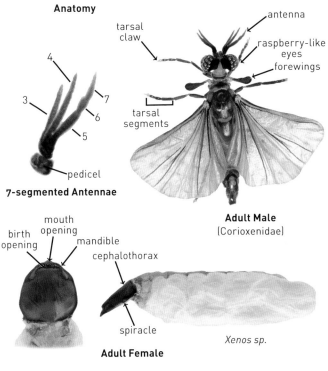

tarsal
claw

antenna

raspberry-like
eyes

forewings

4

3

7

6

5

pedicel

7-segmented Antennae

tarsal
segments

Adult Male
(Corioxenidae)

mouth
opening

birth
opening

mandible

cephalothorax

spiracle

Adult Female

Xenos sp.

SEM of first instar larva of *Stylops* sp.

eye

Male puparium of *Xenos* sp.

CORIOXENIDS Family Corioxenidae*

Found in southern US. Male: Mandibles absent. Tarsi, if 5-segmented, ending in weak claw; **if 4-segmented with** or without **weak claw.** Sensory spots on tarsi. **Antennae 5–7-segmented with third and fourth segments extending out laterally.** Female: Brood canal opening at tip of host. Underside of female toward host. Parasites of Hemiptera (Pentatomidae, Lygaeidae, Cydnidae, Scutelleridae, and Coreidae). World 43, NA 5.

HALICTOPHAGIDS Family Halictophagidae*

Male: **Tarsi 3-segmented without claws. Antennae 6–7-segmented** with segment jutting out laterally off third segmented or third to fourth segmented or third to sixth segmented. Female: Head half or more as long as cephalothorax. Dorsal side not membranous. Abdominal segmented 1–5 with 1 birth organ each. Parasites of Hemiptera (Delphacidae, Eurybrachidae, Fulgoridae, Issidae, Cicadellidae, and Membracidae), Diptera (Tephritidae), and Orthoptera (Tridactylidae). World 132, NA 12.

ELENCHIDS Family Elenchidae*   Not illus.

Male: Tarsi 2-segmented without claws. Antennae 4-segmented third segment jutting out laterally. Female: Cephalothorax without hook-like projection. Abdominal segments 2 and 3 with 1–5 birth organs each. Brood canal opening not crescent shaped. Found in southern US. Parasites of planthoppers (Delphacidae, Eurybrachidae, Fulgoridae, Flatidae, and Dictyopharidae). World 28, NA 2.

MYRMECOLACIDS Family Myrmecolacidae*   Not illus.

Unique in that different sexes parasitize different hosts. Males parasitize ants, and females parasitize mantids and Orthoptera. Male: Narrow, round joints on 7-segmented antennae. Spoon-shaped sclerite under wing (subalare) on metathorax absent. Female: Cephalothorax with hook-like projections behind spiracles. Abdominal segments 2–3 with more than 5 birth organs each (*Caenocholax* with 1 birth organ per segment). Brood canal opening crescent shaped. World 85, NA 2.

STYLOPIDS Family Stylopidae*

Largest family in NA. Male: **Antennae 4–6 broad, flat segments.** Spoon-shaped sclerite under wing (subalare) on metathorax. Female: Dorsally flattened cephalothorax sclerotized and close to host abdominal segment. Brood canal opening slit shaped. Abdominal segments 3–6 with 1 genital aperture each. Parasites of Hymenoptera (Apoidea, Sphecoidea, and Vespoidea). Sometimes split into 2 families: Stylopidae (bee parasites) and Xenidae (parasites of wasps in the families Vespidae, Sphecidae, and Crabronidae). World 167, NA 70.

CORIOXENIDS

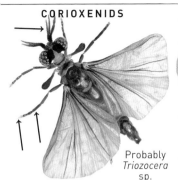

Probably *Triozocera* sp.

HALICTOPHAGIDS

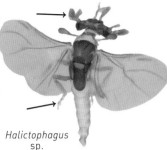

Halictophagus sp.

STYLOPIDS

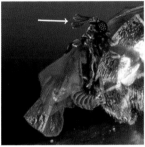

Halictoxenos sp. male mating with female while on host.

Xenos sp. male pupa in Paper Wasp (*Polistes metricus*).

Family Relationships and Hosts

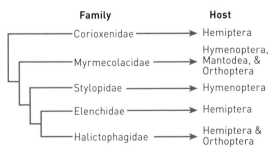

Family	Host
Corioxenidae	Hemiptera
Myrmecolacidae	Hymenoptera, Mantodea, & Orthoptera
Stylopidae	Hymenoptera
Elenchidae	Hemiptera
Halictophagidae	Hemiptera & Orthoptera

Phylogeny based on McMahon et al. (2011).

Wild Olive Tortoise Beetle
(*Physonota alutacea*)
Chrysomelidae

False Bombardier Beetle
larva (*Galerita* sp.)
Carabidae

BEETLES
ORDER COLEOPTERA

Derivation of Name
- Coleoptera: *koleos*, sheath; *pteron*, winged. Referring to first pair of wings forming a hard sheath that covers the second pair of wings when not in flight.

Identification
- 0.2–300 mm TL.
- Mouthparts are usually biting, chewing with distinct mandibles; antennae important for identification, usually 11 segments but as few as 2; FW form elytra, are usually hardened, can be leathery, usually cover abdomen, but can be short, meet in straight line down the back; HW are membranous, usually longer than FW; tarsi are 3–5 segments; abdomen usually 5 visible segments but up to 8.

Classification
- Four suborders: Archostemata, Adephaga, Myxophaga, Polyphaga. Many recent taxonomic changes, but the superfamily arrangement is stable. Some families are now subfamilies, and a few new families have been recognized recently.
- World: 400,000 species; North America: 25,200 species.

Range
- Found on every continent except Antarctica.

Similar orders
- Dermaptera (p. 114): abdomen with pincer-like appendages at tip.
- Hemiptera (p. 174): mouthparts sucking; FW rarely meet in straight line down the back and nearly always overlapping at tip; antennae 1–10-segmented.
- Orthoptera (p. 130): FW with distinct veins; antennae usually long and thread-like, many segmented.

Wedge-shaped Beetle
(*Macrosiagon limbata*)
Ripiphoridae

295

Food
- Adults and larva feed on a variety of living and dead plant and animal tissues.

Behavior
- Adults can be found under bark, leaf litter, rocks, flowers, fungi, and many other habitats.
- Many winged species are good fliers and can fly long distances; others are flightless and may have specialized legs to either run, swim, burrow, or dig.
- Different species can be identified by specific flight patterns.

Life Cycle
- Holometabolous.
- Larvae are quite variable, from grub to wormlike; occupy habitats from water to land; thoracic legs are usually well developed but can be absent.
- Pupa are poorly known.
- Courtship of adults is not well known but they may use antennae in various ways. Pheromones have been reported in a number of groups to attract a mate; stridulation of various body parts has also been reported. Eggs are generally deposited in areas where larval food can be found.

Importance to Humans
- Some species can be pests of crops and stored grains.
- Many are pests of forest trees.
- Beneficial as predators in controlling pests, detritus cycle, fecal material removal, and can be pollinators.

Collecting and Preserving
Immatures
- Collect live larvae.
- Kill with 70% ethanol.
- Permanently store in 70% ethanol.

Adults
- Individuals collected with light, beating vegetation, dung, and carrion.
- Best collected in ethanol. Can be killed with ethyl acetate kill jars, but this technique can degrade DNA.
- Adults pinned through right side of the elytra; small specimens can be pointed.

Resources
- *Beetles of Eastern North America* by Evans, 2014
- *Beetles of Western North America* by Evans, 2021
- *Beetles: The Natural History and Diversity of Coleoptera* by Marshall, 2018
- *American Beetles* by Arnett and Thomas, 2001 (vol. 1) and Arnett et al., 2002 (vol. 2)
- *A Field Guide to the Beetles of North America* by White, 1983
- *How to Know the Beetles* by Arnett et al., 1980
- *Fireflies, Glow-worms, and Lightning Bugs* by Faust, 2017
- *A Field Guide to the Tiger Beetles of the United States and Canada* by Pearson et al., 2006
- International Weevil Community Website—weevil.myspecies.info

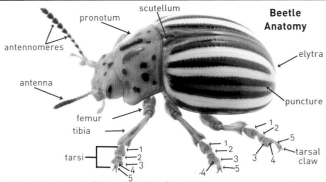

Beetle Anatomy

scutellum
pronotum
antennomeres
antenna
elytra
puncture
femur
tibia
tarsi — 1 2 3 4 5
1 2 3 4 5
1 2 5 3 4 — tarsal claw

For above beetle (Chrysomelidae), tarsi appear 4-segmented on each leg, but they are actually 5-segmented with tarsomere 4 small & hidden between lobes of heart-shaped 3. Tarsal formula would be written 5-5ᵉ5 (5 on front leg, 5 on middle leg, 5 on hind leg).

clypeus
compound eye
frontoclypeal suture
frons
mandible
maxillary palpi
labial palpi

Beetle Head

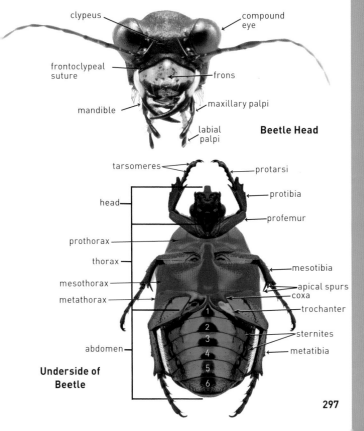

tarsomeres
protarsi
head
protibia
profemur
prothorax
thorax
mesotibia
mesothorax
apical spurs
metathorax
coxa
trochanter
abdomen
sternites
metatibia

Underside of Beetle

297

SILHOUETTES OF BEETLE SUPERFAMILIES

SUBORDER ADEPHAGA

SUBORDER ADEPHAGA
HYDROPHILOIDEA

STAPHYLINOIDEA

SCARABAEOIDEA

SCIRTOIDEA　　　　**DASCILLOIDEA**　　　　**BUPRESTOIDEA**

BYRRHOIDEA

DERODONTOIDEA

BOSTRICHOIDEA

ELATEROIDEA

TENEBRIONOIDEA

CLEROIDEA

COCCINELLOIDEA

CUCUJOIDEA

CHRYSOMELOIDEA

CURCULIONOIDEA

COLEOPTERA

KEY TO BEETLE SUBORDERS

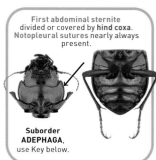

First abdominal sternite divided or covered by **hind coxa**. Notopleural sutures nearly always present.

Suborder ADEPHAGA, use Key below.

Pronotum without notopleural sutures. MOST BEETLES.

Suborder POLYPHAGA, Key 1, p. 302

— hind femur

First abdominal sternite not divided or covered by hind coxa.

KEY TO FAMILIES OF ADEPHAGA

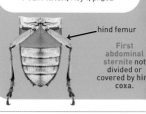

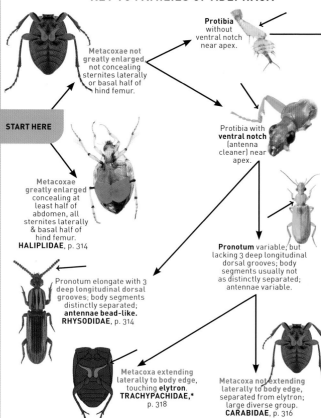

Metacoxae not greatly enlarged, not concealing sternites laterally or basal half of hind femur.

Protibia without ventral notch near apex.

START HERE

Metacoxae greatly enlarged concealing at least half of abdomen, all sternites laterally & basal half of hind femur. **HALIPLIDAE,** p. 314

Protibia with **ventral notch** (antenna cleaner) near apex.

Pronotum variable, but lacking 3 deep longitudinal dorsal grooves; body segments usually not as distinctly separated; antennae variable.

Pronotum elongate with 3 deep longitudinal dorsal grooves; body segments distinctly separated; **antennae bead-like. RHYSODIDAE,** p. 314

Metacoxa extending laterally to body edge, touching **elytron. TRACHYPACHIDAE,*** p. 318

Metacoxa not extending laterally to body edge, separated from elytron; large diverse group. **CARABIDAE,** p. 316

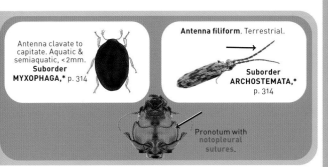

BEETLES

Antenna clavate to capitate. Aquatic & semiaquatic, <2mm.
Suborder MYXOPHAGA,* p. 314

Antenna filiform. Terrestrial.
Suborder ARCHOSTEMATA,* p. 314

Pronotum with notopleural sutures.

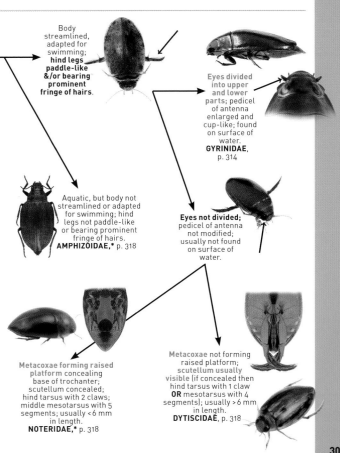

Body streamlined, adapted for swimming; **hind legs paddle-like &/or bearing prominent fringe of hairs.**

Eyes divided into upper and lower parts; pedicel of antenna enlarged and cup-like; found on surface of water.
GYRINIDAE, p. 314

Aquatic, but body not streamlined or adapted for swimming; hind legs not paddle-like or bearing prominent fringe of hairs.
AMPHIZOIDAE,* p. 318

Eyes not divided; pedicel of antenna not modified; usually not found on surface of water.

Metacoxae forming raised platform concealing base of trochanter; scutellum concealed; hind tarsus with 2 claws; middle mesotarsus with 5 segments; usually <6 mm in length.
NOTERIDAE,* p. 318

Metacoxae not forming raised platform; scutellum usually visible (if concealed then hind tarsus with 1 claw **OR** mesotarsus with 4 segments); usually >6 mm in length.
DYTISCIDAE, p. 318

301

KEY 1 TO COMMON FAMILIES OF POLYPHAGA

Pronotum without median groove; ♂ often with **enlarged mandibles**; horn may be present on head. **LUCANIDAE**, p. 324

Pronotum with **median groove;** mandibles not enlarged; **single horn present from middle of head. PASSALIDAE**, p. 324

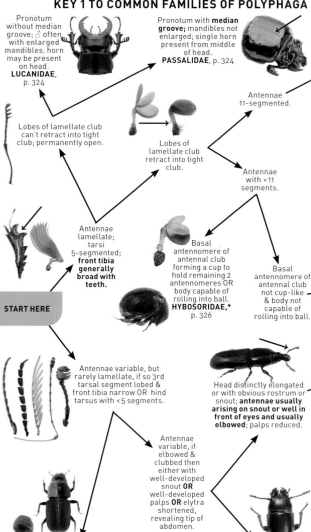

Antennae 11-segmented.

Lobes of lamellate club can't retract into tight club; permanently open.

Lobes of lamellate club retract into tight club.

Antennae with <11 segments.

Antennae lamellate; tarsi 5-segmented; **front tibia generally broad with teeth.**

Basal antennomere of antennal club forming a cup to hold remaining 2 antennomeres OR body capable of rolling into ball. **HYBOSORIDAE,*** p. 326

Basal antennomere of antennal club not cup-like & body not capable of rolling into ball.

START HERE

Antennae variable, but rarely lamellate, if so 3rd tarsal segment lobed & front tibia narrow OR hind tarsus with <5 segments.

Head distinctly elongated or with obvious rostrum or snout; **antennae usually arising on snout or well in front of eyes and usually elbowed**; palps reduced.

Antennae variable, if elbowed & clubbed then either with well-developed snout **OR** well-developed palps **OR** elytra shortened, revealing tip of abdomen.

Antennae short & elbowed with compact club; elytra not reduced, covering entire abdomen; palps greatly reduced. **CURCULIONIDAE** (in part), p. 362

Head not distinctly elongated or with obvious rostrum or snout; antennae variable, but rarely elbowed; palps long & visible. **POLYPHAGA KEY 2**, p. 304

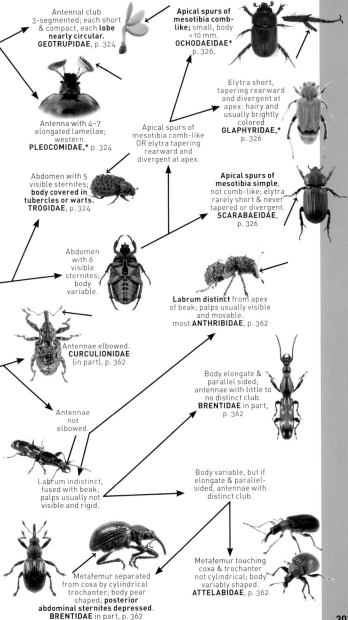

Antennal club 3-segmented; each short & compact, each **lobe nearly circular.** GEOTRUPIDAE, p. 324

Apical spurs of mesotibia comb-like; small, body <10 mm. OCHODAEIDAE,* p. 326,

Antenna with 4–7 elongated lamellae; western. PLEOCOMIDAE,* p. 324

Apical spurs of mesotibia comb-like OR elytra tapering rearward and divergent at apex.

Elytra short, tapering rearward and divergent at apex; hairy and usually brightly colored GLAPHYRIDAE,* p. 326

Abdomen with 5 visible sternites; **body covered in tubercles or warts.** TROGIDAE, p. 324

Apical spurs of mesotibia simple, not comb-like; elytra rarely short & never tapered or divergent. SCARABAEIDAE, p. 326

Abdomen with 6 visible sternites; body variable.

Labrum distinct from apex of beak; palps usually visible and movable. most **ANTHRIBIDAE**, p. 362

Antennae elbowed. **CURCULIONIDAE** (in part), p. 362

Body elongate & parallel sided; antennae with little to no distinct club. **BRENTIDAE** in part, p. 362

Antennae not elbowed.

Labrum indistinct, fused with beak; palps usually not visible and rigid.

Body variable, but if elongate & parallel-sided, antennae with distinct club.

Metafemur separated from coxa by cylindrical trochanter; body pear shaped; **posterior abdominal sternites depressed.** BRENTIDAE in part, p. 362

Metafemur touching coxa & trochanter not cylindrical; body variably shaped. ATTELABIDAE, p. 362

303

KEY 2 TO COMMON FAMILIES OF POLYPHAGA

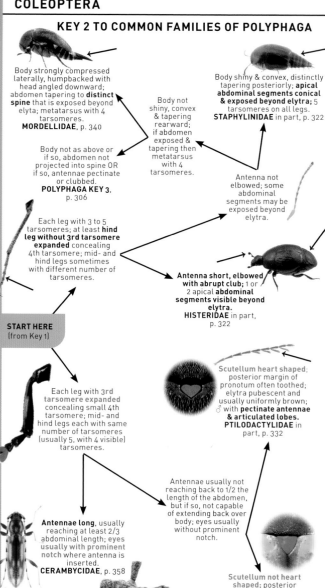

Body strongly compressed laterally, humpbacked with head angled downward; abdomen tapering to **distinct spine** that is exposed beyond elyta; metatarsus with 4 tarsomeres.
MORDELLIDAE, p. 340

Body not shiny, convex & tapering rearward; if abdomen exposed & tapering then metatarsus with 4 tarsomeres.

Body shiny & convex, distinctly tapering posteriorly; **apical abdominal segments conical & exposed beyond elytra;** 5 tarsomeres on all legs.
STAPHYLINIDAE in part, p. 322

Body not as above or if so, abdomen not projected into spine OR if so, antennae pectinate or clubbed.
POLYPHAGA KEY 3, p. 306

Antenna not elbowed; some abdominal segments may be exposed beyond elytra.

Each leg with 3 to 5 tarsomeres; at least **hind leg without 3rd tarsomere expanded** concealing 4th tarsomere; mid- and hind legs sometimes with different number of tarsomeres.

Antenna short, elbowed with abrupt club; 1 or 2 apical **abdominal segments visible beyond elytra.**
HISTERIDAE in part, p. 322

START HERE (from Key 1)

Scutellum heart shaped; posterior margin of pronotum often toothed; elytra pubescent and usually uniformly brown; ♂ with **pectinate antennae & articulated lobes.**
PTILODACTYLIDAE in part, p. 332

Each leg with 3rd tarsomere expanded concealing small 4th tarsomere; mid- and hind legs each with same number of tarsomeres (usually 5, with 4 visible) tarsomeres.

Antennae usually not reaching back to 1/2 the length of the abdomen, but if so, not capable of extending back over body; eyes usually without prominent notch.

Antennae long, usually reaching at least 2/3 abdominal length; **eyes usually with prominent notch** where antenna is inserted.
CERAMBYCIDAE, p. 358

Scutellum not heart shaped; posterior margin of pronotum smooth; antennae rarely pectinate, but if so lacking articulated lobes; elytra variable.

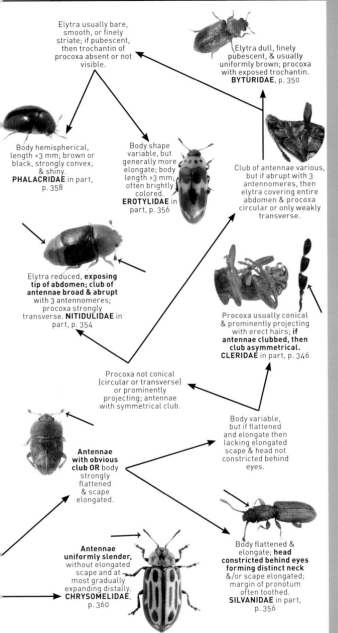

Elytra usually bare, smooth, or finely striate; if pubescent, then trochantin of procoxa absent or not visible.

Elytra dull, finely pubescent, & usually uniformly brown; procoxa with exposed trochantin. **BYTURIDAE**, p. 350

Body hemispherical, length <3 mm; brown or black, strongly convex, & shiny. **PHALACRIDAE** in part, p. 358

Body shape variable, but generally more elongate; body length >3 mm; often brightly colored. **EROTYLIDAE** in part, p. 356

Club of antennae various, but if abrupt with 3 antennomeres, then elytra covering entire abdomen & procoxa circular or only weakly transverse.

Elytra reduced, **exposing tip of abdomen; club of antennae broad & abrupt** with 3 antennomeres; procoxa strongly transverse. **NITIDULIDAE** in part, p. 354

Procoxa usually conical & prominently projecting with erect hairs; **if antennae clubbed, then club asymmetrical. CLERIDAE** in part, p. 346

Procoxa not conical (circular or transverse) or prominently projecting; antennae with symmetrical club.

Body variable, but if flattened and elongate then lacking elongated scape & head not constricted behind eyes.

Antennae with obvious club OR body strongly flattened & scape elongated.

Antennae uniformly slender, without elongated scape and at most gradually expanding distally. **CHRYSOMELIDAE**, p. 360

Body flattened & elongate; **head constricted behind eyes forming distinct neck** &/or scape elongated; margin of pronotum often toothed. **SILVANIDAE** in part, p. 356

305

COLEOPTERA

KEY 3 TO COMMON FAMILIES OF POLYPHAGA

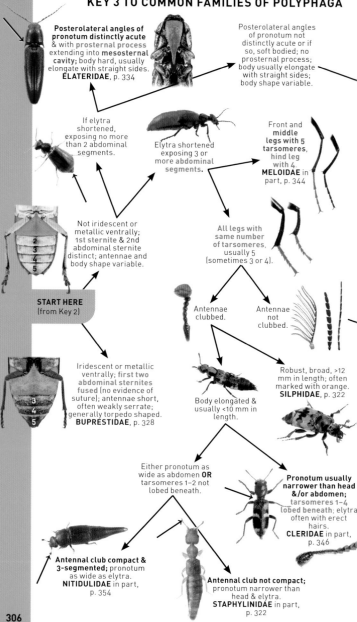

Posterolateral angles of pronotum distinctly acute & with prosternal process extending into mesosternal cavity; body hard, usually elongate with straight sides. **ELATERIDAE**, p. 334

Posterolateral angles of pronotum not distinctly acute or if so, soft bodied; no prosternal process; body usually elongate with straight sides; body shape variable.

If elytra shortened, exposing no more than 2 abdominal segments.

Elytra shortened exposing 3 or more abdominal segments.

Front and middle legs with 5 tarsomeres, hind leg with 4 **MELOIDAE** in part, p. 344

Not iridescent or metallic ventrally; 1st sternite & 2nd abdominal sternite distinct; antennae and body shape variable.

All legs with same number of tarsomeres, usually 5 (sometimes 3 or 4).

START HERE (from Key 2)

Antennae clubbed.

Antennae not clubbed.

Iridescent or metallic ventrally; first two abdominal sternites fused (no evidence of suture); antennae short, often weakly serrate; generally torpedo shaped. **BUPRESTIDAE**, p. 328

Robust, broad, >12 mm in length; often marked with orange. **SILPHIDAE**, p. 322

Body elongated & usually <10 mm in length.

Either pronotum as wide as abdomen **OR** tarsomeres 1–2 not lobed beneath.

Pronotum usually narrower than head &/or abdomen; tarsomeres 1–4 lobed beneath; elytra often with erect hairs. **CLERIDAE** in part, p. 346

Antennal club compact & 3-segmented; pronotum as wide as elytra. **NITIDULIDAE** in part, p. 354

Antennal club not compact; pronotum narrower than head & elytra. **STAPHYLINIDAE** in part, p. 322

306

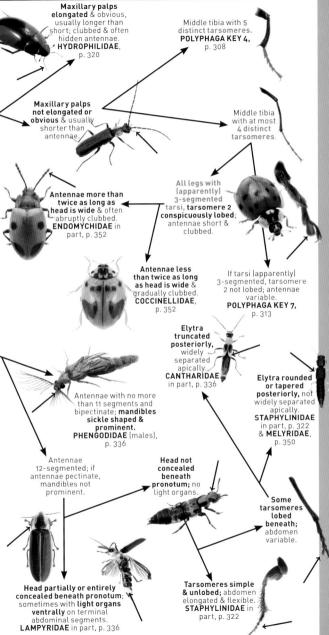

Maxillary palps elongated & obvious, usually longer than short; clubbed & often hidden antennae. **HYDROPHILIDAE**, p. 320

Maxillary palps not elongated or obvious & usually shorter than antennae.

Middle tibia with 5 distinct tarsomeres. **POLYPHAGA KEY 4,** p. 308

Middle tibia with at most **4 distinct tarsomeres**.

Antennae more than twice as long as head is wide & often abruptly clubbed. **ENDOMYCHIDAE** in part, p. 352

All legs with (apparently) 3-segmented tarsi, **tarsomere 2 conspicuously lobed**; antennae short & clubbed.

Antennae less than twice as long as head is wide & gradually clubbed. **COCCINELLIDAE**, p. 352

If tarsi (apparently) 3-segmented, tarsomere 2 not lobed; antennae variable. **POLYPHAGA KEY 7,** p. 313

Elytra truncated posteriorly, widely separated apically. **CANTHARIDAE** in part, p. 336

Elytra rounded or tapered posteriorly, not widely separated apically. **STAPHYLINIDAE** in part, p. 322 & **MELYRIDAE**, p. 350

Antennae with no more than 11 segments and bipectinate; **mandibles sickle shaped & prominent. PHENGODIDAE** (males), p. 336

Antennae 12-segmented; if antennae pectinate, mandibles not prominent.

Head not concealed beneath pronotum; no light organs.

Some tarsomeres lobed beneath; abdomen variable.

Head partially or entirely concealed beneath pronotum; sometimes with **light organs ventrally** on terminal abdominal segments. **LAMPYRIDAE** in part, p. 336

Tarsomeres simple & unlobed; abdomen elongated & flexible. **STAPHYLINIDAE** in part, p. 322

COLEOPTERA

KEY 4 TO COMMON FAMILIES OF POLYPHAGA

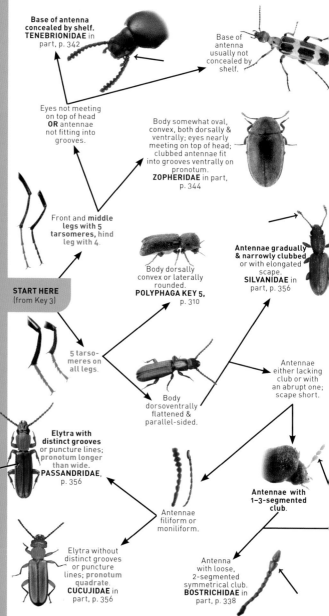

Base of antenna concealed by shelf.
TENEBRIONIDAE in part, p. 342

Base of antenna usually not concealed by shelf.

Eyes not meeting on top of head **OR** antennae not fitting into grooves.

Body somewhat oval, convex, both dorsally & ventrally; eyes nearly meeting on top of head; clubbed antennae fit into grooves ventrally on pronotum.
ZOPHERIDAE in part, p. 344

Front and **middle legs with 5 tarsomeres,** hind leg with 4.

Body dorsally convex or laterally rounded.
POLYPHAGA KEY 5, p. 310

Antennae gradually & narrowly clubbed or with elongated scape.
SILVANIDAE in part, p. 356

START HERE (from Key 3)

5 tarso- meres on all legs.

Body dorsoventrally flattened & parallel-sided.

Antennae either lacking club or with an abrupt one; scape short.

Elytra with distinct grooves or puncture lines; pronotum longer than wide.
PASSANDRIDAE, p. 356

Antennae with 1-3-segmented club.

Antennae filiform or moniliform.

Elytra without distinct grooves or puncture lines; **pronotum quadrate.**
CUCUJIDAE in part, p. 356

Antenna with loose, 2-segmented symmetrical club.
BOSTRICHIDAE in part, p. 338

308

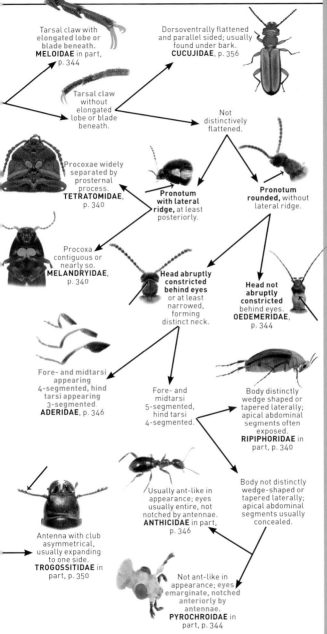

Tarsal claw with elongated lobe or blade beneath.
MELOIDAE in part, p. 344

Dorsoventrally flattened and parallel sided; usually found under bark.
CUCUJIDAE, p. 356

Tarsal claw without elongated lobe or blade beneath.

Not distinctly flattened.

Procoxae widely separated by prosternal process.
TETRATOMIDAE, p. 340

Pronotum with lateral ridge, at least posteriorly.

Pronotum rounded, without lateral ridge.

Procoxa contiguous or nearly so.
MELANDRYIDAE, p. 340

Head abruptly constricted behind eyes or at least narrowed, forming distinct neck.

Head not abruptly constricted behind eyes.
OEDEMERIDAE, p. 344

Fore- and midtarsi appearing 4-segmented, hind tarsi appearing 3-segmented.
ADERIDAE, p. 346

Fore- and midtarsi 5-segmented, hind tarsi 4-segmented.

Body distinctly wedge shaped or tapered laterally; apical abdominal segments often exposed.
RIPIPHORIDAE in part, p. 340

Usually ant-like in appearance; eyes usually entire, not notched by antennae.
ANTHICIDAE in part, p. 346

Body not distinctly wedge-shaped or tapered laterally; apical abdominal segments usually concealed.

Antenna with club asymmetrical, usually expanding to one side.
TROGOSSITIDAE in part, p. 350

Not ant-like in appearance; eyes emarginate, notched anteriorly by antennae.
PYROCHROIDAE in part, p. 344

KEY 5 TO COMMON FAMILIES OF POLYPHAGA

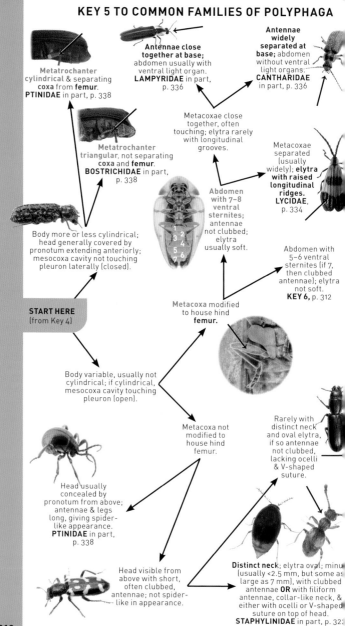

Metatrochanter cylindrical & separating coxa from femur.
PTINIDAE in part, p. 338

Antennae close together at base; abdomen usually with ventral light organ.
LAMPYRIDAE in part, p. 336

Antennae widely separated at base; abdomen without ventral light organs.
CANTHARIDAE in part, p. 336

Metatrochanter triangular, not separating **coxa** and **femur**.
BOSTRICHIDAE in part, p. 338

Metacoxae close together, often touching; elytra rarely with longitudinal grooves.

Metacoxae separated (usually widely); **elytra with raised longitudinal ridges.**
LYCIDAE, p. 334

Abdomen with 7–8 ventral sternites; antennae not clubbed; elytra usually soft.

Body more or less cylindrical; head generally covered by pronotum extending anteriorly; mesocoxa cavity not touching pleuron laterally (closed).

Abdomen with 5–6 ventral sternites (if 7, then clubbed antennae); elytra not soft.
KEY 6, p. 312

START HERE (from Key 4)

Metacoxa modified to house hind **femur.**

Body variable, usually not cylindrical; if cylindrical, mesocoxa cavity touching pleuron (open).

Metacoxa not modified to house hind femur.

Rarely with distinct neck and oval elytra, if so antennae not clubbed, lacking ocelli & V-shaped suture.

Head usually concealed by pronotum from above; antennae & legs long, giving spider-like appearance.
PTINIDAE in part, p. 338

Head visible from above with short, often clubbed, antennae; not spider-like in appearance.

Distinct neck; elytra oval; minut (usually <2.5 mm, but some as large as 7 mm), with clubbed antennae **OR** with filiform antennae, collar-like neck, & either with ocelli or V-shaped suture on top of head.
STAPHYLINIDAE in part, p. 32

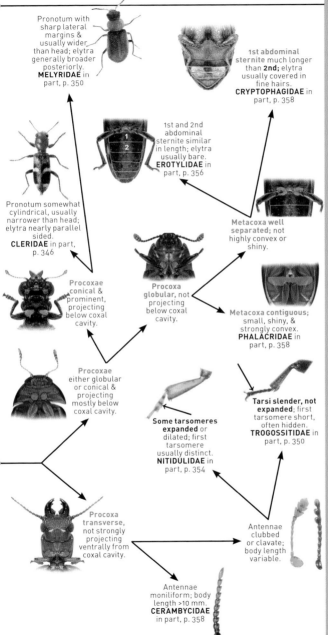

Pronotum with sharp lateral margins & usually wider than head; elytra generally broader posteriorly.
MELYRIDAE in part, p. 350

1st abdominal sternite much longer than **2nd**; elytra usually covered in fine hairs.
CRYPTOPHAGIDAE in part, p. 358

1st and 2nd abdominal sternite similar in length; elytra usually bare.
EROTYLIDAE in part, p. 356

Pronotum somewhat cylindrical, usually narrower than head; elytra nearly parallel sided.
CLERIDAE in part, p. 346

Metacoxa well separated; not highly convex or shiny.

Procoxae conical & prominent, projecting below coxal cavity.

Procoxa globular, not projecting below coxal cavity.

Metacoxa contiguous; small, shiny, & strongly convex.
PHALACRIDAE in part, p. 358

Procoxae either globular or conical & projecting mostly below coxal cavity.

Some tarsomeres expanded or dilated; first tarsomere usually distinct.
NITIDULIDAE in part, p. 354

Tarsi slender, not expanded; first tarsomere short, often hidden.
TROGOSSITIDAE in part, p. 350

Procoxa transverse, not strongly projecting ventrally from coxal cavity.

Antennae clubbed or clavate; body length variable.

Antennae moniliform; body length >10 mm.
CERAMBYCIDAE in part, p. 358

KEY 6 TO COMMON FAMILIES OF POLYPHAGA

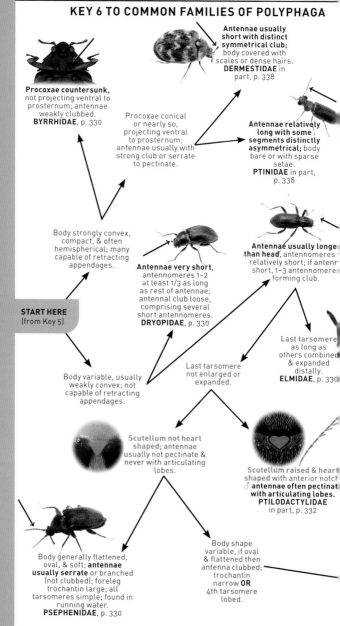

Antennae usually short with distinct symmetrical club; body covered with scales or dense hairs.
DERMESTIDAE in part, p. 338

Procoxae countersunk, not projecting ventral to prosternum; antennae weakly clubbed.
BYRRHIDAE, p. 330

Procoxae conical or nearly so, projecting ventral to prosternum; antennae usually with strong club or serrate to pectinate.

Antennae relatively long with some segments distinctly asymmetrical; body bare or with sparse setae.
PTINIDAE in part, p. 338

Body strongly convex, compact, & often hemispherical; many capable of retracting appendages.

Antennae very short, antennomeres 1–2 at least 1/3 as long as rest of antennae; antennal club loose, comprising several short antennomeres.
DRYOPIDAE, p. 330

Antennae usually longer than head, antennomeres relatively short; if antenna short, 1–3 antennomeres forming club.

START HERE [from Key 5]

Last tarsomere as long as others combined & expanded distally.
ELMIDAE, p. 330

Body variable, usually weakly convex; not capable of retracting appendages.

Last tarsomere not enlarged or expanded.

Scutellum not heart shaped; antennae usually not pectinate & never with articulating lobes.

Scutellum raised & heart shaped with anterior notch ♂ **antennae often pectinate with articulating lobes.**
PTILODACTYLIDAE in part, p. 332

Body generally flattened, oval, & soft; **antennae usually serrate** or branched (not clubbed); foreleg trochantin large; all tarsomeres simple; found in running water.
PSEPHENIDAE, p. 330

Body shape variable, if oval & flattened then antenna clubbed; trochantin narrow **OR** 4th tarsomere lobed.

KEY 7 TO COMMON FAMILIES OF POLYPHAGA

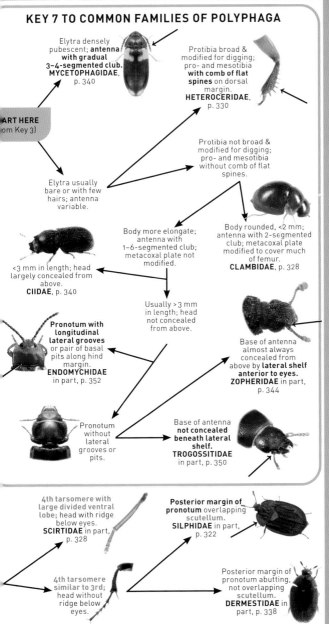

Elytra densely pubescent; **antenna with gradual 3–4-segmented club.** MYCETOPHAGIDAE, p. 340

Protibia broad & modified for digging; pro- and mesotibia **with comb of flat spines** on dorsal margin. **HETEROCERIDAE,** p. 330

ART HERE om Key 3)

Elytra usually bare or with few hairs; antenna variable.

Protibia not broad & modified for digging; pro- and mesotibia without comb of flat spines.

Body more elongate; antenna with 1–6-segmented club; metacoxal plate not modified.

Body rounded, <2 mm; antenna with 2-segmented club; metacoxal plate modified to cover much of femur. **CLAMBIDAE,** p. 328

<3 mm in length; head largely concealed from above. **CIIDAE,** p. 340

Usually >3 mm in length; head not concealed from above.

Pronotum with longitudinal lateral grooves or pair of basal pits along hind margin. **ENDOMYCHIDAE** in part, p. 352

Base of antenna almost always concealed from above by **lateral shelf anterior to eyes.** **ZOPHERIDAE** in part, p. 344

Pronotum without lateral grooves or pits.

Base of antenna **not concealed beneath lateral shelf.** **TROGOSSITIDAE** in part, p. 350

4th tarsomere with large divided ventral lobe; head with ridge below eyes. **SCIRTIDAE** in part, p. 328

Posterior margin of pronotum overlapping scutellum. **SILPHIDAE** in part, p. 322

4th tarsomere similar to 3rd; head without ridge below eyes.

Posterior margin of pronotum abutting, not overlapping scutellum. **DERMESTIDAE** in part, p. 338

313

COLEOPTERA

Suborder Archostemata
Somewhat flattened, elongated, and parallel sided with apex of HW rolle
Antennae filiform. Tarsi 5-5-5. Uncommon to rare.

RETICULATED BEETLES Family Cupedidae*
Brown, gray, or black with forward-projecting antennae. Densely scaled
elytra reticulate with closely spaced square punctures. Pronotum
extends posteriorly into groove of mesosternum. Found in oak, chestnut
and pine logs. TL 7–25 mm; World 30, NA 4.

TELEPHONE-POLE BEETLE Family Micromalthidae*   Not illus.
Single species, *Micromalthus debilis,* found in decaying oak and
chestnut logs. Dark, shiny, soft bodied with large head. Legs and bead-
like antennae yellowish. Complex life history with larvae capable of
parthenogenetic reproduction. TL 1.5–2.5 mm; World 1, NA 1.

Suborder Myxophaga
Found in wet or moist environments feeding on filamentous algae.
Antennae clavate to capitate. Tarsi 3-3-3. Uncommon.

MINUTE BOG BEETLES Family Sphaeriusidae*   Not illus.
Minute, oval, convex, shiny, dark beetles. Prominent head. Antennae
yellowish and capitate with long setae terminally. Large hind coxal plate
Abdominal sternites of unequal width. Found along streams in mud, mos
and under stones. TL 0.5–1.2 mm; World 20, NA 3.

SKIFF BEETLES Family Hydroscaphidae*
Small, tan or brown, with **shortened elytra revealing several tapering
abdominal segments.** Wings fringed with long setae. Antennae
8-segmented with **terminal segment developed into club.** Live within
algae on rocks and in streams. TL 1.0–2.0 mm; World 25, NA 2.

Suborder Adephaga
Hind coxae divide first abdominal sternite. Usually with filiform antenna
Tarsi 5-5-5. Notopleural sutures present. Predaceous.

WRINKLED BARK BEETLES Family Rhysodidae
Slender, reddish-brown or gray, elongate with **3 deep longitudinal dorsa
grooves on elongated pronotum. Antennae bead-like and 11-segmente**
Body segments distinctly separated. Elytra distinctly striate. Found unde
bark. Sometimes treated as Carabidae. TL 5.0–8.0 mm; World >350, NA

WHIRLIGIG BEETLES Family Gyrinidae
Black, shiny, oval and streamlined body. **Eyes divided laterally. Antenna
short and clubbed.** Front legs long; hind legs modified for swimming. Em
odor when handled. TL 2.5–15 mm; World 900, NA 60.

CRAWLING WATER BEETLES Family Haliplidae
Yellow or brown, oval, convex with spots. Elytra with distinct punctures
darkened in most species. **Hind coxae enlarged and plate-like.** Found in
vegetation in or near water. TL 1.5–5.0 mm; World 220, NA 70.

RETICULATED BEETLES

Cupes capitatus

Tenomerga cinerea

SKIFF BEETLES

Hydroscapha natans

WRINKLED BARK BEETLES

Omoglymmius hamatus

WHIRLIGIG BEETLES

Dineutus sp.

Gyrinus sp.

CRAWLING WATER BEETLES

Haliplus sp.

Peltodytes sexmaculatus

COLEOPTERA

Suborder Adephaga cont.

GROUND BEETLES Family Carabidae

Large, variable family; many are dark, somewhat flattened, and quick runners. **Antennae filiform and pubescent. Front tibia with a prominent distal notch for cleaning antennae.** Hind trochanter elongate; at one-third length of femur. TL 0.7–85 mm; World 34,000, NA 2,500.

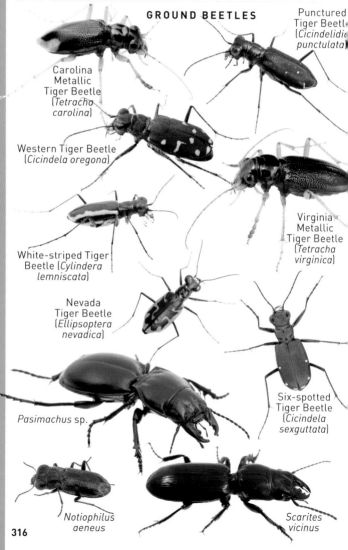

GROUND BEETLES

Punctured Tiger Beetle (*Cicindelidia punctulata*)

Carolina Metallic Tiger Beetle (*Tetracha carolina*)

Western Tiger Beetle (*Cicindela oregona*)

Virginia Metallic Tiger Beetle (*Tetracha virginica*)

White-striped Tiger Beetle (*Cylindera lemniscata*)

Nevada Tiger Beetle (*Ellipsoptera nevadica*)

Pasimachus sp.

Six-spotted Tiger Beetle (*Cicindela sexguttata*)

Notiophilus aeneus

Scarites vicinus

CARABIDAE

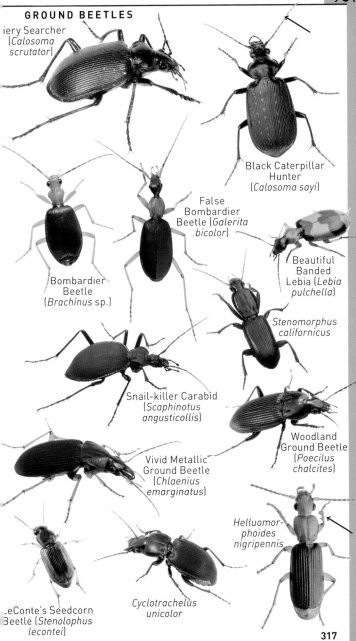

GROUND BEETLES

iery Searcher
(*Calosoma scrutator*)

Black Caterpillar Hunter
(*Calosoma sayi*)

False Bombardier Beetle (*Galerita bicolor*)

Bombardier Beetle
(*Brachinus* sp.)

Beautiful Banded Lebia (*Lebia pulchella*)

Stenomorphus californicus

Snail-killer Carabid
(*Scaphinotus angusticollis*)

Woodland Ground Beetle
(*Poecilus chalcites*)

Vivid Metallic Ground Beetle
(*Chlaenius emarginatus*)

Helluomorphoides nigripennis

eConte's Seedcorn Beetle (*Stenolophus lecontei*)

Cyclotrachelus unicolor

317

COLEOPTERA

FALSE GROUND BEETLES Family Trachypachidae*
Uniformly dark with metallic luster; very similar to Carabidae, but antennae not densely pubescent and hind coxae extend laterally to elytral margin. Pronotum with lateral margins. Western; single genus, *Trachypachus*, in NA. TL 4–7 mm; World 6, NA 3.

BURROWING WATER BEETLES Family Noteridae*
Similar to Dytiscidae, but scutellum not visible. Hind tarsus with 2 subequal claws. Body broadly oval and streamlined; dark, usually with reddish head and pronotum. Protibia fits in groove on ventral margin of femur. Hind coxal plates large. TL 1–6 mm; World 250, NA 14.

TROUT-STREAM BEETLES Family Amphizoidae*
Elongate, oval, somewhat convex, and dark. Largely devoid of setae. Somewhat resemble tenbrionids. Found in cold mountain streams of western NA. Single genus, *Amphizoa*. TL 11–16 mm; World 6, NA 3.

PREDACEOUS DIVING BEETLES Family Dytiscidae
Smooth, oval body, generally streamlined. Generally dark. Elytra generally with punctures along stria. **Antennae with 11 antennomeres; first spherical (appearing as 2 segments); second antennomere elongate; remaining segments filiform.** Scutellum present. **Hind tibiae with long hairs for swimming;** hind legs move together vs. alternately (Hydrophilidae). Aquatic, found in a variety of habitats. TL 1–40 mm; World 4,300, NA 500.

Predaceous Diving Beetle
(*Cybister fimbriolatus*)

ADEPHAGA

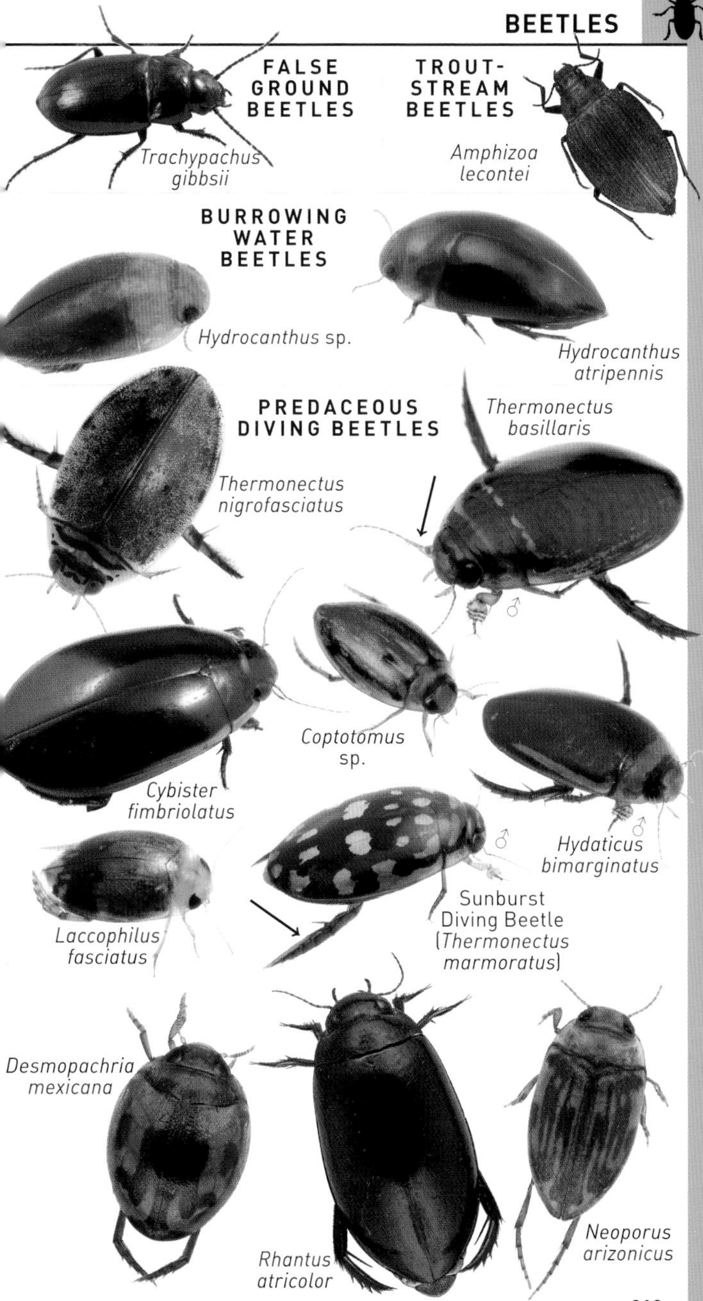

FALSE GROUND BEETLES

Trachypachus gibbsii

TROUT-STREAM BEETLES

Amphizoa lecontei

BURROWING WATER BEETLES

Hydrocanthus sp.

Hydrocanthus atripennis

PREDACEOUS DIVING BEETLES

Thermonectus basillaris

Thermonectus nigrofasciatus

Coptotomus sp.

♂

Cybister fimbriolatus

♂ *Hydaticus bimarginatus*

Laccophilus fasciatus

Sunburst Diving Beetle (*Thermonectus marmoratus*)

Desmopachria mexicana

Rhantus atricolor

Neoporus arizonicus

319

Suborder Polyphaga

Hind margin of first abdominal sternite entire. Antennae variable. Tarsi often 5-5-5, but variable. Notopleural suture absent.

Superfamily Hydrophiloidea

WATER SCAVENGER BEETLE Family Hydrophilidae

Generally broadly oval, convex dorsally, flat or concave ventrally, and dark or dull in color. **Antennae clavate (6–10 antennomeres) with first elongate; often in groove along lower margin of eye. Maxillary palps long, extending beyond head, appearing as antennae. Clypeus large, usually fused with frons, but separated from labrum.** Metasternum prolonged posteriorly as sharp spine in some. Hind legs move alternately vs. together (Dytiscidae). Five abdominal sterna visible. Most aquatic; some terrestrial. TL 1–40 mm; World 3,400, NA 260.

HELOPHORID BEETLES Family Helophoridae

Body elongate with 5–7 longitudinal grooves on pronotum. Underside of body pubescent with fine microsculpture. Found on or around emergent vegetation in standing or slow-moving water; adults often attracted to lights. Single genus, *Helophorus*, in NA, with most species found in the north or west. Traditionally considered a subfamily of Hydrophilidae. TL 2–8 mm; World 200, NA 43.

EPIMETOPID BEETLES Family Epimetopidae*

Pronotum with central projection forming shelf above head. Four sternites visible ventrally. Found on edges of streams and shallow ponds. Single genus in NA, *Epimetopus*, restricted to southern, mostly southwestern, states. TL 1.0–3.5 mm; World <100, NA 4.

MINUTE MUD-LOVING BEETLES Family Georissidae*   Not illus.

Reddish-brown to black, broadly oval and small with head and pronotum rouglyly sculptured. Anterior coxae and trochanters fused. Basal sternite enlarged. Found in or near wet soil and often heavily covered with sand or mud. Single genus in NA, *Georissus*. TL 1–3 mm; World 75, NA 2.

HYDROCHID BEETLES Family Hydrochidae

Elongate with heavily punctured body; pronotum narrowing posteriorly, narrower than base of elytra. Antennae 7-segmented; terminal 3 segments forming club; cup-like basal segment. Found in slow-moving streams and standing water. Single genus in NA, *Hydrochus*. Sometimes placed within Hydrophilidae. TL 1.5–5.5 mm; World 180, NA 31.

FALSE CLOWN BEETLES Family Sphaeritidae*   Not illus.

Black with metallic blue luster. Similar to histerids, but antennae not elbowed and tibia less expanded, lacking teeth externally. Single abdominal segment exposed beyond elytra dorsally. Found in carrion, manure, and decaying fungi. Single species, *Sphaerites politus*, in western NA. TL 3.5–5.6 mm; World 4, NA 1.

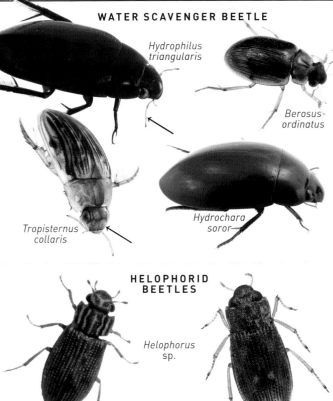

WATER SCAVENGER BEETLE

Hydrophilus triangularis

Berosus ordinatus

Tropisternus collaris

Hydrochara soror

HELOPHORID BEETLES

Helophorus sp.

HYDROCHID BEETLES

Hydrochus sp.

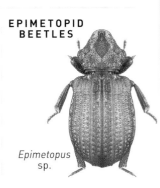

EPIMETOPID BEETLES

Epimetopus sp.

COLEOPTERA

Superfamily Hydrophiloidea cont.

CLOWN BEETLES Family Histeridae
Broadly oval, compact, generally shiny black and strongly convex, with **elytra squared off apically, revealing 1–2 segments. Antennae elbowed and clubbed.** Tibia dilated; protibia usually toothed or spined. TL 0.5–18 mm; World 4,000, NA 440.

Superfamily Staphylinoidea

MINUTE MOSS BEETLES Family Hydraenidae*   Not illus.
Dark and elongate or oval. Antennae with last 5 antennomeres clubbed. Six or seven abdominal sterna visible. Semiaquatic, found along margins of streams and in nearby sphagnum moss. TL 0.5–3 mm; World 1,400, NA 70.

FEATHERWINGED BEETLES Family Ptiliidae
Minute and oval; HW with long fringe of hairs, often extending from under elytra. Antennae with whorls of long hairs. Many polymorphic: "normal" morph with well-developed eyes, wings, and pigmentation; "vestigial" morph with eyes, wings, and body pigmentation reduced or lacking. TL 0.4–1.2 mm; World 550, NA 120.

PRIMITIVE CARRION BEETLE Family Agyrtidae*   Not illus.
Elongate or oblong and somewhat flattened in shape; light to dark brown and usually shiny. Elytra with 9 or 10 striae and not reduced. Antennae filiform to clavate. Five visible abdominal sterna. Found in carrion and decaying vegetation. TL 4–14 mm; World 70, NA 11.

ROUND FUNGUS BEETLES Family Leiodidae*
Body strongly convex, broadly ovate to elongate, shiny black or brown, and glabrous. **Eigth antennomere usually smaller than seventh or ninth.** Thorax and elytra frequently with transverse striae; others smoothed and glossy. Many capable of rolling into ball. TL 1.5–6.5 mm; World 3,500, NA 382.

CARRION BEETLES Family Silphidae
Large, dark, often with red, yellow, and orange maculation. **Antennae with 11 antennomeres; clavate or capitate, occasionally elbowed.** Pronotum with distinct lateral edges, sometimes expanded. **Elytra often truncate.** Males usually with broadly expanded protarsi. Associated with decaying animals and some fungi. TL 7–45 mm (usually 12–20 mm); World 200, NA 30.

ROVE BEETLES Family Staphylinidae
Elongate, with **shortened elytra, typically exposing 3–6 abdominal segments,** and active; others oval, strongly convex, and shiny. Antennae thread-like or clubbed. Tarsal formula usually 5-5-5, but variable. Common under rocks and logs. TL 1–25 mm (usually 2–10 mm); World 64,000, NA 4,400.

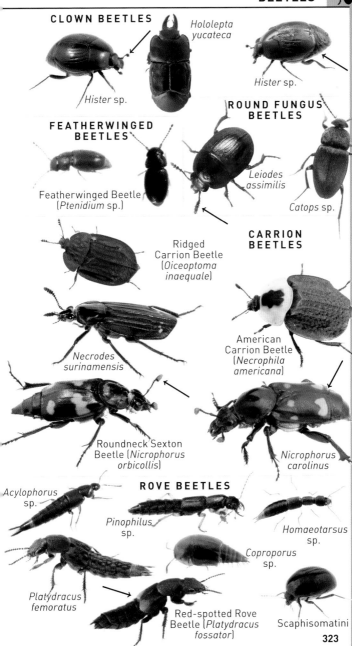

CLOWN BEETLES

Hololepta yucateca

Hister sp.

Hister sp.

FEATHERWINGED BEETLES

Featherwinged Beetle (*Ptenidium* sp.)

ROUND FUNGUS BEETLES

Leiodes assimilis

Catops sp.

Ridged Carrion Beetle (*Oiceoptoma inaequale*)

CARRION BEETLES

Necrodes surinamensis

American Carrion Beetle (*Necrophila americana*)

Roundneck Sexton Beetle (*Nicrophorus orbicollis*)

Nicrophorus carolinus

ROVE BEETLES

Acylophorus sp.

Pinophilus sp.

Homaeotarsus sp.

Coproporus sp.

Platydracus femoratus

Red-spotted Rove Beetle (*Platydracus fossator*)

Scaphisomatini

323

COLEOPTERA

SCARABAEOIDEA

Superfamily Scarabaeoidea

STAG BEETLES Family Lucanidae

Medium to large, brown to black; males often with **enlarged mandibles.**
Antennae with 10 antennomeres; **club comprised of 3–7 antennomeres;**
first antennomere often as long as remaining antennomeres. Mandibles
extend beyond apex of labrum. TL 8–60 mm; World 1,500, NA 38.

FALSE STAG BEETLES Family Diphyllostomatidae*   Not illus.

Brown, elongate, with heavily setose legs and body. Seven abdominal
sternites (as opposed to five in other Scarabaeoidea) visible. Lack tibial
spurs. Females have reduced eyes and vestigial wings. Single genus,
Diphyllostomata, limited to CA. TL 5–9 mm; World 3, NA 3.

BESS BEETLES Family Passalidae

Large, robust, shiny and black. Distinct neck between pronotum and
abdomen. **Pronotum grooved in middle; elytra deeply grooved.** Large,
robust, toothed mandibles. Antennae with 10 antennomeres; **3 (usually)
to 6 forming a club.** Found communally within galleries of rotting logs,
where they are subsocial. TL 30–40 mm; World 500, NA 2.

ENIGMATIC SCARAB BEETLES Family Glaresidae*

Small, light to dark brown, oblong-oval and convex. Moderately dense
short setae dorsally. **Head deflexed and eyes divided by canthus.**
Antennae with 10 antennomeres; last 3 forming club. Hind femora and
tibiae enlarged, covering most of abdomen. Single genus, *Glaresis*,
typically found in arid areas at lights. TL 2.5–6 mm; World 80, NA 25.

HIDE BEETLES Family Trogidae

Oblong, convex, gray to dark brown; appearing warty dorsally and often
encrusted with dirt. Head deflexed. Antenna with 10 antennomeres; **last
3 forming club.** Basal antennomere robust; second arising before apex
of first. Often with moderately **dense gray or brown setae dorsally.**
Scavenge dead animal remains. TL 3–20 mm; World 300, NA 40.

RAIN BEETLES Family Pleocomidae*

Stout bodied, dark, large and densely covered with setae ventrally, on
the legs and at margins of elytra and thorax. Glossy dorsally; reddish-
brown to black. **Clypeal process deeply bifurcated.** Vertex with conical
tubercle or erect horn medially. Antennae with 11 antennomeres;
antennomere 2 short and moniliform; **club comprised of 4 to 8
antennomeres** (female shorter and stouter than male). Adults live in
the ground, usually emerging only at dusk or after a rain. Single genus,
Plecoma, restricted to western NA. TL 17–44 mm; World 26, NA 26.

EARTH-BORING SCARAB BEETLES Family Geotrupidae

Stout bodied, strongly convex, oval, and usually black or brown.
Antennae with 11 antennomeres. **Elytra strongly grooved or striate.**
Tarsi long and slender. **Protibiae broadened; scalloped along outer
edge.** Most found beneath dung. TL 5–45 mm; World 620, NA 55.

324

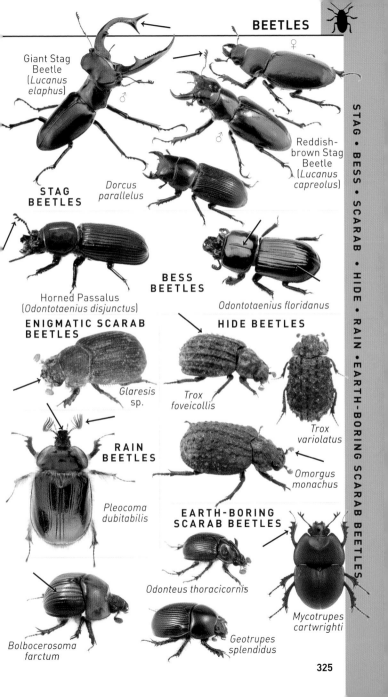

BEETLES

Giant Stag Beetle (*Lucanus elaphus*)

♀

♂

♂

Reddish-brown Stag Beetle (*Lucanus capreolus*)

Dorcus parallelus

STAG BEETLES

Horned Passalus (*Odontotaenius disjunctus*)

BESS BEETLES

Odontotaenius floridanus

ENIGMATIC SCARAB BEETLES

Glaresis sp.

HIDE BEETLES

Trox foveicollis

Trox variolatus

RAIN BEETLES

Pleocoma dubitabilis

Omorgus monachus

EARTH-BORING SCARAB BEETLES

Odonteus thoracicornis

Mycotrupes cartwrighti

Bolbocerosoma farctum

Geotrupes splendidus

325

COLEOPTERA

Superfamily Scarabaeoidea cont.

SAND-LOVING SCARAB BEETLES Family Ochodaeidae*
Small, elongate, convex, yellow to reddish-brown. Antennae with 9 or 10 antennomeres; **last 3 forming club.** Labrum and mandibles produced beyond labrum. Hind tibia with long apical spur pectinate along one edge. Elytra with or without striae. Some can form a complete ball when disturbed. TL 3–10 mm; World 80, NA 22.

SCAVENGER SCARAB BEETLES Family Hybosoridae*
Small, dark, oval, often glossy and dorsally convex. Mandibles and labrum project forward beyond clypeus. Antennae with 10 antennomeres; **last 3 forming club;** basal antennomere of club excavated to receive 2 apical (tomentose) antennomeres. **Pronotum convex, wider at base than elytra.** TL 3–7 mm; World 600, NA 5.

BUMBLE BEE SCARAB BEETLES Family Glaphyridae*
Often brightly colored, hairy, and fly during the day visiting flowers. Elongate head deflexed. **Elytra often narrowed toward tip.** Single genus, *Lichnanthe*, in NA. TL 6–20 mm; World 80, NA 8.

SCARAB BEETLES Family Scarabaeidae
Diverse, but generally heavy bodied, oval or elongate, and usually convex. Antennae with 8–11 antennomeres; **last 3–5 generally forming club.** Pronotum variable, with or without horns or tubercles. **Front tibia often expanded, with outer edge scalloped.** Abdomen with 6 visible sternites. TL 2–60 mm; World 28,000, NA 1,700.

SCARAB BEETLES

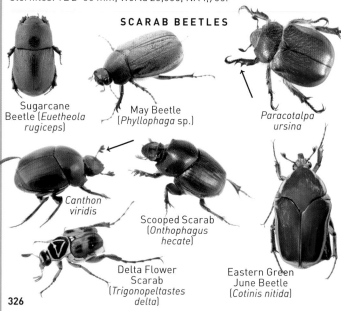

Sugarcane Beetle (*Euetheola rugiceps*)

May Beetle (*Phyllophaga* sp.)

Paracotalpa ursina

Canthon viridis

Scooped Scarab (*Onthophagus hecate*)

Delta Flower Scarab (*Trigonopeltastes delta*)

Eastern Green June Beetle (*Cotinis nitida*)

BEETLES

SAND-LOVING SCARAB BEETLES

Neochodaeus praesidii

SCAVENGER SCARAB BEETLES

Hybosorus illigeri

Pill Scarab Beetle
(*Germarostes globosus*)

BUMBLE BEE SCARAB BEETLES

Lichnanthe rathvoni

SCARAB BEETLES

Eastern Hercules Beetle
(*Dynastes tityus*)

♂

Ox Beetle
(*Strategus antaeus*)

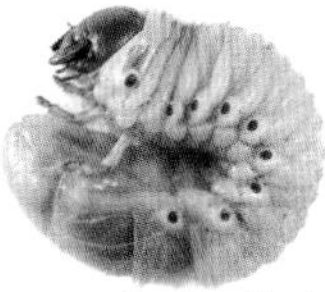

♀   ♂

Ox Beetle larva
(*Strategus* sp.)

Rainbow Scarab
(*Phanaeus difformis*)

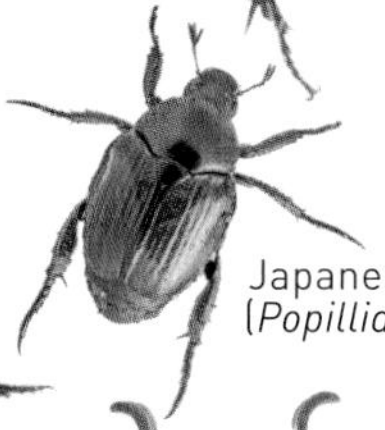

Japanese Beetle
(*Popillia japonica*)

Grapevine Beetle
(*Pelidnota punctata*)

Bumble Flower Beetle
(*Euphoria inda*)

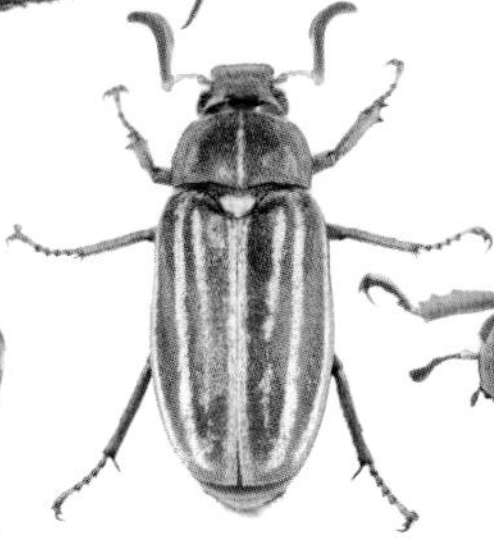

Lined June Beetle
(*Polyphylla occidentalis*)

Glorious Scarab
(*Chrysina gloriosa*) **327**

SCARAB BEETLES

Superfamily Scirtoidea

PLATE-THIGH BEETLES Family Eucinetidae* — Not illus.

Small, compact, elliptical, brown or black, and capable of jumping. Head small, narrowed anteriorly and deflexed. Antennae with 11 antennomeres, slightly serrate, expanding distally. Pronotum short, narrowed anteriorly. Tarsi 5-5-5. TL 0.8–4 mm; World >50, NA 11.

MINUTE BEETLES Family Clambidae*

Minute, oval, dark, convex with head strongly deflexed; most capable of rolling into a ball. Eyes partially or completely divided. Antennae with 8 or 10 antennomeres. Hind coxae dilated into broad plates extending to elytral edge. Tarsi 4-4-4. TL 0.7–2 mm; World 170, NA 12.

MARSH BEETLES Family Scirtidae

Oval, oblong, somewhat flattened body; pale yellow to black, sometimes spotted. Head relatively large, strongly deflexed. Antennae filiform with 11 antennomeres; **first antennomere large, inserted between eyes.** Prothoracic coxae conical. Larvae with multi-segmented antennae. TL 1.5–12 mm; World 800, NA 50.

Superfamily Dascilloidea

SOFT-BODIED PLANT BEETLES Family Dascillidae* Not illus.

Elongate, slightly convex, mottled brown to gray with dense setae covering body. Head usually visible from above; sometimes with conspicuous mandibles. Prosternal keel weakly developed. Antennae with 11 antennomeres; generally serrate. Tarsi 5-5-5. Found on vegetation near water; western. TL 8–18 mm; World 80, NA 5.

CICADA PARASITE BEETLES Family Rhipiceridae*

Elongate, brownish with orange antennae and prominent mandibles. **Antennae arising from prominent tubercles at base of mandibles;** flabellate in male, serrate to pectinate in female. Lack prosternal process. Membranous lobes on tarsomeres 1–4. Larvae are ectoparasitoids of cicada nymphs. Adults rarely seen, but aggregate. Single genus, *Sandalus*, in NA. TL 11–25 mm; World 70, NA 5.

Superfamily Buprestoidea

FALSE JEWEL BEETLES Family Schizopodidae — Not illus.

Similar to buprestids, but differ in having a wide metepisternum and deeply bilobed fourth tarsomere. Body stout and strongly convex; color varies from tan to green and copper. Some species sexually dimorphic. Restricted to southwest in NA. TL 6–18 mm; World 7, NA 7.

METALLIC WOOD-BORING BEETLES Family Buprestidae

Cylindrical to flattened, usually **tapering to tip;** hard bodied, generally iridescent. Head projecting anteriorly. **Antennae generally serrate.** five abdominal sternites; first two fused. Transverse metasternal suture present. TL 3–33 mm (usually <20 mm); World 15,000, NA 788.

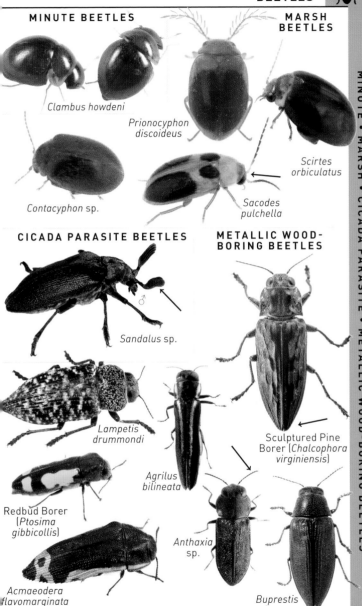

MINUTE BEETLES

Clambus howdeni

Prionocyphon discoideus

Contacyphon sp.

MARSH BEETLES

Scirtes orbiculatus

Sacodes pulchella

CICADA PARASITE BEETLES

Sandalus sp. ♂

Lampetis drummondi

Redbud Borer (*Ptosima gibbicollis*)

Acmaeodera flavomarginata

METALLIC WOOD-BORING BEETLES

Sculptured Pine Borer (*Chalcophora virginiensis*)

Agrilus bilineata

Anthaxia sp.

Buprestis decora

COLEOPTERA

Superfamily Byrrhoidea

PILL BEETLES Family Byrrhidae
Small, oval, convex, compact, brown to black body. Head deflexed, concealed from above. Antennae with 11 antennomeres; clavate, capitate, or filiform. Hind coxae large, extending to elytral margin. Legs and antennae capable of being retracted. Tarsi 4-4-4 or 5-5-5. Associated with moss. TL 1–10 mm; World 430, NA 35.

RIFFLE BEETLES Family Elmidae
Ovate to elongate body. Antennae slender with 7–11 antennomeres; filiform. Long thin unmodified legs with **large claws.** Head deflexed, somewhat concealed by pronotum. Elytra often with yellow or red coloration and rugose. Tarsi 5-5-5. Aquatic or semiaquatic. TL 1–8 mm; World 1,500, NA 100.

LONG-TOED WATER BEETLES Family Dryopidae
Elongate-oval, brown to gray, with head somewhat obscured beneath pronotum. Some species covered in fine pubescence. Antennae with 8–11 antennomeres, but very **short with most antennomeres wider than long and concealed** beneath prosternal lobe. Tarsi 5-5-5. Cling to rocks and logs in streams. TL 1–8 mm; World 300, NA 13.

TRAVERTINE BEETLES Family Lutrochidae
Body ovate, strongly convex, and covered in dense pubescence. Antennae with 11 antennomeres and short; length of 3–11 subequal to 1–2. Tarsi 5-5-5. Aquatic, eat algae, found only on travertine deposits. Single genus, *Lutrochus*, in NA. TL 2–6 mm; World 11, NA 3.

MINUTE MARSH-LOVING BEETLES Family Limnichidae*
Minute, oval to elongate, convex, and generally brown, often covered with colorful pubescence. Antennae with 11 antennomeres, short, clavate with most antennomeres broader than long. Anterior trochantin exposed. Tarsi 5-5-5. Aquatic. TL 1–2 mm; World 400, NA 28.

VARIEGATED MUD-LOVING BEETLES Family Heteroceridae
Oblong, somewhat flattened with **pronounced mandibles**. Color tan to black with variable patterning. **Pro- and mesotibiae with comb of flattened spines** on dorsal margin. First visible abdominal sternite with paired, arching stridulatory files. Tarsi 4-4-4. Associated with moist sand and muddy riparian habitats. TL 1–8 mm; World 500, NA 40.

WATER PENNY BEETLES Family Psephenidae
Dark, soft bodied, oval to ovate body with very short, fine, dense pubescence. Antennae with 11 antennomeres, moniliform, serrate, or flabellate, inserted beneath frontal ridges. Tarsi 5-5-5. Adults typically found, sometimes in aggregation, on tops of rocks emerging from streams. Larvae, oval and flat, found adhering to stones and other objects submerged in streams. TL 3–7 mm; World 300, NA 16.

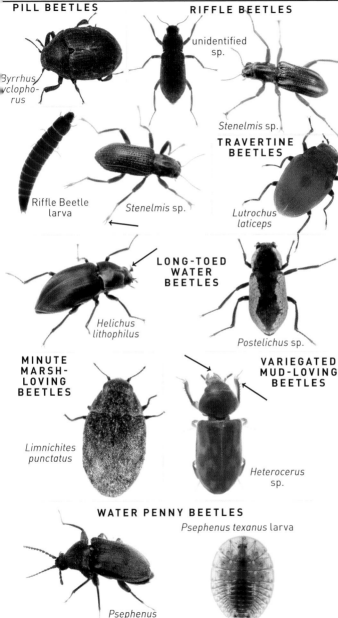

PILL BEETLES

Byrrhus cyclophorus

RIFFLE BEETLES

unidentified sp.

Stenelmis sp.

Riffle Beetle larva

Stenelmis sp.

TRAVERTINE BEETLES

Lutrochus laticeps

LONG-TOED WATER BEETLES

Helichus lithophilus

Postelichus sp.

MINUTE MARSH-LOVING BEETLES

Limnichites punctatus

VARIEGATED MUD-LOVING BEETLES

Heterocerus sp.

WATER PENNY BEETLES

Psephenus texanus larva

Psephenus texanus

COLEOPTERA

Superfamily Byrrhoidea cont.

TOE-WINGED BEETLES Family Ptilodactylidae

Elongate, oval, and brownish with head generally concealed from above. Antennae with 11 antennomeres; **pectinate in males, serrate in females. Pronotum anteriorly narrowed and rounded;** base crenulate. Tarsi 5-5-5 with third tarsomere lobed beneath and fourth often reduced. Live on vegetation. TL 2–16 mm; World 500, NA 25.

TURTLE BEETLES Family Chelonariidae*

Oval, compact, moderately convex, and glossy with patches of white scale-like setae on elytra. Antennae with 11 antennomeres; **1–3 elongated and flattened. Anterior and lateral portions of pronotum with sharply defined carina;** crenulate basally. Can retract head and appendages into grooves on body. Adults on vegetation; larvae feed on rotting vegetation. Until recently, single NA species, *Chelonarum lecontei*. An apparently introduced and undescribed species has recently been discovered in FL.
TL 4–5 mm; World 250, NA 2.

FOREST STREAM BEETLES Family Eulichadidae*   Not illus.

Body elongate, elaterid-like, with white setae on scutellum. Body brown. Antennae with 11 antennomeres; serrate (strongly so in males). Single NA species, *Stenocolus scutellaris*, found in mountains of northern CA. TL 10–30 mm; World 40, NA 1.

CALLIRHIPID CEDAR BEETLES

Family Callirhipidae*   Not illus.

Elongate, slightly convex, brown to black body. Antennae with 11 antennomeres arising on frontal prominence; pectinate to flabellate and more modified in male. Pronotum with well-developed basal interlocking mechanism. Single species, *Zenoa picea*, in NA. TL 9–23 mm; World 150, NA 1.

Superfamily Elateroidea

SOFT-BODIED PLANT BEETLES

Family Artematopodidae*   Not illus.

Elongate to ovate, convex, tan to brown, generally pubescent body. Head deflexed. Antennae with 11 antennomeres; long and filiform. Tarsi 5-5-5; tarsomeres 2–4 or 3–4 lobed beneath. Four to five abdominal sternites visible. TL 3–10 mm; World 45, NA 8.

TEXAS BEETLES Brachypsectridae*   Not illus.

Oblong, somewhat depressed with acute, carinate hind pronotal angles. Antennae with 11 antennomeres; 5–11 serrate. Presternum elongated below head. Five visible abdominal sternites. Single species, *Brachypsectra fulva*, in southwestern NA. Larvae found under bark or rocks. TL 4–8 mm; World 5, NA 1.

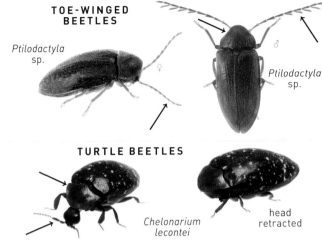

TOE-WINGED BEETLES

Ptilodactyla sp.
♀

Ptilodactyla sp.
♂

TURTLE BEETLES

Chelonarium lecontei

head retracted

Superfamily Elateroidea cont.

RARE CLICK BEETLES Family Cerophytidae*   Not illus.
Elongate-oblong, somewhat flattened, brown to black bodies. Antennae with 11 antennomeres; males pectinate, females serrate. Frons protruding forward. Hind trochanters nearly as long as femora. Adults can propel themselves similar to click beetles. Single genus, *Cerophytum*, in NA. TL 5–9 mm; World 20, NA 2.

FALSE CLICK BEETLES Family Eucnemidae*   Not illus.
Similar to elaterids, but head is strongly deflexed and second antennomere offset from first. Elongate, convex, and brownish. Antennae with 11 antennomeres. Labrum membranous and mostly hidden. Anterior margin of prosternum straight, not lobed. Abdominal segments connate. TL 3–25 mm; World 2,000, NA >100.

FALSE METALLIC WOOD-BORING BEETLES
Family Throscidae*   Not illus.
Small, oblong-oval, somewhat flattened and brownish to black. Head deflexed. Pronotum tightly compressed against elytra. Antennae with 11 antennomeres; clavate to capitate. Prosternum lobed anteriorly, nearly concealing mouthparts. Found on vegetation and in leaf litter. TL 1–6 mm; World 150, NA 20.

FALSE SOLDIER BEETLES Family Omethidae*   Not illus.
Small, soft bodied, similar to cantharids and lampyrids. Antennae with 11 antennomeres and widely separated; filiform, serrate, or pectinate. Tarsi 5-5-5; 3 and 4 with bifid lobes below. Abdomen with 7–8 ventral sternites visible. Telegeusinae males with elongated and compressed maxillary palps. TL 3–12 mm; World >40, NA 12.

Superfamily Elateroidea cont.

CLICK BEETLES Family Elateridae

Elongate with **elytra typically rounded and narrowing posteriorly.**
Prothorax large and articulating so that they can "jump" or
flip themselves. **Antennae serrate to pectinate.** Prosternal
spine fits into groove on mesosternum. **Posterior corners of
pronotum usually projecting backward in sharp spines or points.**
Phytophagous; larvae slender and called wireworms.
TL 1–60 mm (usually 12–30 mm); World 10,000, NA 970.

NET-WINGED BEETLES Family Lycidae

Soft bodied, often colorful, with head partially obscured by
pronotum. Antennae usually with 11 antennomeres; filiform, **serrate,
pectinate** to flabellate. Mesocoxae distinctly separated. **Elytra
with network of raised veins;** longitudinal ones more distinct
than transverse. **Head produced into snout** in some species. Eitht
abdominal sternites visible in males; seven in females. Found on
foliage; typically in wooded areas. TL 2–22 mm; World 4,600, NA 80.

NET-WINGED BEETLES

Caenia dimidiata

Calopteron sp. larva

Plateros sp.

Calopteron terminale

Lycus sanguineus

larva

Calopteron discrepans

Eros humeralis

ELATEROIDEA

CLICK BEETLES

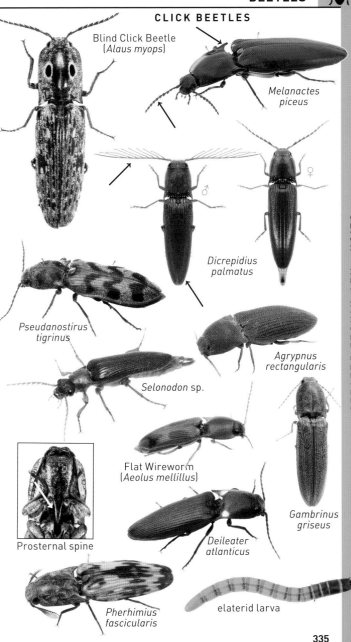

Blind Click Beetle
(*Alaus myops*)

Melanactes piceus

♂

♀

Dicrepidius palmatus

Pseudanostirus tigrinus

Agrypnus rectangularis

Selonodon sp.

Flat Wireworm
(*Aeolus mellillus*)

Prosternal spine

Gambrinus griseus

Deileater atlanticus

Pherhimius fascicularis

elaterid larva

335

Superfamily Elateroidea cont.

GLOWWORM BEETLES Family Phengodidae
Adult males with **short, leathery elytra generally narrowing posteriorly,** pronounced eyes. **Antennae bipectinate** with 12 antennomeres. Adult females larviform and possess bioluminescent organs. Males frequent lights. TL 6–30 mm; World 250, NA 23.

FIREFLIES Family Lampyridae
Soft bodied, **head concealed below pronotum** and last 2 abdominal sternites often modified as bioluminescent organ. Antennae usually with 11 **antennomeres; filiform to serrate.** Elytra soft, loosely meeting and generally flattened. Most found on vegetation during the day, but active at night; others, lacking bioluminescent organ, are active during the day. Larvae are predaceous, feeding on smaller insects and snails. TL 4–20 mm; World 2,200, NA 170.

SOLDIER BEETLES Family Cantharidae
Elongate, soft bodied with **head exposed beyond pronotum,** and often aposematically colored. Antennae with 11 antennomeres; filiform, serrate, pectinate, or flabellate. Labrum generally membranous, concealed beneath clypeus. Seven or eight abdominal sternites exposed. Common on vegetation. TL 1.2–18 mm; World 5,100, NA 470.

Superfamily Derodontoidea

JACOBSON'S BEETLES Family Jacobsoniidae*   Not illus.
Minute, elongate, lack scutellum. Antennae 10 to 11 antennomeres with apical club comprised of 2 antennomeres. Metasternum as long or longer than 5 visible abdominal sternites combined. Single genus, *Derolathrus*, in NA. TL <1 mm; World 23, NA 2.

GLOWWORM BEETLES

Phengodes fusciceps
♂

Phengodes sp.
♂

Distremocephalus texanus
♂

♀

Phengodes sp.

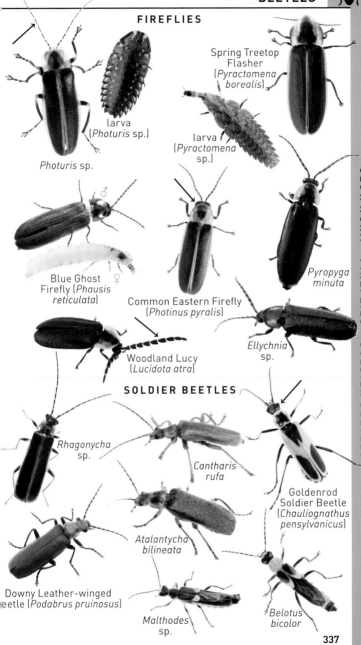

FIREFLIES

larva
(*Photuris* sp.)

Photuris sp.

Spring Treetop
Flasher
(*Pyractomena
borealis*)

larva
(*Pyractomena*
sp.)

Blue Ghost
Firefly (*Phausis
reticulata*) ♀

♂

Pyropyga
minuta

Common Eastern Firefly
(*Photinus pyralis*)

Ellychnia
sp.

Woodland Lucy
(*Lucidota atra*)

SOLDIER BEETLES

Rhagonycha
sp.

*Cantharis
rufa*

Goldenrod
Soldier Beetle
(*Chauliognathus
pensylvanicus*)

*Atalantycha
bilineata*

Downy Leather-winged
Beetle (*Podabrus pruinosus*)

Malthodes
sp.

*Belotus
bicolor*

GLOWWORM • FIREFLIES • SOLDIER BEETLES

337

Superfamily Derodontoidea cont.

TOOTH-NECKED BEETLES Family Derodontidae*
Small, elongate, dorsally convex, brownish body. Pair of ocelli present along inner margin of compound eyes. Pronotum with strongly toothed lateral margins and generally rugose dorsally. Elytra with many rows of large square punctures or polished dark spots. Found in woody fungi and under bark of rotting trees. Adults are active in late winter through early spring. TL 2–3 mm; World 37, NA 9.

Superfamily Nosodendroidea

WOUNDED-TREE BEETLES Family Nosodendridae*
Oval, convex, glossy black body capable of retracting appendages. **Prolegs flattened;** tibiae held anterior to femora at rest. Antennae with 11 antennomeres; capitate, with club protected in cavities between prolegs. Single genus, *Nosodendron*, in NA. Adults and larvae found in slime flux on trees and at oozing wounds. TL 5–6 mm; World 50, NA 2.

Superfamily Bostrichoidea

CARPET BEETLES Family Dermestidae
Oval or obovate, compact, strongly convex, often with pattern of variously colored scales on body. Antennae with 5 to 11 antennomeres; **clubbed,** fitting within shallow or deep groove. Median ocellus frequently present. Hind coxae excavated for receiving femora. Found on dry animal carcasses, in homes, stored products, and animal nests. TL 2–12 mm; World 1,000, NA 124.

HORNED POWDER-POST BEETLES Family Bostrichidae
Stout, cylindrical, black to brown, with **head strongly deflexed and concealed beneath the pronotum** except in the subfamily Lyctinae. Antennae with 8–11 antennomeres; straight with **loose club of 3–4 antennomeres. Pronotum often roughly sculptured.** Most species wood-boring, attacking living trees. TL 2–52 mm (usually < 24 mm); World 570, NA 70.

SPIDER & DEATH-WATCH BEETLES Family Ptinidae
Small, compact, variable body. Some species capable of contracting head and appendages within hooded prothorax (death-watch beetles). Other species with elytra strongly convex, long legs and antennae (spider beetles). Antennae generally filiform to serrate and 11 antennomeres; some with 1–2-segmented club, others with **terminal 3 antennomeres elongated.** TL 1–9 mm (most <5 mm); World 2,200, NA 400.

ENDECATOMID BEETLES Family Endecatomidae*   Not illus.
Elongate, gray to brown, similar to bostrichids, but lateral margins of pronotum with row of projecting straight or recurved setae. Elytra with microtubercles, bearing a single seta, arranged in irregular, reticulate pattern. Single genus, *Endecatomus*, in NA. TL 4–5 mm; World 4, NA 2.

DERODONTOIDEA • BOSTRICHOIDEA

TOOTH-NECKED BEETLES

Laricobius rubidus

WOUNDED-TREE BEETLES

Slime Flux Beetle (*Nosodendron unicolor*)

CARPET BEETLES

Black Larder Beetle (*Dermestes ater*)

larva

Cryptorhopalum sp.

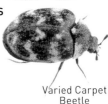

Varied Carpet Beetle (*Anthrenus verbasci*)

HORNED POWDER-POST BEETLES

Red-shouldered Bostrichid (*Xylobiops basilaris*)

Apple Twig Borer (*Amphicerus bicaudatus*)

Horned Powder-post Beetle (*Lichenophanes bicornis*)

Stout's Hardwood Borer (*Polycaon stouti*)

Powder-post Bostrichid (*Amphicerus cornutus*)

SPIDER & DEATH-WATCH BEETLES

Smooth Spider Beetle (*Gibbium aequinoctiale*)

Drugstore Beetle (*Stegobium paniceum*)

Ernobius sp.

Superfamily Tenebrionoidea

SHIP-TIMBER BEETLES Family Lymexylidae*   Not illus.
Elongate, cylindrical ,or somewhat flattened. Maxillary palps
expanded and fan-like (especially in male). Antennae with 11
antennomeres, relatively short; filiform or serrate. Prothoracic coxae
cylindrical and projecting. Elytra long, loosely fitting, and lacking
grooves. Tarsi 5-5-5. Attracted to lights.
TL 10–15 mm; World 70, NA 2.

HAIRY FUNGUS BEETLES Family Mycetophagidae
Oblong to ovate, somewhat flattened, brown, orange, or yellow,
covered in pubescence. Antennae with 11 antennomeres; **clubbed.**
Tarsi 4-4-4, some 3-4-4 (males). Abdomen with 5 visible sternites.
Found under bark feeding on fungus.
TL 1.5–5.5 mm; World 130, NA 26.

MINUTE TREE-FUNGUS BEETLES Family Ciidae
Elongate to oval, cylindrical, convex, brown to black. Head somewhat
concealed from above. Antennae short, 8–10 antennomeres; last 2–3
forming club. Tarsi 4-4-4. TL 0.5–6 mm; World 650, NA 84.

POLYPORE FUNGUS BEETLES Family Tetratomidae*
Oblong to ovate, somewhat flattened, and pubescent. Eyes notched.
Antennae **clavate with 11 antennomeres;** apical 3–4 forming loose
club in some. Front coxae separated by prosternal process. Tarsi 5-5-
4. Five visible abdominal sternites. TL 5–15 mm; World 155, NA 26.

FALSE DARKLING BEETLES Family Melandryidae
Variable morphology; most slender-elongate or broad-oval. Tip of
maxillary palp often large and hatchet or knife shaped. Antennae
moniliform or filiform to serrate with 11 antennomeres; apical 3–5
forming club in some. Most with first hind tarsomere elongated. **Hind
tibial spurs distinct.** Tarsi 5-5-4. TL 3–20 mm; World 430, NA 60.

TUMBLING FLOWER BEETLES Family Mordellidae
Long, narrow, compact, and distinctly wedge shaped, with **pointed
abdomen extending beyond elytra. Head large and triangular** in
front; sharply constricted behind eyes. Antennae serrate, clavate, or
filiform with 11 antennomeres. Scutellum visible. Tarsi 5-5-4. Exhibit
distinctive jumping behavior when disturbed.
TL 1.5–15 mm (usually 3–8); World 1,500, NA 200.

WEDGE-SHAPED BEETLES Family Ripiphoridae
Elongate and most often distinctly wedge shaped with black, red,
orange, or yellow coloration. Some species resemble flies or wasps.
Head deflexed. Antennae with 10 or usually 11 antennomeres;
flabellate, pectinate, or serrate (more elaborate in males). Pronotum
large, narrowest behind head. Tarsi 5-5-4. Exhibit complex life
history as parasitoids. TL 3–15 mm; World 400, NA 50.

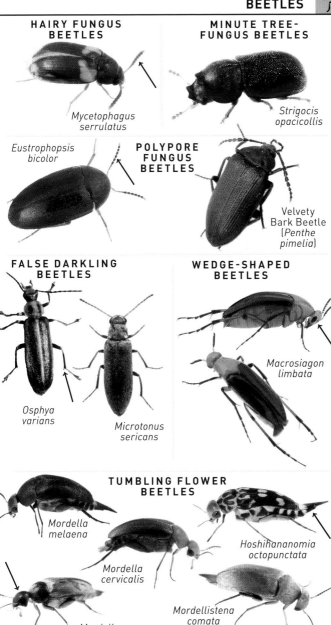

HAIRY FUNGUS BEETLES

Mycetophagus serrulatus

MINUTE TREE-FUNGUS BEETLES

Strigocis opacicollis

Eustrophopsis bicolor

POLYPORE FUNGUS BEETLES

Velvety Bark Beetle (*Penthe pimelia*)

FALSE DARKLING BEETLES

Osphya varians

Microtonus sericans

WEDGE-SHAPED BEETLES

Macrosiagon limbata

TUMBLING FLOWER BEETLES

Mordella melaena

Mordella cervicalis

Mordella sp.

Hoshihananomia octopunctata

Mordellistena comata

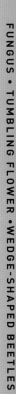

TENEBRIONOIDEA

Superfamily Tenebrionoidea cont.

DARKLING BEETLES Family Tenebrionidae
Extremely diverse. Elongate and cylindrical to oblong, oval, and somewhat flattened. Eyes strongly notched. Antennae moniliform, serrate, or clavate with bases concealed from above; 11 antennomeres. Procoxal cavities closed. Elytra completely cover abdomen; sometimes fused together and variously pitted, ridged, grooved, or smooth. Tarsi 5-5-4. Five abdominal sternites visible; one through three fused. Found in numerous habitats, including sand, decaying wood, fungi, and mammal nests. They are most commonly found feeding on fungi, lichen, algae, stored products, and flowers. In the arid southwest, they replace the ecological niche occupied by carabids in more verdant areas. TL 1–80 mm; World 20,000, NA 1,200.

CRYPTIC FUNGUS BEETLES Family Archeocrypticidae*   Not illu
Small, oval, strongly convex, tan to black, and finely clothed with recumbent setae. Antennae with 11 antennomeres; 9–11 forming gradual club. Prothorax with ventral process extending posteriorly between coxae. Elytra with rows of fine punctures. Tarsi 5-5-4. Abdomen with 5 visible sternites. TL 1.5–2 mm; World 53, NA 2.

FALSE LONGHORN BEETLES Family Stenotrachelidae* Not illus
Elongate, soft bodied, convex, and narrowing posteriorly. Head elongate, narrowing posteriorly. Eyes notched. Filiform antennae with 11 antennomeres. Pronotum elongate, narrowing anteriorly. Tarsi 5-5-4; pretarsal claws with ventral lobe. TL 6–20 mm; World 20, NA 10.

SYNCHROA BARK BEETLES Family Synchroidae*   Not illu
Elongate, narrow, somewhat flattened and tapering posteriorly. Similar to elaterids, but lack prosternal clicking mechanism, have 5-5-4 tarsal formula, and antennal bases are hidden from above. Tibial spurs finely notched. TL 7–13 mm; World 8, NA 2.

PALM AND FLOWER BEETLES Family Mycteridae*   Not illus
Variable in appearance. Head may be elongate with short rostrum or no Body elongate and/or somewhat flattened or stout and convex. Antenna with 11 antennomeres; variable, but with terminal antennomere constricted near its middle. Sulcus on underside of prothorax in front of legs. TL 3–9 mm; World 160, NA 12.

JUGULAR-HORNED BEETLES Family Prostomidae*   Not illus
Elongate, parallel sided and flattened with large prognathic mandibles. Body color is reddish-brown. Antennae with 11 antennomeres and weak club formed of apical 3. Tarsi 4-4-4. Larvae and adults typically found in dead logs. Single species, *Prostomis americanus*, in western NA. TL 6 mm; World 30, NA 1.

DARKLING BEETLES

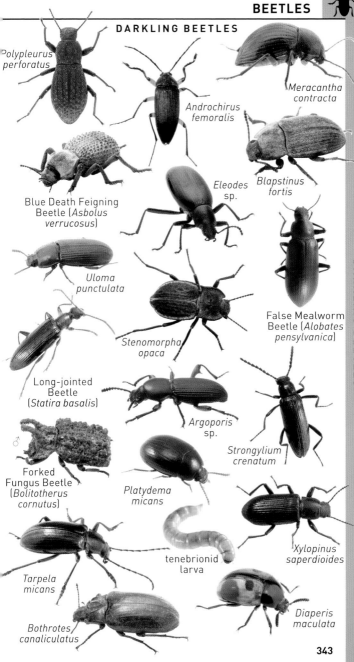

Polypleurus perforatus

Androchirus femoralis

Meracantha contracta

Blapstinus fortis

Eleodes sp.

Blue Death Feigning Beetle (*Asbolus verrucosus*)

Uloma punctulata

False Mealworm Beetle (*Alobates pensylvanica*)

Long-jointed Beetle (*Statira basalis*)

Stenomorpha opaca

♂

Argoporis sp.

Strongylium crenatum

Forked Fungus Beetle (*Bolitotherus cornutus*)

Platydema micans

Xylopinus saperdioides

Tarpela micans

tenebrionid larva

Bothrotes canaliculatus

Diaperis maculata

343

TENEBRIONOIDEA

Superfamily Tenebrionoidea cont.

IRONCLAD BEETLES Family Zopheridae
Diverse in form. Elongate and cylindrical or sometimes flattened and oval; usually brown or black. Eyes notched. **Antennae short and stout with 9–11 antennomeres;** base concealed from above, and apical 2–3 antennomeres form club. Pronotum may be elongate or transverse; **margins often expanded,** sometimes elaborately toothed. Elytra with raised surface sculpturing. Tarsi 4-4-4 or 5-5-4.
TL 2–34 mm; World 1,700, NA 110.

FALSE BLISTER BEETLES Family Oedemeridae
Slender, elongate, soft bodied, variously colored black, brown, tan, or gray often with yellow, orange, or red pronotum. **Pronotum usually longer than wide with anterior margin slightly concealing head.** Antennae filiform; 11 antennomeres. Tarsi 5-5-4; penultimate tarsomere wide and setose. TL 5–20 mm; World 1,500, NA 90.

BLISTER BEETLES Family Meloidae
Soft bodied, elongate, somewhat cylindrical with exposed head. Filiform antennae with 11 antennomeres; middle antennomeres modified in some males. **Eyes notched**. Pronotum generally narrower than head and base of elytra; rounded laterally. **Elytra somewhat rolled over abdomen laterally.** Tarsi 5-5-4; claws with ventral blade. Many with complex life histories, including hypermetamorphosis. TL 3–70 mm (usually 10–20 mm); World 3,000, NA 410.

CONIFER BARK BEETLES Family Boridae*   Not illus.
Elongate, dark, parallel sided, distinctly punctate and somewhat convex dorsally. Antennae with 11 antennomeres; bases hidden from above; last 3 antennomeres forming club. Pronotum with distinct lateral carinae. Tarsi 5-5-4. Under bark of live and dead pines. TL 11–23 mm; World 4, NA 2.

DEAD LOG BEETLES Family Pythidae
Elongate, subcylindrical to depressed with distinct punctation of various depths over body. **Head nearly square, with bulging eyes** and mandibles projecting forward. Antennae largely moniliform with 11 antennomeres; third elongated, terminal 3 somewhat enlarged. Pronotum rounded or subquadrate. Tarsi 5-5-4. Found beneath loose bark on logs. TL 6–22 mm; World 23, NA 7.

FIRE-COLORED BEETLES Family Pyrochroidae
Elongate, slightly to moderately flattened, with conspicuous "neck" constriction behind notched eyes. Generally yellowish-brown to black with red, yellow, or orange pronotum. Antennae with 11 antennomeres; filiform, serrate, to pectinate (more complex in males). Tarsi 5-5-4; claws simple or distinctly toothed at base. Larvae under bark of dead trees; some adults on vegetation. TL 4–20 mm; World 170, NA 50.

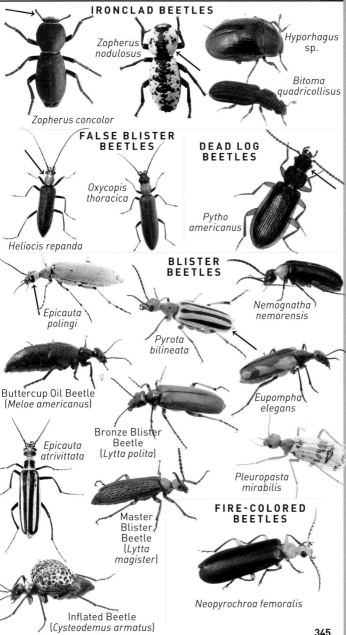

BEETLES

IRONCLAD BEETLES

Zopherus nodulosus

Zopherus concolor

Hyporhagus sp.

Bitoma quadricollisus

FALSE BLISTER BEETLES

Oxycopis thoracica

Heliocis repanda

DEAD LOG BEETLES

Pytho americanus

BLISTER BEETLES

Epicauta polingi

Pyrota bilineata

Nemognatha nemorensis

Buttercup Oil Beetle (*Meloe americanus*) ♀

Bronze Blister Beetle (*Lytta polita*)

Eupompha elegans

Epicauta atrivittata

Master Blister Beetle (*Lytta magister*)

Pleuropasta mirabilis

FIRE-COLORED BEETLES

Neopyrochroa femoralis

Inflated Beetle (*Cysteodemus armatus*)

Superfamily Tenebrionoidea cont.

NARROW-WAISTED BARK BEETLES Family Salpingidae

Elongate, convex or flat, brownish to black, variously glossy. Head may be produced into rostrum. Eyes usually present. Antennae with 11 antennomeres; filiform or clavate. Elytra completely covering abdomen or exposing 3–4 segments. Tarsi 5-5-4. Found on foliage and flowers or under logs. TL 3–4 mm; World 300, NA 20.

ANTLIKE FLOWER BEETLES Family Anthicidae

Many ant-like in appearance, in part because of **abruptly constricted head.** Antennae with 11 antennomeres; filiform, serrate, or weakly clubbed. Pronotum narrowed or constricted at base. Elytra covered in short hairs. Tarsi 5-5-4; penultimate tarsomere narrowly lobed beneath. TL 2–12 mm; World 3,500, NA 230.

BROAD-HIPPED FLOWER BEETLES

Family Ischaliidae* Not illus.
Elongate, somewhat convex. Pronotum with 3 angles posteriorly. Distinct ridges laterally on pronotum and elytra. HW reduced. Antennae filiform. Tarsi 5-5-4. Single genus, *Eupleurida*, in NA; sometimes included within Anthicidae. TL 4–6.5 mm; World 34, NA 3.

ANT-LIKE LEAF BEETLES Family Aderidae*

Elongate with broad, deflexed **head constricted posteriorly.** Eyes notched, hairy, and coarsely faceted. Antennae with 11 antennomeres; filiform, clavate, or pectinate. Tarsi appear 4-4-3 (actually 5-5-4); tarsomere 1 long, penultimate short. TL 1–4 mm; World 900, NA 50.

FALSE FLOWER BEETLES Family Scraptiidae

Soft bodied and variable. Elongate to ovate, flattened to moderately convex. Some clothed in fine setae with deeply notched eyes. Head not retracted within pronotum. Antennae with 11 antennomeres; filiform to moniliform, some with distinct club. Tarsi 5-5-4; penultimate tarsomere sometimes with fleshy lobes. Typically found on foliage and flowers; some in logs. TL 1–13 mm; World 500, NA 45.

Superfamily Cleroidea

CHECKERED BEETLES Family Cleridae

Elongate, narrow, typically covered in bristly setae and variously colored red, orange, yellow, or blue. **Pronotum somewhat cylindrical, narrower than elytra.** Antennae with 9–11 antennomeres; **filiform, serrate, pectinate, or flabellate and variously clubbed.** Tarsi 5-5-5; pad with seta between claws. TL 2–24 mm; World 3,600, NA 300.

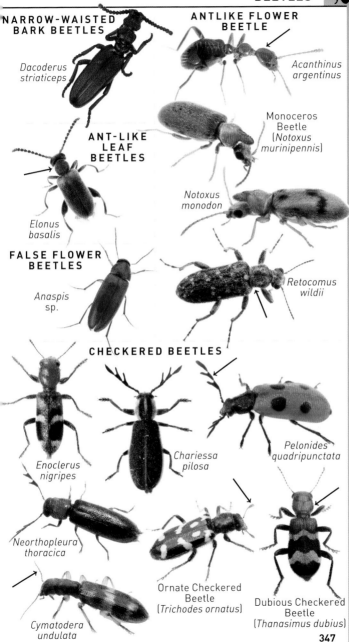

NARROW-WAISTED BARK BEETLES

Dacoderus striaticeps

ANTLIKE FLOWER BEETLE

Acanthinus argentinus

ANT-LIKE LEAF BEETLES

Elonus basalis

Monoceros Beetle (*Notoxus murinipennis*)

Notoxus monodon

FALSE FLOWER BEETLES

Anaspis sp.

Retocomus wildii

CHECKERED BEETLES

Enoclerus nigripes

Chariessa pilosa

Pelonides quadripunctata

Neorthopleura thoracica

Ornate Checkered Beetle (*Trichodes ornatus*)

Dubious Checkered Beetle (*Thanasimus dubius*)

Cymatodera undulata

Superfamily Cleroidea cont.

MAURONISCID BEETLES Family Mauroniscidae*   Not illus.

Elongate, tan to brown, somewhat convex and clothed in setae. Head elongagte; mouthparts often elongate and highly modified. Antennae with 11 antennomeres; filiform to slightly serrate with loose club formed by terminal 5 or 6 antennomeres. Pronotum convex with lateral margins carinate. Tarsi 5-5-5; claws unarmed. Formerly considered Melyridae. TL 2–5 mm; World 26, NA 8.

SHIELD BEETLES Family Peltidae

Dorsoventrally flattened and superficially resembling Nitidulidae. Small, broadly oval, generally glabrous or with elytral tubercles setose. Head projecting forward, not concealed beneath pronotum with deeply emarginate anterior margins. Antennae 11-segmented with terminal 3 segments forming loose club. Lateral margins or pronotum and elytra broadly explanate. Found under bark of dead pine and spruce trees. Single genus, *Peltis*, in NA; formerly included within Trogossitidae. TL 5–12 mm; World 10, NA 3.

LOPHOCATERID BEETLES Family Lophocateridae

Oblong to elongate and somewhat flattened; dorsal surface may be pubescent or glabrous. Head small, not concealed by pronotum. Terminal antennomeres (9–11) forming loose asymmetrical or symmetrical club. Pronotum broad with anterior margins emarginate or truncate; lateral margins explanate. Formerly included within Trogossitidae. TL 3–11 mm; World 120, NA 8.

THYMALID BEETLES Family Thymalidae

Ovoid, somewhat convex, with lateral margins of elytra explanate. Antennae 11-segmented with terminal 3 segments forming loose symmetrical club. Pronotum narrowed and rounded laterally. Found on bracket fungus. Single species, *Thymalus marginicollis*, found in eastern NA can retract legs and head. Formerly included within Trogossitidae. TL 7 mm; World 10, NA 1.

RHADALID BEETLES Family Rhadalidae*

Elongate, convex, and uniformly clothed in dense, erect pubescence. Head partially concealed under pronotum; eyes coarsely faceted or weakly emarginate. Antennae serrate. Pronotum with lateral margins explanate. Most of our species found in the southwestern US; formerly included within Melyridae. TL 3.5–7 mm; World 300, NA 13.

THANEROCLERID BEETLES Family Thanerocleridae

Elongate with weakly clubbed antennae. Found in eastern NA, either under bark infested with bark beetles (*Zendosus sanguineus*) or in stored products preying on insects (*Thaneroclerus buquet*). Formerly considered a subfamily of Cleridae. TL 3–7 mm; World 10, NA 2.

SHIELD BEETLES

LOPHOCATERID BEETLES

Peltis pippings-koeldi

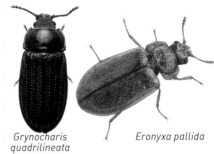

Grynocharis quadrilineata

Eronyxa pallida

THYMALID BEETLES

Thymalus marginicollis

RHADALID BEETLES

Rhadalus testaceus

Semijulistus flavipes

THANEROCLERID BEETLES

Zenodosus sanguineus

Thaneroclerus buquet

COLEOPTERA

Superfamily Cleroidea cont.

SOFT-WINGED FLOWER BEETLES Family Melyridae

Head broad with distinct clypeus and bulging eyes; soft, loose-fitting elytra often widening posteriorly. Antennae with 11 antennomeres or appearing as 10 with second antennomere hidden under distal end of first. **Pronotum wider than long; sides margined or keeled.** Tarsi generally 5-5-5; claws simple or toothed.
TL 2–7 mm; World 6,000, NA 520.

BARK-GNAWING BEETLES Family Trogossitidae

Elongate, rigid, parallel sided, somewhat convex to oblong-oval and flattened to convex. Antennae generally with 11 antennomeres; **last 1 to 3 forming club. Pronotum wider than head, either squarish** or wider than long. Tarsi 5-5-5; first very small; claws subequal and simple. Many species found beneath bark or crawling on dead branches. TL 2–22 mm (most 5–15 mm); World 600, NA 60.

FRUITWORM BEETLES Family Byturidae

Oblong, convex, tan with fine, dense pubescence. Antennae with 11 antennomeres; clavate, last 3 antennomeres forming club. Pronotum as wide as elytra, narrowing anteriorly. Tarsi 5-5-5; tarsomeres 2–3 with broad, plate-like lobes. Found on foliage of flowering plants like blackberries. TL 2.5–5.5 mm; World 24, NA 2.

FALSE SKIN BEETLES Family Biphyllidae*   Not illus.

Tan to dark brown; oblong-oval, somewhat convex; covered with setae lying both erect and flat. Antennae with 11 antennomeres; last 3 antennomeres forming club. Pronotum wider than head with pair of ridges on slightly rounded sides. Scutellum partially exposed. Tarsi 5-5-5; tarsomeres 2 and 3 with slender, pubescent lobes. Can occur on standing wood snags. TL 2–4 mm; World 200, NA 3.

Superfamily Coccinelloidea

MINUTE BARK BEETLES Family Cerylonidae

Elongate or oval, robust, somewhat flattened, generally smooth and glossy. All larvae and many adults with piercing-sucking mouthparts. Antennae usually with 10 antennomeres; last 1 or 2 forming club. Elytra usually grooved with rows of punctures. Tarsi usually 4-4-4; rarely 3-3-3. Found in decaying wood and leaf litter. TL 1–5 mm (usually <2 mm); World 300, NA 12.

WELL POLISHED BEETLES Family Euxestidae*   Not illus.

Oblong, oval, shiny reddish-brown to black. Pronotum sparsely punctate; posterior margin wavy. Elytra narrowing posteriorly and rounded at tips; punctures coarse, sparse, and irregular. Tarsi 4-4-4. Found under bark of hardwoods and on vegetation. Formerly considered Cerylonidae. TL 2–3 mm; World 70, NA 2.

SOFT-WINGED FLOWER BEETLES

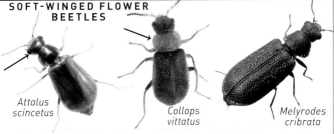

Attalus scincetus

Collops vittatus

Melyrodes cribrata

BARK-GNAWING BEETLES

Temnoscheila sp.

Tenebroides laticollis

MINUTE BARK BEETLES

FRUITWORM BEETLES

Cerylon castaneum

Raspberry Fruitworm (*Byturus unicolor*)

Superfamily Coccinelloidea cont.

MURMIDIID BEETLES Family Murmidiidae*   Not illus.
Brown, oval, strongly convex, covered with short, fine, inconspicuous setae. Pronotum twice as long as wide; lateral cavities to hold antennae visible in some. Antennae with 10 antennomeres; first 2 form club. Elytra with rows of distinct punctures. Tarsi 4-4-4. Found on decomposing vegetation and in stored products like seeds and rice. Formerly considered Cerylonidae. TL 1–2 mm; World 5, NA 4.

EUPSILOBIID BEETLES Family Eupsilobiidae*   Not illus.
Oval, strongly convex, shiny brown. Head twice as wide as long with frontoclypeal suture. Antennae with 11 antennomeres; insertion point not visible from above; last 2 antennomeres form club. Pronotum wider than head; explanate anteriorly. Tarsi 4-4-4. Single species, *Eidoreus politus*, known from FL. Formerly considered Endomychidae. TL 1 mm; World 15, NA 1.

Superfamily Coccinelloidea cont.

HANDSOME FUNGUS BEETLES Family Endomychidae

Usually elongate-oval, broadly rounded or flattened, often with orange or red coloration. Frontoclypeal suture present. Antennae with 11 antennomeres; **last 3 antennomeres usually forming loose club. Pair of distinct basal grooves dorsally on pronotum.** Elytra with irregular punctures. Tarsi appear 3-3-3 (actually 4-4-4); first 2 tarsomeres bilobed beneath. TL 4–10 mm; World 1,800, NA 45.

ANAMORPHID BEETLES Family Anamorphidae*   Not illus.

Most nearly hemispherical, brownish to black, shiny, densely clothed with fine setae. Head largely obscured by rounded pronotum. Antennae with 11 antennomeres; last 3 forming club. Tarsi 4-4-4. Formerly considered Endomychidae. TL 1–2 mm; World 170, NA 11.

DRY BARK BEETLES Family Bothrideridae*   Not illus.

Elongate, narrow, cylindrical, somewhat flattened or oblong and convex. Antennae with 10 to 11 antennomeres; 1–3 antennomeres forming club and basal insertions exposed. Pronotum longer than wide. Scutellum visible. Tarsi 4-4-4. Found under bark of conifers and hardwoods. TL 1.5–13 mm; World 300, NA 18.

TEREDID BEETLES Family Teredidae*   Not illus.

Elongate, convex; pronotum generally rounded laterally and often narrowing posteriorly. Eyes may be absent (*Anommatus*). In others, eyes present. Antennae with 10–11 antennomeres; last antennomeres forming club and basal insertions exposed. Trochanters large and distinct. Tarsi 4-4-4. Found under bark of conifers and hardwoods. Formerly considered Bothrideridae.
TL 3–4 mm; World 120, NA 4.

MYCETAEID BEETLES Family Mycetaeidae*   Not illus.

Elongate, convex, reddish-brown, shiny, clothed in yellowish setae. Antennae setose with 11 antennomeres; last 3 antennomeres forming loose club. Tarsi appear 3-3-3 (actually 4-4-4); first 2 tarsomeres bilobed beneath. Single adventive species, *Mycetaea subterranea*, in NA. Formerly considered Endomychidae. TL 2 mm; World 5, NA 1.

LADY BEETLES Family Coccinellidae

Hemispherical or oval and convex and often red or orange and black. Head somewhat or deeply inserted within prothorax. Antennae with 7–11 antennomeres; **terminating in club. Pronotum strongly convex, wider than long, distinctly margined or keeled laterally.** Scutellum small and triangular. Elytra smoothly rounded. Tarsi 4-4-4 with third tarsomere minute; sometimes appearing 3-3-3. Abdomen with 5–7 abdominal sternites visible. Found on flowers and plants, where they feed on aphids. TL 0.8–11 mm; World 6,000, NA 500.

HANDSOME FUNGUS BEETLES

Aphorista vittata

Endomychus biguttatus

LADY BEETLES

larva

Polished Lady Beetle (*Cycloneda munda*)

Spotted Lady Beetle (*Coleomegilla maculata*)

Ashy Gray Lady Beetle (*Olla v-nigrum*)

Multicolored Asian Lady Beetle (*Harmonia axyridis*)

Seven-spotted Lady Beetle (*Coccinella septempunctata*)

Twice-stabbed Lady Beetle (*Chilocorus* sp.)

Barber's Lady Beetle (*Brachiacantha barberi*)

V-marked Lady Beetle (*Neoharmonia venusta*)

Cactus Lady Beetle (*Chilocorus cacti*)

Mealybug Destroyer (*Cryptolaemus montrouzieri*)

Streaked Lady Beetle (*Myzia pullata*)

Dusky Lady Beetle (*Scymnus louisianae*)

Twenty-spotted Lady Beetle (*Psyllobora vigintimaculata*)

Superfamily Coccinelloidea cont.

MINUTE HOODED BEETLES Family Corylophidae
Tiny, oval or circular, with head generally concealed from above by pronotum. Antennae with 9–11 antennomeres; **last 3 forming a club.** Base of forelegs surrounded entirely by thoracic plates. Tarsi 4-4-4. Six abdominal sternites visible. TL 0.5–2.5 mm; World 200, NA 60.

AKALYPTOISCHIID SCAVENGER BEETLES
Family Akalyptoischiidae*   Not illus.
Elongate-oval to broadly oval, hard bodied, and somewhat convex. Antennae with 11 antennomeres; 2–7 forming club. Prothoracic coxal cavities closed. Tarsi 5-5-4 (rarely 4-4-4); tarsomeres 1–2 or 1–3 densely setose ventrally. Elytra coarsely punctate. Five abdominal sternites visible; first two fused. Single genus, *Akalyptoischion*; limited to western NA. TL 1.5–4 mm; World 24, NA 24.

MINUTE BROWN SCAVENGER BEETLES Family Latridiidae
Elongate, somewhat convex, **widest at middle,** and usually brownish. Body either glabrous or finely pubescent. Antennae with 10–11 antennomeres; **terminal 2–3 forming club.** Pronotum usually narrower than elytral base. Elytra rounded anteriorly and grooved. Tarsi 3-3-3. Five to six abdominal sternites visible. Most common during rainy season; found in leaf litter. TL 0.8–3 mm; World 1,000, NA 140.

Superfamily Cucujoidea

CRYPTIC SLIME MOLD BEETLES Family Sphindidae
Cylindrical, parallel sided, convex, and brownish with head partially visible from above. Antennae with 10–11 antennomeres; 2–3 terminal antennomeres forming club; **scape and pedicel enlarged.** Prothorax same width as elytra. Tarsi 5-5-5; 2–3 broadly lobed ventrally. Five visible abdominal sternites. TL 1.5–3.5 mm; World 60, NA 9.

PALMETTO BEETLES Family Smicripidae*   Not illus.
Elongate, parallel sided, and somewhat flattened. Head as wide as pronotum. Antennae with 11 antennomeres; 9–11 forming club. Maxilla with single lobe. Elytra short, exposing last 2 abdominal segments and pygidium. Last abdominal sternite equal in width to preceding 4. Tarsi 4-4-4 or 5-5-5. Single genus, *Smicrips*, in NA. TL 1–2 mm; World 6, NA 2.

SAP-FEEDING BEETLES Family Nitidulidae
Elongate and robust, nearly hemispherical or slightly flattened. Antennae with 11 antennomeres; **last 3 forming club.** Pronotum wider than long; anterior margin slightly or deeply notched; **sides strongly keeled.** Elytra generally **smooth with rounded tips; occasionally short.** Tarsi 5-5-5; tarsomeres broad with antennomere 4 small. Five abdominal sternites visible. Found on flowers, in leaf litter, under loose bark, at sap, and at carrion. TL 1–15 mm; World 4,500, NA 165.

BEETLES

MINUTE HOODED BEETLES

Holopsis marginicollis and larva

CRYPTIC SLIME MOLD BEETLES

Sphindus americanus

MINUTE BROWN SCAVENGER BEETLES

Melanophthalma sp.

SAP-FEEDING BEETLES

Epuraea sp.

Colopterus unicolor

Pallodes pallidus

Lobiopa undulata

Carpophilus melanopterus

Stelidota octomaculata

Small Hive Beetle (*Aethina tumida*)

Amphotis schwarzi

Cryptarcha ampla

Thalycra sp.

355

Superfamily Cucujoidea cont.

SHORT-WINGED FLOWER BEETLES Family Kateretidae

Most elongate-oval, dark with **shortened elytra exposing pygidium** and weakly clubbed antennae (made up of 3 antennomeres). Maxilla with 2 distinct lobes. Pronotum slightly narrower than elytra. Scutellum large and triangular. Tarsi 5-5-5. TL 1.5–6 mm; World 100, NA 13.

MINUTE CLUBBED BEETLES Family Monotomidae

Elongate, parallel sided, brownish to black with **terminal abdominal segment exposed.** Antennae with 10 antennomeres; **last 1 or 2 forming club.** Pronotum subquadrate or elongate; lateral margins smooth or toothed. Tarsi usually 5-5-5 (females) and 5-5-4 (males). Five visible abdominal sternites. TL 1.5–6 mm; World 250, NA 55.

SILVANID FLAT BARK BEETLES Family Silvanidae

Elongate, parallel sided to ovate, and somewhat flattened. Antennae with 11 antennomeres; **filiform or loosely moniliform terminating in club. Pronotum longer than wide with distinct margins;** either smooth or waved. Elytra covering abdomen; typically with punctures or sculpturing. Tarsi 5-5-5 (some 4-4-4). Five visible abdominal sternites. Found under bark, decaying vegetation, and fruit. TL 2–15 mm; World 500, NA 32.

PARASITIC FLAT BARK BEETLES Family Passandridae

Elongate, reddish-brown, either nearly flattened or somewhat cylindrical. Head, pronotum, and elytra heavily sculptured with grooves and carinae. Antennae with 11 antennomeres; moniliform and pubescent. Scutellum small. Tarsi 5-5-5; simple claws. TL 4–13 mm; World 36, NA 3.

FLAT BARK BEETLES Family Cucujidae

Elongate, parallel sided, red to reddish-brown, and strongly flattened. Head oblong and triangular. Antennae with 11 antennomeres; moniliform. Pronotum with subequal sides or shorter than wide. Scutellum small. Elytra finely punctured. Tarsi 5-5-4 (male) and 5-5-5 (female); claws simple. Found under bark. TL 6–25 mm; World 47, NA 7.

LINED FLAT BARK BEETLES Family Laemophloeidae

Elongate, flattened, or somewhat cylindrical, brown to reddish-brown. Head and **pronotum with pair of fine lateral lines or grooves.** Antennae with 10 antennomeres; usually filiform. Scutellum broad to triangular. Elytra generally with complete lateral ridge. Tarsi generally 5-5-4 in males and 5-5-5 in females; tarsomere 1 short, claws simple. Found under bark. TL 1–5 mm (generally <3 mm); World 430, NA 52.

PLEASING FUNGUS BEETLES Family Erotylidae

Elongate-oval or long and parallel sided, often variously colored brown or black with red and orange. Antennae with 11 antennomeres; **terminal 3, 4, or 5 forming club. Pronotum with lateral margins.** Scutellum visible. Elytra smooth, sometimes with pits. Tarsi 5-5-5; sometimes appearing 4-4-4. TL 2–22 mm; World 3,500, NA 80.

BEETLES

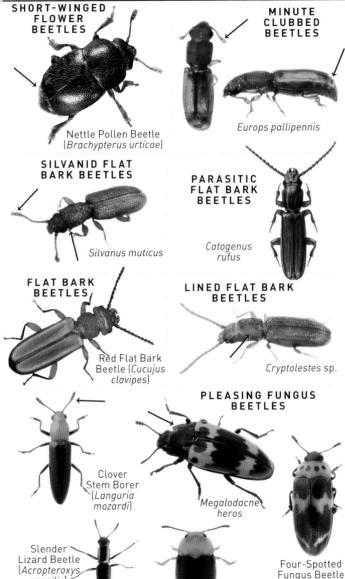

SHORT-WINGED FLOWER BEETLES

Nettle Pollen Beetle (*Brachypterus urticae*)

MINUTE CLUBBED BEETLES

Europs pallipennis

SILVANID FLAT BARK BEETLES

Silvanus muticus

PARASITIC FLAT BARK BEETLES

Catogenus rufus

FLAT BARK BEETLES

Red Flat Bark Beetle (*Cucujus clavipes*)

LINED FLAT BARK BEETLES

Cryptolestes sp.

PLEASING FUNGUS BEETLES

Clover Stem Borer (*Languria mozardi*)

Megalodacne heros

Slender Lizard Beetle (*Acropteroxys gracilis*)

Triplax thoracica

Four-Spotted Fungus Beetle (*Ischyrus quadripunctatus*)

Superfamily Cucujoidea cont.

SHINING FLOWER BEETLES Family Phalacridae
Broadly oval and convex to nearly hemispherical, shiny brown or
black, and glabrous. Antennae with 11 antennomeres; **last 1 to 3
forming elongated club.** Scutellum relatively large and triangular.
Tarsi 5-5-5; 1–3 broad and hairy ventrally, 4 tiny and obscure, claws
with tooth or broadly expanded basally. TL 1–3 mm; World 630, NA 122.

SILKEN FUNGUS BEETLES Family Cryptophagidae
Elongate-oval to oval, robust, brown to black, and often clothed with
long silky setae. Antennae with 11 antennomeres; last 3 forming
loose club. Pronotum often with pair of depressions along base
and sometimes with sculptured or toothed lateral margins. Elytra
with irregular punctures. Tarsi 5-5-5; sometimes 5-5-4 in males. In
decaying matter with fungi. TL 0.8–5 mm; World 600, NA 150.

CYBOCEPHALID BEETLES Family Cybocephalidae   Not illus.
Dark, ovoid, highly convex, and shiny. Can roll into ball-like position
with head facing downward. Tarsomeres lobed beneath; 4 tarsal
segments on all legs. Single genus, *Cybocephalus*, in NA; sometimes
considered subfamily of Nitidulidae. TL 0.5–2.5 mm; World 220, NA 5.

Superfamily Chrysomeloidea
Tarsi 5-5-5; appearing 4-4-4 with fourth tarsomere small, enveloped
beneath lobes of 3.

DISTENIID BEETLES Family Disteniidae
Superficially similar to Cerambycidae. Elongate, dark brown to
black, with dense pubescence and tapering elytra. Clypeus and frons
distinct; not in same plane. Antennae with 11 antennomeres; filiform,
usually with clusters of long setae below. Antennomere 2 inflated and
long; 1 and 3 subequal. Single species, *Elytrimitatrix undata*, found in
NA. Found on dead hardwoods. TL 16–26 mm; World 340, NA 1.

LONG-HORNED BEETLES Family Cerambycidae
Large family with variably shaped species. Typically elongate,
broadest across elytra, robust, with **filiform antennae at least half
as long as body** (often times much longer). **Eyes generally markedly
notched around antennal bases.** Found in a variety of situations,
including at flowers, trees, and at lights. Larvae in dead or damaged
wood. TL 3–60 mm; World 10,300, NA 1,000.

**SHINING FLOWER
BEETLES**

Acylomus
sp.

**SILKEN FUNGUS
BEETLES**

Cryptophagus
sp.

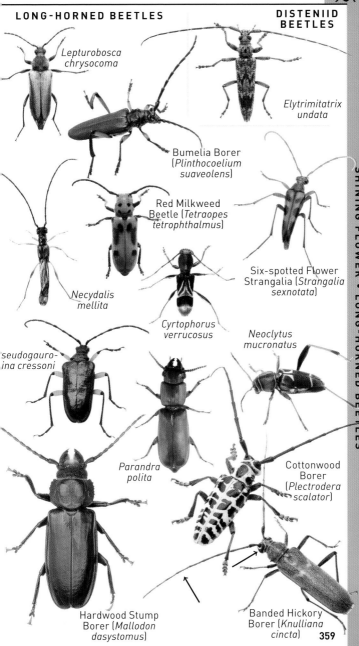

BEETLES

LONG-HORNED BEETLES

Lepturobosca chrysocoma

Bumelia Borer (*Plinthocoelium suaveolens*)

Red Milkweed Beetle (*Tetraopes tetrophthalmus*)

Necydalis mellita

Cyrtophorus verrucosus

seudogauro-ina cressoni

Parandra polita

Hardwood Stump Borer (*Mallodon dasystomus*)

DISTENIID BEETLES

Elytrimitatrix undata

Six-spotted Flower Strangalia (*Strangalia sexnotata*)

Neoclytus mucronatus

Cottonwood Borer (*Plectrodera scalator*)

Banded Hickory Borer (*Knulliana cincta*)

359

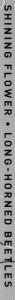

Superfamily Chrysomeloidea cont.

RAVENOUS LEAF BEETLES Family Orsodacnidae

Elongate, somewhat convex. Head with subquadrate labrum. Antennae with 11 antennomeres; filiform and inserted laterally between eyes and base of mandibles. **Pronotum longer than wide, rounded laterally** without a keel. All tibia with pair of spurs apically. Elytra cover abdomen; rounded apically. Found on blooming shrubs and trees. TL 4–9 mm; World 40, NA 4.

MEGALOPODID LEAF BEETLES

Family Megalopodidae   Not illus.

Elongate-oblong, convex, often shiny and coarsely punctured with pubescence. Antennae with 11 antennomeres; short, attaching low on head between mandibles and eyes; antennomeres 5–11 becoming serrate. Pronotum expanded medially on sides. TL 3–5 mm; World 350, NA 9.

LEAF BEETLES Family Chrysomelidae

Large, diverse family. Typically elongate, somewhat cylindrical, occasionally hemispherical or compact and flattened. Antennae filiform, sometimes serrate, pectinate, flabellate, or clavate. **Pronotum broader than head;** usually subquadrate or wider than long and often keeled laterally. Elytra may or may not conceal abdomen. Hind legs modified for jumping in some. Found on foliage and flowers and some in seeds (Bruchinae). TL 1–17 mm; World 35,000, NA 1,900.

Superfamily Curculionoidea

CIMBERID WEEVILS Family Cimberidae*   Not illus.

Elongate body with elongated rostrum becoming wider at tip, distinct labrum, and straight, filiform antennae inserted near tip of rostrum. Elytra lacking grooves or rows of punctures, but often clothed in setae. Tarsi 5-5-5, but appear 4-4-4 with fourth small; second lobed at middle. Hind tibia with 2 apical spurs. TL 3–5.5 mm; World 70, NA 14.

PINE FLOWER SNOUT BEETLES

Family Nemonychidae*   Not illus.

Similar to Cimberidae, but second tarsomere truncate at middle and not projected over base of third and hind tibia with single apical spur. Single species, *Atopomacer ites,* known from pines. TL 3–5.5 mm; World 4 NA 1.

CYCAD WEEVILS Family Belidae*   Not illus.

Brown to reddish-brown with rostrum and enlarged front femora (in males). Antennae with 11 antennomeres; straight with 9–11 forming club and inserted basally on rostrum. Pronotum with sharply keeled sides. Elytra short, exposing pygidium. Tarsi 5-5-5; appearing 4-4-4 with tarsomeres 2–3 broadly lobed. Single genus, *Rhopalotria*, found on *Zamia* cycads in FL. TL 3–6 mm; World 380, NA 3.

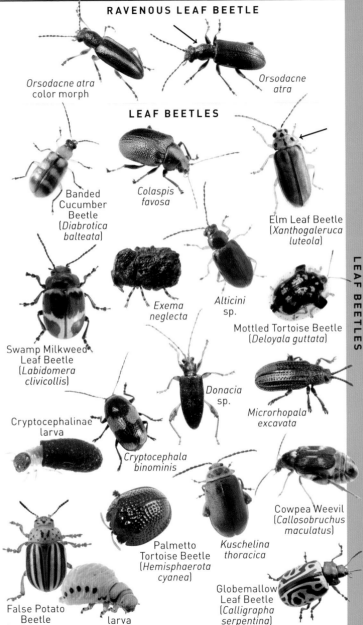

RAVENOUS LEAF BEETLE

Orsodacne atra
color morph

Orsodacne atra

LEAF BEETLES

Banded
Cucumber
Beetle
(*Diabrotica
balteata*)

*Colaspis
favosa*

Elm Leaf Beetle
(*Xanthogaleruca
luteola*)

*Exema
neglecta*

Alticini
sp.

Mottled Tortoise Beetle
(*Deloyala guttata*)

Swamp Milkweed
Leaf Beetle
(*Labidomera
clivicollis*)

Donacia
sp.

*Microrhopala
excavata*

Cryptocephalinae
larva

*Cryptocephala
binominis*

Cowpea Weevil
(*Callosobruchus
maculatus*)

Palmetto
Tortoise Beetle
(*Hemisphaerota
cyanea*)

*Kuschelina
thoracica*

Globemallow
Leaf Beetle
(*Calligrapha
serpentina*)

False Potato
Beetle
(*Leptinotarsa
juncta*)

larva

LEAF BEETLES

361

Superfamily Curculionoidea cont.

FUNGUS WEEVILS Family Anthribidae

Elongate, brown to black, and often mottled with broad, **flat rostrum** and grooves underneath. Antennae straight and clubbed. **Pronotum with transverse ridge basally.** Elytra not covering pygidium. Tarsi 5-5-5, but appearing 4-4-4 with fourth small and hidden by spongy-pubescent third. Abdomen with 5 visible sternites; 1–4 fused. Found on shelf fungi, dead and diseased branches. TL 0.4–16 mm; World 3,900, NA 88.

LEAF ROLLING WEEVILS Family Attelabidae

Elongate, oval or stout and compact, variously convex with short broad or long and **slender rostrum widening at tip.** Antennae with 11 antennomeres; straight with antennomeres 9–11 forming loose club. **Pronotum narrower than base of elytra.** Elytra cover entire abdomen. Tarsi 5-5-5, but appear 4-4-4 with fourth small and hidden beneath lobes of 3. TL 1.8–7.5 mm; World 360, NA 51.

STRAIGHT-SNOUTED WEEVILS Family Brentidae

Diverse family; either long, slender, and parallel sided; stout and pear shaped; ant-like (p. 303); or large and robust (e.g., *Ithycerus noveboracensis*). Rostrum short and broad to long and slender. Body usually without scales or setae. Antennae usually straight, but may be elbowed, attaching at middle to sides of rostrum. Tarsi 5-5-5. Five abdominal sternites; first two fused and longer than three and four. TL 1–40 mm; World 4,000, NA 150.

SNOUT AND BARK BEETLES Family Curculionidae

Very large and diverse family. Typically broadly oval, long and cylindrical, to strongly humpbacked. **Rostrum, if present, long and slender or short and broad.** Body sculptured or smooth, sometimes with scales or variously metallic in color. Antennae with 11 antennomeres; **clubbed and elbowed with 9–11 forming compact club.** Pronotum slightly wider than head and usually without sharp margins or keels laterally. Scutellum small or absent. Tarsi usually 5-5-5, but appearing 4-4-4 with fourth very small and hidden beneath lobes of third. TL 1–35 mm (usually 5–15 mm); World 50,000, NA 2,500.

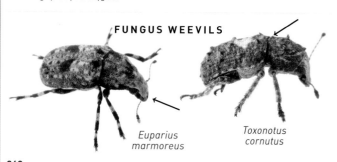

FUNGUS WEEVILS

Euparius marmoreus

Toxonotus cornutus

LEAF ROLLING WEEVILS

Eugnamptus angustatus

Synolabus bipustulatus

Homoeolabus analis

Pterocolus ovatus

SNOUT AND BARK BEETLES

STRAIGHT-SNOUTED WEEVILS

Oak Timberworm (*Arrhenodes minutus*)

Rice Weevil (*Sitophilus oryzae*)

Plum Curculio (*Conotrachelus nenuphar*)

Curculio sp.

Southern Blue-green Citrus Root Weevil (*Pachnaeus litus*)

Ironweed Curculio (*Rhodobaenus tredecimpunctatus*)

Black Turpentine Beetle (*Dendroctonus terebrans*)

Ash Bark Beetle (*Hylesinus mexicanus*)

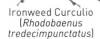

Antlike Weevil (*Myrmex* sp.)

Ambrosia Beetle (*Myoplatypus flavicornis*)

Asian Ambrosia Beetle (*Xylosandrus crassiusculus*)

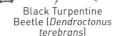

Northern Caddisfly
(*Pycnopsyche* sp.)
Limnephilidae

October Caddis
(*Dicosmoecus* sp.)
Limnephilidae

CADDISFLIES
ORDER TRICHOPTERA

Derivation of Name
- Trichoptera: *tricho*, hair; *ptera*, winged—referring to the hairy wings. The name caddisfly is derived from *cadyss*, meaning cotton or silk and refers to the larvae of this group, most of which use silk to make cases or retreats.

Identification
- 2–40 mm TL.
- Similar to moths in general appearance, but with wings held roof-like over abdomen and covered with hairs. FW usually dark, sometimes colorful with patterns; HW usually clear and smaller than FW.
- Mouthparts are chewing, with reduced mandibles but well-developed palps; feed on liquids.
- Antennae are relatively long and filiform.
- Larvae and pupae are aquatic; most build silken retreats or cases made of substrate and vegetative materials connected by silk.

Classification
- Two suborders (Annulipalpia and Integripalpia; some recognize the paraphyletic Spicipalpia as a third suborder) and 28 families in NA. Suborders are divided based on the adult mouthparts.
- Families are largely identified using characters found on the head and thorax, including the presence and shape of warts, the ocelli, the maxillary palps, and spurs and spines on the legs.
- World: 14,000 species; North America: 1,500 species.

Range
- Found on every continent except Antarctica.

Similar orders
- Lepidoptera (p. 378): scales on wings; coiled proboscis for mouthpart.
- Neuroptera (p. 280): FW and HW similar in shape and size and lacking dense hairs.
- Hymenoptera (p. 236): no hairs on wings; variable antennae.

KEY TO CADDISFLY FAMILIES

Rarely encountered families not included in the key. See text for descriptions.

Beraeidae, p. 374
Dipseudopsidae, p. 370
Ecnomidae, p. 370
Goeridae, p. 376
Hydrobiosidae, p. 374

Ptilocolepidae, p. 372
Rossianidae, p. 376
Sericostomatidae, p. 376
Uenoidae, p. 376
Xiphocentronidae, p. 370

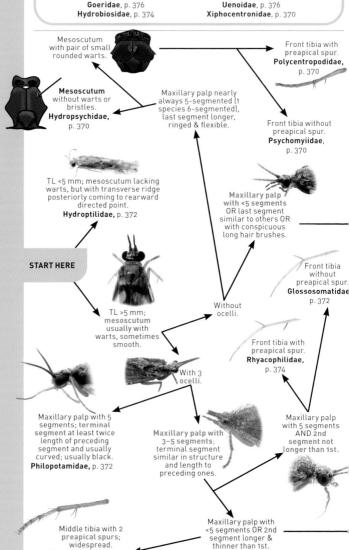

Mesoscutum with pair of small rounded warts.

Mesoscutum without warts or bristles. **Hydropsychidae**, p. 370

Front tibia with **preapical spur. Polycentropodidae,** p. 370

Maxillary palp nearly always 5-segmented (1 species 6-segmented), last segment longer, ringed & flexible.

Front tibia without preapical spur. **Psychomyiidae,** p. 370

TL <5 mm; mesoscutum lacking warts, but with transverse ridge posteriorly coming to rearward directed point. **Hydroptilidae,** p. 372

Maxillary palp with <5 segments OR last segment similar to others OR with conspicuous long hair brushes.

START HERE

Front tibia without preapical spur. **Glossosomatidae** p. 372

Without ocelli.

TL >5 mm; mesoscutum usually with warts, sometimes smooth.

Front tibia with preapical spur. **Rhyacophilidae,** p. 374

With 3 ocelli.

Maxillary palp with 5 segments; terminal segment at least twice length of preceding segment and usually curved; usually black. **Philopotamidae,** p. 372

Maxillary palp with 3–5 segments; terminal segment similar in structure and length to preceding ones.

Maxillary palp with 5 segments AND 2nd segment not longer than 1st.

Middle tibia with 2 **preapical spurs**; widespread. **Phryganeidae,** p. 377

Maxillary palp with <5 segments OR 2nd segment longer & thinner than 1st.

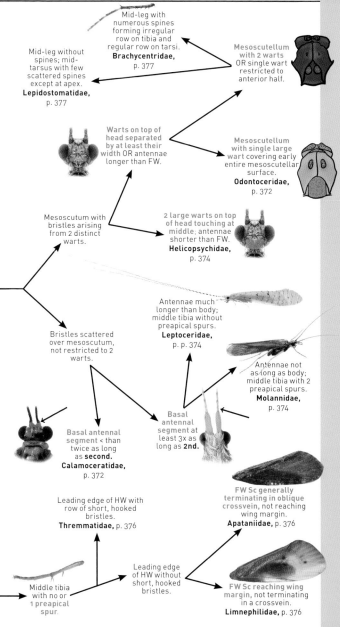

Mid-leg with numerous spines forming irregular row on tibia and regular row on tarsi. **Brachycentridae,** p. 377

Mid-leg without spines; mid-tarsus with few scattered spines except at apex. **Lepidostomatidae,** p. 377

Mesoscutellum with 2 warts OR single wart restricted to anterior half.

Warts on top of head separated by at least their width OR antennae longer than FW.

Mesoscutellum with single large wart covering early entire mesoscutellar surface. **Odontoceridae,** p. 372

Mesoscutum with bristles arising from 2 distinct warts.

2 large warts on top of head touching at middle; antennae shorter than FW. **Helicopsychidae,** p. 374

Antennae much longer than body; middle tibia without preapical spurs. **Leptoceridae,** p. p. 374

Bristles scattered over mesoscutum, not restricted to 2 warts.

Antennae not as long as body; middle tibia with 2 preapical spurs. **Molannidae,** p. 374

Basal antennal segment < than twice as long as **second.** **Calamoceratidae,** p. 372

Basal antennal segment at least 3x as long as **2nd.**

FW Sc generally terminating in oblique crossvein, not reaching wing margin. **Apataniidae,** p. 376

Leading edge of HW with row of short, hooked bristles. **Thremmatidae,** p. 376

Leading edge of HW without short, hooked bristles.

Middle tibia with no or 1 preapical spur.

FW Sc reaching wing margin, not terminating in a crossvein. **Limnephilidae,** p. 376

TRICHOPTERA

Food
- Adults feed on nectar and other fluids or not at all.
- Most larvae feed on dead or living plant material; some are predatory.
- Larvae display a variety of feeding strategies, including leaf shredding, algal grazing, collector filtering, and predation.

Behavior
- Adults are generally moderately strong fliers and active at night.
- Larvae live in a variety of aquatic habitats, where they may build cases from substrate materials, construct net retreats, or be free-living.

Life Cycle
- Holometabolous.
- Larvae are aquatic, most with abdominal gills, cylindrical in shape, and typically undergo 5 instars.
- Pupae are aquatic, with gills, swimming legs, and functional mouthparts to chew their way out of the pupal case.
- Variable number of generations per year(s); overwinter as larva or pupa.
- Adults short-lived, generally only a few weeks.

Importance to Humans
- Larvae serve as food source for many fish and other aquatic insects.
- Adults and some larvae (Hydropsychidae) are used as models for artificial flies in fly-fishing, and larvae are used as bait.
- Larvae are useful indicators of water quality.

Collecting and Preserving
Immatures
- Collect using dipnet or scoop, searching submerged rocks, logs, and leaf packs.
- Initially fix using Kahle's fluid (11% formalin [3–4% concentration], 28% ethanol [95% concentration], 2% glacial acetic acid, 59% water) for 24 hours or less, then permanently store in 80% ethanol. Kahle's may destroy DNA, so alternatively initially fix in 95% ethanol.

Adults
- Easiest collected at lights, beating sheet or with aspirator off vegetation, bridge pilings, etc.
- Store in 80% ethanol; some groups best preserved pinned.

Resources
- World Checklist, entweb.sites.clemson.edu/database/trichopt
- *Caddisflies* by LaFontaine, 1981
- *Caddisflies: The Underwater Architects* by Wiggins, 2004
- *Larvae of the North American Caddisfly Genera, 2nd ed.* by Wiggins, 1996

Maxillary Palps

Polycentropus sp. (Polycentropodidae); fixed terminal, ringed segment.

Ceraclea sp. (Leptoceridae); unringed terminal segment.

Forewing Venation

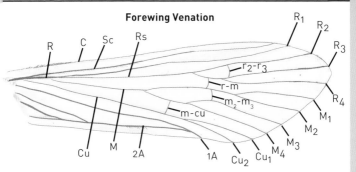

Pro- and Mesonota

Beraea gorteba
Beraeidae

Brachycentrus numerosus
Brachycentridae

Heteroplectron americanum
Calamoceratidae

Goera calcarata
Goeridae

Helicopsyche borealis
Helicopsychidae

Hydropsyche simulans
Hydropsychidae

Theilopsyche sp.
Lepidostomatidae

Ceraclea tarsipunctata
Leptoceridae

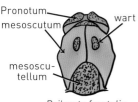

Psiloreta frontalis
Odontoceridae

Psychomyia flavida
Psychomyiidae

Agarodes crassicornis
Sericostomatidae

Pro- and mesonota of Trichoptera. Modified from Ross/Illinois Natural History Survey illustrations.

Suborder Annulipalpia

Characterized by the fixed terminal segment of the adult maxillary palps, which are often ringed. Larvae make fixed-retreats.

Superfamily Hydropsychoidea

DIPSEUDID CADDISFLIES Family Dipseudopsidae*   Not illus.

Light brown. Ocelli absent. Scutum possesses warts. Wings with R_2 branching from R_3 at the radial crossvein. Larvae are burrowers in sandy substrate of streams. Single genus, *Phylocentropus*, restricted to central and eastern NA. TL 7–10 mm; World 110, NA 5.

ECNOMIDAE CADDISFLIES Family Ecnomidae*   Not illus.

Light brown. Mesoscutum with pair of small, rounded warts. R_1 in FW branched; R_{2+3} unbranched. Single species in NA, *Austrotinodes texensis*, restricted to central TX. TL 2–10 mm; World 470, NA 1.

TUBE MAKER CADDISFLIES Family Polycentropodidae

Brownish bodies, generally with mottled wings, and lacking ocelli. Segment 5 of maxillary palps elongated, **usually twice as long as segment 4, and ringed.** Mesoscutum with pair of small warts. **Foreleg tibia generally with preapical spur.** Fork of R_2 and R_3 in FW originating well beyond radial sector crossvein. Larvae found in various lentic and lotic habitats with slight to moderate current, where they make trumpet retreats. TL 7–13 mm; World 720, NA 75.

NET TUBE CADDISFLIES Family Psychomyiidae   Not illus.

Brown or tan bodies, lacking ocelli. Segment 5 of maxillary palps elongated, usually twice as long as segment 4, and ringed; segments 2 and 3 subequal in length. Scutum possesses warts. Foreleg tibia lacks preapical spur. Larvae live in streams, constructing tube-like retreats, often camouflaged with detritus.
TL 2–5 mm; World 450, NA 14.

XIPHOCENTRONID CADDISFLIES

Family Xiphocentronidae*   Not illus.

Brown or tan bodies lacking ocelli. Scutum with warts. Front tibia without preapical spur. Larvae live in streams, constructing tube-like retreats, often camouflaged with fine sand.
TL 2.5–5 mm; World 145, NA 3.

NETSPINNING CADDISFLIES Family Hydropsychidae

Most are brownish with some **mottling in wings.** Segment 5 of maxillary palps elongated with cross striations. Ocelli absent. Mesoscutum and mesoscutellum lack warts. Female with middle leg often flattened and dilated, and both sexes lack preapical spurs on front tibiae. Larvae live in fast-flowing water, where they construct case-like net retreats. TL 5–19 mm; World 1,600, NA 158.

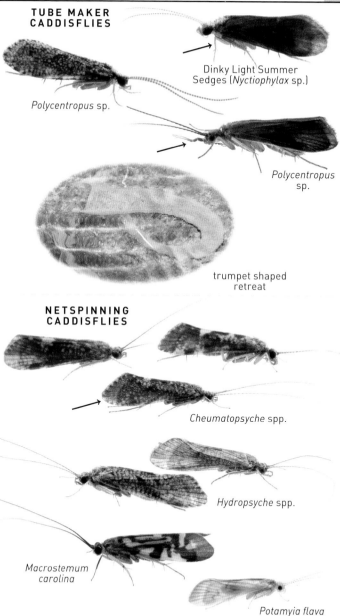

TUBE MAKER CADDISFLIES

Dinky Light Summer Sedges (*Nyctiophylax* sp.)

Polycentropus sp.

Polycentropus sp.

trumpet shaped retreat

NETSPINNING CADDISFLIES

Cheumatopsyche spp.

Hydropsyche spp.

Macrostemum carolina

Potamyia flava

Superfamily Philopotamoidea

FINGERNET CADDISFLIES Family Philopotamidae

Dark, brownish to black bodies with gray wings. Ocelli present. Maxillary palps 5-segmented; **segment 5 twice or more length of segment 4.** Females of *Dolophilodes distinctus* can be winged (summer) or wingless (winter). Larvae restricted to riffles of streams, where they make finger-shaped retreats. TL 5–10 mm; World 1,170, NA 51.

Suborder Integripalpia
Terminal segment of maxillary palps unringed. Larvae build portable cases out of sediment and debris.

Superfamily Glossosomatoidea

LITTLE BLACK CADDISFLIES Family Glossosomatidae

Dark bodied with wings more or less mottled with short antennae. Maxillary palps 5-segmented. Ocelli present. Both sexes lack apical, preapical spurs or both on the front tibiae. Larvae generally occur in cold-water streams with considerable current, where they build a case resembling a tortoiseshell. TL 2–13 mm; World 1,800, NA 80.

Superfamily Hydroptiloidea

MICROCADDISFLIES Family Hydroptilidae

Very small with very hairy, mottled wings, and antennae shorter than FW. Wings narrow with long fringes of hairs; some clubbed. Mesoscutum lacks warts; mesoscutellum is a narrowly triangular ridge posteriorly. Larvae make laterally compressed purse-shaped cases only in the final instar and are found in lentic and lotic habitats with variable flows. TL 1.5–5 mm; World 2,000, NA 271.

PTILOCOLEPID CADDISFLIES Family Ptilocolepidae*   Not illus.

Adults like microcaddisflies. Pronotum with mesal warts set close together. Larvae found in cold springs and seeps. Single genus in NA, *Palaegapetus*. TL 1–5 mm; World 11, NA 3.

Superfamily Leptoceroidea

CALAMOCERA TID CADDISFLIES Family Calamoceratidae

Orange-black or brownish-black bodies and wings. **Antennae are 2–3 times length of body. Maxillary palps 5- or 6-segmented.** Mesoscutum elongated with simple longitudinal row of small warts on each side. Larvae build cases out of large leaf pieces or hollowed twigs in springs and spring runs. TL 12–20 mm; World 120, NA 4.

MORTARJOINT CASEMAKERS Family Odontoceridae

Blackish body and gray wings with light dots. Ocelli absent. Scutellum dome-like with single large wart occupying most of surface. Larvae construct cylindrical, sometimes curved, cases of sand; found in riffles of swift streams. TL ~12 mm; World 100, NA 13.

PHILOPOTAMOIDEA • GLOSSOSOMATOIDEA • HYDROPTILOIDEA

FINGERNET CADDISFLIES

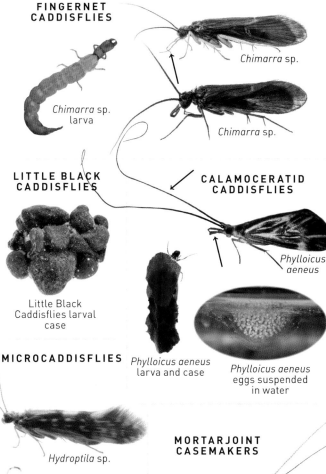

Chimarra sp.

Chimarra sp. larva

Chimarra sp.

LITTLE BLACK CADDISFLIES

CALAMOCERATID CADDISFLIES

Little Black Caddisflies larval case

Phylloicus aeneus

MICROCADDISFLIES

Phylloicus aeneus larva and case

Phylloicus aeneus eggs suspended in water

Hydroptila sp.

MORTARJOINT CASEMAKERS

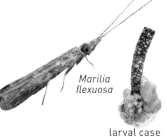

Hydroptila sp.

Marilia flexuosa

unknown species

larval case

LONG-HORNED CADDISFLIES Family Leptoceridae
Elongated, slender, and often pale with **long antennae** (1 1/2 to 3 times as long as wings) and slender. **Maxillary palps 5-segmented.** Scutum with warts scattered in wide rows on either side. Larvae make elongated cases out of various materials and are found in a variety of habitats. TL 5–18 mm; World 2,000, NA 116.

HOOD CASEMAKERS Family Molannidae
Brownish-gray wings that may be mottled. Maxillary palps stout, hairy, and 5-segmented with the first 2 segments short. Middle legs with pair of preapical spurs. Rest with wings curled around body, which is held at an angle above the substrate. Larvae make shield-shaped cases of sand grains and are usually found in small springs with sandy substrate. TL 10–16 mm; World 40, NA 7.

Superfamily Rhyacophiloidea

FREE-LIVING CADDISFLIES Family Rhyacophilidae   Not illus.
Dark bodies and wings mottled, with short antennae. Ocelli present. Tibia of forelegs with apical and preapical spurs. Larvae don't make cases and are predaceous; found in cool, clean streams. TL 7–25 mm; World 753, NA 127.

HYDROBIOSID CADDISFLIES Family Hydrobiosidae* Not illus.
Similar to Rhyacophilidae, representing the southern hemisphere Gondwanan equivalent. Larvae are free-living and predaceous. Single genus in NA, *Atopsyche*, found in the southwest. TL 8–14 mm; World 400, NA 3.

Superfamily Sericostomatoidea

BERAEID CADDISFLIES Family Beraeidae*   Not illus.
Small, brown; scutum lacks warts. Middle and hind tarsi with crown of black apical spines; first tarsal segment with preapical spines. Larvae construct smooth, curved sand-grain cases; found in seepage habitats. Single genus in NA, *Beraea*, found in the northeast south to GA. TL ~5 mm; World 50, NA 3.

SNAIL-CASE CADDISFLIES Family Helicopsychidae
Straw colored with wings mottled brown. Scutellum with pair of narrow transverse warts. Anterior margin of HW with row of modified hairs in basal half and slight concavity in distal half. Larvae occur in a variety of habitats, where they use small rock particles to construct case that resembles snail shell. Single genus, *Helicopsyche*, in NA. TL 5–8 mm; World 230, NA 4.

LEPTOCEROIDEA • RHYACOPHILOIDEA • SERICOSTOMATOIDEA

LONG-HORNED CADDISFLIES

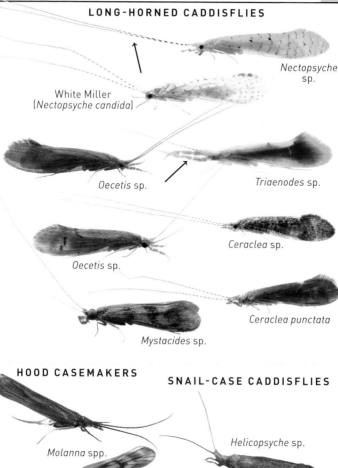

Nectopsyche sp.

White Miller
(*Nectopsyche candida*)

Oecetis sp.

Triaenodes sp.

Ceraclea sp.

Oecetis sp.

Ceraclea punctata

Mystacides sp.

HOOD CASEMAKERS

SNAIL-CASE CADDISFLIES

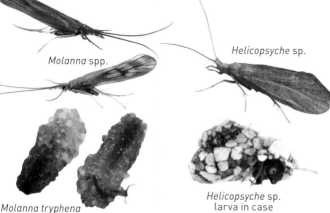

Helicopsyche sp.

Molanna spp.

Molanna tryphena
larva in case

Helicopsyche sp.
larva in case

BUSHTAILED CADDISFLIES

Family Sericostomatidae*   Not illus.

Mesonotum with deep medial groove. Single pair of pronotal and mesoscutal warts. Larvae are found in lakes and streams, where they construct tapered, short, curved cases of sand grains. TL ~15 mm; World 90, NA 16.

Superfamily Limnephiloidea

EARLY SMOKY WING SEDGES Family Apataniidae   Not illus.

Closely related to limnephilids. Maxillary palps in male 3-segmented; females are 5-segmented. Larvae distinctive in that the third thoracic segment lacks plates but has row of transverse bristles. TL 10–25 mm; World 180, NA 34.

GOERID CADDISFLIES Family Goeridae*   Not illus.

Closely related to limnephilids. Maxillary palps in male 3-segmented; females are 5-segmented. Larvae with second thoracic segment distinctive, bearing 4 or more plates. TL 10–25 mm; World 160, NA 12.

NORTHERN CADDISFLIES Family Limnephilidae

Brownish or tan with narrow FW that are often mottled or patterned. Ocelli present or absent. Maxillary palps in male 3-segmented; females are 5-segmented. Forelegs with no or 1 tibial spur. Larvae construct cases from a variety of materials, including plants, minerals, and snail shells. TL 10–31 mm; World 826, NA 234.

ROSSIANID CADDISFLIES Family Rossianidae*   Not illus.

Restricted to western NA. Larvae found in organic muck of spring seeps or stream gravel deposits under moss. Build stout, slightly curved cases out of rock fragments. TL ~7 mm; World 2, NA 2.

STONECASE CADDISFLIES Family Uenoidae*   Not illus.

Three ocelli present. Maxillary palps in male 3-segmented; females are 5-segmented; segment 2 slender and longer than 1. Middle tibia with 1 or no preapical spurs. Hindwing with row of stout, apically hooked setae along anterior margin. Larvae construct stout cases of rock particles. Restricted to western NA and emerge in the fall. TL 7–25 mm; World 80, NA 18.

THREMMATID CADDISFLIES Family Thremmatidae   Not illus.

Similar to the Uenoidae. Adults are yellow and brown, often with a pattern of large pale diamonds along middle. Emerge in the fall. Larvae construct stout cases of rock particles. TL 7–25 mm; World 50, NA 33.

BIZARRE CADDISFLIES

Lepidostoma sp.

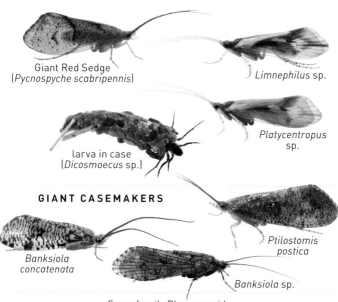

Giant Red Sedge
(*Pycnospyche scabripennis*)

Limnephilus sp.

Platycentropus sp.

larva in case
(*Dicosmoecus* sp.)

GIANT CASEMAKERS

Banksiola concatenata

Ptilostomis postica

Banksiola sp.

Superfamily Phryganeoidea

HUMPLESS CASEMAKER CADDISFLIES

Family Brachycentridae   Not illus.
Ocelli absent. Maxillary palps usually 3-segmented in males and pressed up against head; 5-segmented in females. Scutum with pair of moderately separated warts; scutellum with pair of larger warts. Tibia of middle leg with irregular row of spines. Larvae occur in a variety of lotic habitats and generally construct 4-sided logcabin–like cases. TL 5–10 mm; World 100, NA 34.

BIZARRE CADDISFLIES Family Lepidostomatidae

Similar to the Brachycentridae, but maxillary palps sometimes appearing 1-segmented and tibia of middle leg lacking row of spines. Larvae occur in small, cold streams and occasionally along lakeshores where they construct a diverse array of cases.
TL 7–25 mm; World 136, NA 72.

GIANT CASEMAKERS Family Phryganeidae

Large, often with patterned gray or brown wings. Ocelli are present. Maxillary palps in male are 4-segmented; female are 5-segmented. Tibia of foreleg bears 2 or more spurs. Larvae are common around marshes, ponds, and lakes but found in flowing water as well; cases are made of strips of vegetation glued together longitudinally. TL 10–30 mm; World 36, NA 28.

377

Gulf Fritillary
(*Agraulis vanillae*)
Nymphalidae

Imperial Moth
Caterpillar
(*Eacles imperialis*)
Saturniidae

BUTTERFLIES & MOTHS
ORDER LEPIDOPTERA

Derivation of Name
- Lepidoptera: *lepidos*, scaled; *pteron*, winged. Referring to the scales on the wings.

Identification
- 2–255 mm Wing Span (WS).
- Mouthparts form a proboscis, antennae of butterflies usually long and slender or clubbed at end; antennae of moths is usually filiform, setaceous, or plumose; HW and FW covered in scales; HW smaller than FW.

Classification
- Butterflies and moths have historically been separated into 2 suborders. However, in the light of recent molecular evidence, the family tree for Lepidoptera has placed butterflies in the middle of moths, suggesting they are day-flying moths.
- Some authors refer to butterflies as Rhopalocera, referring to the clubbed antennae.
- For convenience, moths are often artificially broken into Macrolepidoptera (large moths) and Microlepidoptera (small moths).
- World: 182,500 species (moths: 165,000; butterflies: 17,500); North America: 12,000 species (moths: 11,000; butterflies: 866).

Range
- Found on every continent except Antarctica.

Similar orders
- Trichoptera (p. 364): few or no scales on wings although can have hairs; no coiled proboscis.
- Hymenoptera (p. 236): no scales on wings; has chewing mouthparts.

Vagabond Crambus
(*Agriphila vulgivagellus*)
Crambidae

Food
- Adults feed on nectar and other fluids.
- Caterpillars feed on various host plants.

Behavior
- Caterpillars feed in various ways: some feed on edge of leaf, some skeletonize the leaf, and some eat holes in leaves; others are borers in stems and wood. Some create silk from modified salivary glands.
- Adults of different species can be identified by specific flight patterns.

Life Cycle
- Holometabolous.
- Larvae or caterpillars typically with 5 instars. Have long and cylindrical bodies with 13 segments (3 thoracic + 10 abdominal); have 3 pairs of legs on the thorax near their well-developed head; prolegs generally on S3–6 and S10.
- The pupal stage is a chrysalis in butterflies, and in moths the larva often spins a cocoon around itself, then sheds its skin, transforming into a pupa inside the cocoon.
- Generations per year(s) vary; often overwinter as larva or pupa.

Importance to Humans
- Some species can be pests of crops, stored grains, or fabrics.
- Most species are pollinators.
- Lepidoptera produce natural silk.

Collecting and Preserving
Adults
- Butterflies collected with a lightweight aerial net and at baited traps.
- Many moths collected at lights; others collected on flowers.
- For larger specimens, inject moth or butterfly with alcohol or ethyl acetate in thorax, being careful not to damage scales on wings, or place live specimen in envelope and then place envelope in kill jar, or place in freezer for several hours. When movement has stopped, carefully remove specimen and spread using spreading board; string or wax paper can be used to hold wings in place to dry (see p. 36).
- Micromoths can also be spread (see https://tinyurl.com/curatingmicros)

Immatures
- Collect live caterpillars. Bring water to a boil, remove from heat, and drop live caterpillar in water. Remove from water after 1 minute. Permanently store in 80% ethanol.

Resources
- Butterflies and Moths of North America—butterfliesandmoths.org
- Moth Photographers Group—mothphotographersgroup.msstate.edu
- *Peterson Field Guide to Moths of Northeastern North America* by Beadle and Leckie, 2012
- *Peterson Field Guide to Moths of Southeastern North America* by Leckie and Beadle, 2018
- *Butterflies of North America* by Brock and Kaufman, 2003
- *Caterpillars of Eastern North America: A Guide to Identification and Natural History* by Wagner, 2005
- *Key to Canadian Families* by J. Dombroskie—doi:10.3752/cjai.2011.17

Head Structure

Wing Coupling

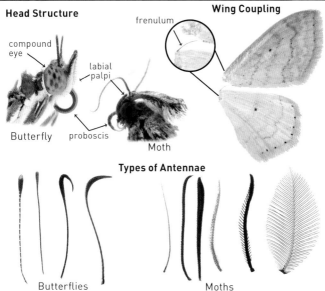

compound eye

labial palpi

frenulum

Butterfly

proboscis

Moth

Types of Antennae

Butterflies

Moths

Butterflies usually have clubbed and hooked antennae, while moths have filiform, bipectinate, & variously modified antennae.

Wing Venation

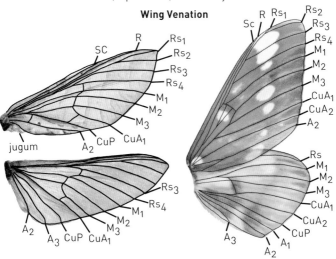

SC

R

Rs_1

Rs_2

Rs_3

Rs_4

M_1

M_2

M_3

CuP CuA_1

A_2

jugum

Rs_3

Rs_4

M_1

M_2

M_3 CuP CuA_1

A_2

A_3

Sc   R Rs_1 Rs_2

Rs_3

Rs_4

M_1

M_2

M_3

CuA_1

CuA_2

A_2

Rs

M_1

M_2

M_3

CuA_1

CuA_2

CuP

A_3 A_1

A_2

Primitive homoneurous venation (similar in both wings) of *Phassus giganteus* (Hepialidae) on the left and more derived heteroneurous venation, differing between wings, of *Citheronia regalis* (Saturniidae) on the right.

381

SILHOUETTES OF COMMON MOTH FAMILIES

Tiger & Lichen Moths
(Erebidae: Arctiinae)

Tussock Moths
(Erebidae: Lymantriinae)

Crambid Snout Moths
(Crambidae)

Twirler Moths
(Gelechiidae)

Slug Caterpillar Moths
(Limacodidae)

Tent Caterpillar &
Lappet Moths
(Lasiocampidae)

Geometrid Moths
(Geometridae)

Owlet Moths
(Noctuidae)

Fairy Moths
(Adelidae)

Casebearer Moths
(Coleophoridae)

Prominent Moths
(Notodontidae)

Pyralid Moths
(Pyralidae)

Metalmark Moths
(Choreutidae)

Tortricid Moths
(Tortricidae)

Fringe-tufted Moths
(Epermeniidae)

Clearwing Moths
(Sesiidae)

Leaf Skeletonizer
Moths (Zygaenidae)

Many-plume Moths
(Alucitidae)

American Silkworm Moths
(Apatelodidae)

Hooktip Moths
(Drepanidae)

Plume Moths
(Pterophoridae)

Silk Moths
(Saturniidae)

Sphinx Moths
(Sphingidae)

Leaf Blotch Miner Moths
(Gracillariidae)

Sack-bearer Moths
(Mimallonidae)

Euteliid Moths
(Euteliidae)

Swallowtail Moths
(Uraniidae)

KEY 1 TO BUTTERFLY & MOTH SUPERFAMILIES

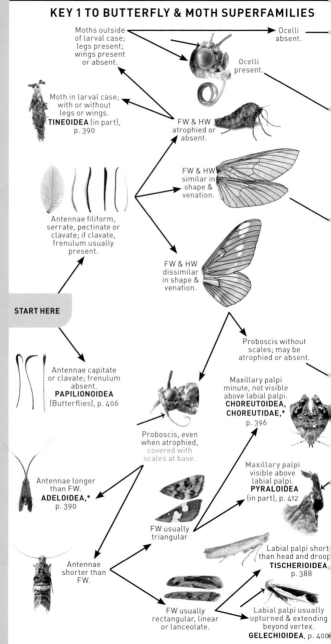

Moths outside of larval case; legs present; wings present or absent.

Ocelli absent.

Ocelli present.

Moth in larval case; with or without legs or wings. **TINEOIDEA** (in part), p. 390

FW & HW atrophied or absent.

FW & HW similar in shape & venation.

Antennae filiform, serrate, pectinate or clavate; if clavate, frenulum usually present.

FW & HW dissimilar in shape & venation.

START HERE

Proboscis without scales; may be atrophied or absent.

Antennae capitate or clavate; frenulum absent. **PAPILIONOIDEA** (Butterflies), p. 406

Maxillary palpi minute, not visible above labial palpi. **CHOREUTOIDEA, CHOREUTIDAE,** * p. 396

Proboscis, even when atrophied, covered with scales at base.

Maxillary palpi visible above labial palpi. **PYRALOIDEA** (in part), p. 412

Antennae longer than FW. **ADELOIDEA,** * p. 390

FW usually triangular

Labial palpi short than head and droop **TISCHERIOIDEA** p. 388

Antennae shorter than FW.

FW usually rectangular, linear or lanceolate.

Labial palpi usually upturned & extending beyond vertex. **GELECHIOIDEA**, p. 400

384

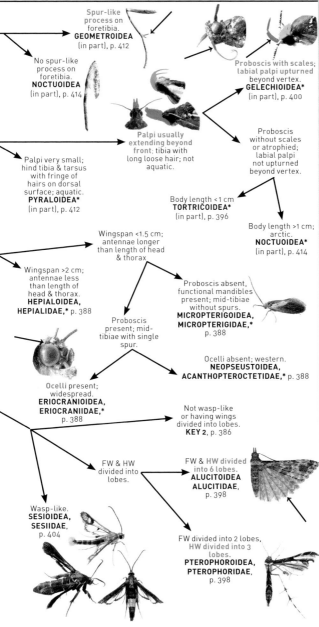

Spur-like process on foretibia. **GEOMETROIDEA** (in part), p. 412

No spur-like process on foretibia. **NOCTUOIDEA** (in part), p. 414

Proboscis with scales; labial palpi upturned beyond vertex. **GELECHIOIDEA*** (in part), p. 400

Palpi usually extending beyond front; tibia with long loose hair; not aquatic.

Proboscis without scales or atrophied; labial palpi not upturned beyond vertex.

Palpi very small; hind tibia & tarsus with fringe of hairs on dorsal surface; aquatic. **PYRALOIDEA*** (in part), p. 412

Body length <1 cm **TORTRICOIDEA*** (in part), p. 396

Body length >1 cm; arctic. **NOCTUOIDEA*** (in part), p. 414

Wingspan <1.5 cm; antennae longer than length of head & thorax

Wingspan >2 cm; antennae less than length of head & thorax. **HEPIALOIDEA, HEPIALIDAE,*** p. 388

Proboscis absent, functional mandibles present; mid-tibiae without spurs. **MICROPTERIGOIDEA, MICROPTERIGIDAE,*** p. 388

Proboscis present; mid-tibiae with single spur.

Ocelli absent; western. **NEOPSEUSTOIDEA, ACANTHOPTEROCTETIDAE,*** p. 388

Ocelli present; widespread. **ERIOCRANIOIDEA, ERIOCRANIIDAE,*** p. 388

Not wasp-like or having wings divided into lobes. **KEY 2,** p. 386

FW & HW divided into lobes.

FW & HW divided into 6 lobes. **ALUCITOIDEA ALUCITIDAE,** p. 398

Wasp-like. **SESIOIDEA, SESIIDAE,** p. 404

FW divided into 2 lobes, HW divided into 3 lobes. **PTEROPHOROIDEA, PTEROPHORIDAE,** p. 398

KEY 2 TO BUTTERFLY & MOTH SUPERFAMILIES

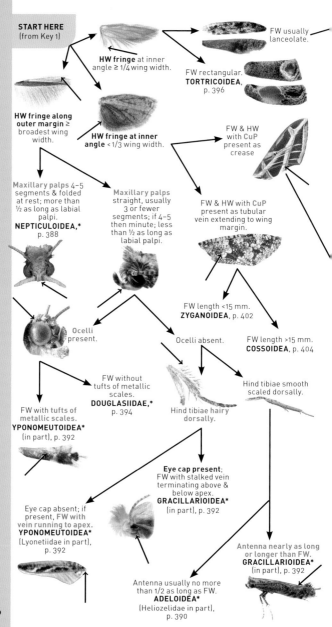

START HERE
(from Key 1)

HW fringe at inner angle ≥ 1/4 wing width.

FW usually lanceolate.

FW rectangular.
TORTRICOIDEA,
p. 396

HW fringe along outer margin ≥ broadest wing width.

HW fringe at inner angle < 1/3 wing width.

FW & HW with CuP present as crease

Maxillary palps 4–5 segments & folded at rest; more than ½ as long as labial palpi.
NEPTICULOIDEA,*
p. 388

Maxillary palps straight, usually 3 or fewer segments; if 4–5 then minute; less than ½ as long as labial palpi.

FW & HW with CuP present as tubular vein extending to wing margin.

FW length <15 mm.
ZYGANOIDEA, p. 402

Ocelli present.

Ocelli absent.

FW length >15 mm.
COSSOIDEA, p. 404

FW without tufts of metallic scales.
DOUGLASIIDAE,*
p. 394

Hind tibiae smooth scaled dorsally.

FW with tufts of metallic scales.
YPONOMEUTOIDEA*
(in part), p. 392

Hind tibiae hairy dorsally.

Eye cap present; FW with stalked vein terminating above & below apex.
GRACILLARIOIDEA*
(in part), p. 392

Eye cap absent; if present, **FW with vein running to apex.**
YPONOMEUTOIDEA*
(Lyonetiidae in part), p. 392

Antenna nearly as long or longer than FW.
GRACILLARIOIDEA*
(in part), p. 392

Antenna usually no more than 1/2 as long as FW.
ADELOIDEA*
(Heliozelidae in part), p. 390

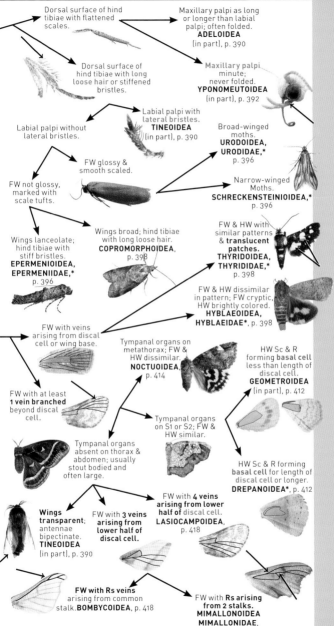

Dorsal surface of hind tibiae with flattened scales.

Maxillary palpi as long or longer than labial palpi; often folded.
ADELOIDEA
(in part), p. 390

Dorsal surface of hind tibiae with long loose hair or stiffened bristles.

Maxillary palpi minute; never folded.
YPONOMEUTOIDEA
(in part), p. 392

Labial palpi with lateral bristles.
TINEOIDEA
(in part), p. 390

Labial palpi without lateral bristles.

Broad-winged moths.
URODOIDEA, URODIDAE,*
p. 396

FW glossy & smooth scaled.

FW not glossy, marked with scale tufts.

Narrow-winged Moths.
SCHRECKENSTEINIOIDEA,*
p. 396

Wings lanceolate; hind tibiae with stiff bristles.
EPERMENIOIDEA, EPERMENIIDAE,*
p. 396

Wings broad; hind tibiae with long loose hair.
COPROMORPHOIDEA,
p. 398

FW & HW with similar patterns & **translucent patches.**
THYRIDOIDEA, THYRIDIDAE,*
p. 398

FW & HW dissimilar in pattern; FW cryptic, HW brightly colored.
HYBLAEOIDEA, HYBLAEIDAE*, p. 398

FW with veins arising from discal cell or wing base.

Tympanal organs on metathorax; FW & HW dissimilar.
NOCTUOIDEA,
p. 414

HW Sc & R forming **basal cell** less than length of discal cell.
GEOMETROIDEA
(in part), p. 412

FW with at least **1 vein branched** beyond discal cell.

Tympanal organs on S1 or S2; FW & HW similar.

Tympanal organs absent on thorax & abdomen; usually stout bodied and often large.

HW Sc & R forming **basal cell** for length of discal cell or longer.
DREPANOIDEA*, p. 412

Wings transparent; antennae bipectinate.
TINEOIDEA
(in part), p. 390

FW with 3 veins arising from lower half of discal cell.

FW with 4 veins arising from lower half of discal cell.
LASIOCAMPOIDEA,
p. 418

FW with Rs veins arising from common stalk. **BOMBYCOIDEA,** p. 418

FW with Rs arising from 2 stalks.
MIMALLONOIDEA MIMALLONIDAE,
p. 418

387

ERIOCRANIOIDEA • NEOPSEUSTOIDEA • HEPIALIOIDEA • NEPTICULOIDEA

Superfamily Micropterigoidea

MANDIBULATE ARCHAIC MOTHS Family Micropterigidae*
Small and dark with **fringe on wings and often having metallic scales.** Mandibulate mouthparts instead of proboscis. Mid-tibia lacks spurs. FW Sc forked near middle. Adults feed on pollen; larvae feed on moss and liverworts. WS 2–13 mm; World 110, NA 3.

Superfamily Eriocranioidea

ERIOCRANIID MOTHS Family Eriocraniidae*
Similar to clothes moths but have metallic scales on wings. Have a single spur on middle tibia. Ocelli present, M_1 in HW and FW not stalked with $R_{4,5}$. Anal veins in FW fused near far edge of wing. Females with horny, piercing ovipositor. Larvae are leaf miners. WS 5–16 mm; World 25, NA 13.

Superfamily Neopseustoidea

ARCHAIC SUN MOTHS Family Acanthopteroctetidae* Not illus.
This group is small and iridescent. They lack ocelli. Anal veins in FW are separated and not fused; M_1 stalked with $R_{4,5}$. Found in western NA. WS 5 mm; World 6, NA 4.

Superfamily Hepialoidea

GHOST MOTHS Family Hepialidae*
FW and HW similar shape, united by jugum. Rs with 3–4 branches. Most gray or brown with silver spots in wing. Adults lack mouthparts and do not feed. WS 23–100 mm; World 500, NA 21.

Superfamily Nepticuloidea
Minute with base of antennae enlarged, forming cap over eye. Larvae are leaf and stem miners.

PIGMY MOTHS Family Nepticulidae*
Eyecap present. Wing venation is reduced; most veins in FW branching off of R. FW is broad with minute spines under scales. WS 1–8 mm; World 862, NA 97.

WHITE EYECAP MOTHS Family Opostegidae*
Large eyecaps. FW venation reduced and unbranched. HW linear and broadly fringed. White with sparse dark markings. WS 2–8 mm; World 100, NA 11.

Superfamily Tischerioidea

TRUMPET LEAF MINER MOTHS Family Tischeriidae*
Wings lanceolate. FW with single R_3 very long extending halfway to base. **Hind tibia with large mass of hair.** Larvae are pests on oak and apple trees in the east. WS 5–13 mm; World 80, NA 50.

MANDIBULATE ARCHAIC MOTHS

Goldcap Moss-eater Moth
(*Epimartyria auricrinella*)

ERIOCRANIID MOTHS

Purplish Birch-miner Moth
(*Eriocrania semipurpurella*)

Chinquapin Leaf-miner Moth
(*Dyseriocrania griseocapitella*)

GHOST MOTHS

Four-spotted Ghost
Moth (*Sthenopis
purpurascens*)

PIGMY MOTHS

*Etainia
sericopeza*

*Ectoedemia
rubifoliella*

WHITE EYECAP MOTHS

*Pseudopostega
cretea*

TRUMPET LEAF MINER MOTHS

*Coptotriche
citrinipennella*

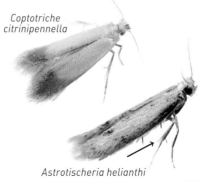

Astrotischeria helianthi

389

Superfamily Adeloidea

Lack ocelli. Ovipositor elongated and piercing. HW with large frenulum. Short proboscis with scales near base.

SHIELD BEARERS Family Heliozelidae*
Maxillary palps 5-segmented. Lanceolate wings. HW stem of Cu not branched; no discal cell. FW dark gray at base with bright yellow, brown or silver markings on distal portions of wings.
WS 2–10 mm; World 123, NA 31.

FAIRY MOTHS Family Adelidae
Small, day-flying moths. **Male antennae can be 2–3 times as long as body.** Basal half of antennae hairy in female. Palps 2–4-seg. Often with iridescent markings. Larvae leaf miners when young, becoming case makers. Wingspread 2–28 mm; World 294, NA 18.

YUCCA MOTHS Family Prodoxidae
Often with white wings. Antennae simple and half the length of rounded FW. Pollinate yuccas by collecting pollen with the palps (see yellow on photos of Prodoxinae), stuffing the pollen ball into the stigma. WS 5–33 mm; World 98, NA 56.

LEAFCUTTER MOTHS Family Incurvariidae
Small dark moths. Antennae half the length of the body. **Rough scales on head.** Folded part of maxillary palps about half as long as width of head. Wings with little reduction in venation. Piercing ovipositor.
WS 5–18 mm; World 51, NA 6.

TRIDENTAFORMID MOTHS Family Tridentaformidae* Not illus.
Males with grappling hook at tip of abdomen bearing rows of broad spines. Previously placed in Incurvariidae or Prodoxidae, but erected as family in 2015. Single species in the family, *Tridentaforma fuscoleuca*, found in western NA. WS 2–10 mm; World 1, NA 1.

Superfamily Tineoidea

Scales on front of the head erect. Lateral bristles on labial palps. Female with pair of slender, ventral pseudapophyses in S10. Most species small, brown and gray.

CLOTHES MOTHS Family Tineidae
Most are dark colors with **tufts of rough scales on head.** Generalized wing venation; R_5 terminating at costa. Many species with whorls of scales on each segment. Larvae feed on fungi and natural fibers, including clothes and carpets. WS 2–25 mm; World 3,000, NA 190.

BAGWORMS Family Psychidae
Larvae have characteristic cases that are easily seen when leaves fall off trees. Adult females are wingless, legless, and worm-like, remaining in larval case. Males usually dark with **pectinate antennae.**
WS 12–38 mm; World 1,350, NA 28.

SHIELD BEARERS

Antispila ampelopsifoliella

YUCCA MOTHS

Prodoxinae

Prodoxus decipiens

CLOTHES MOTHS

Clemens' Grass Tubeworm Moth
(*Acrolophus popeanella*)

Dark-collared Tinea
(*Tinea apicimaculella*)

Clothes Moth
(*Homosetia* sp.)

Evergreen Bagworm Moth
(*Thyridopteryx ephemeraeformis*)

FAIRY MOTHS

*Cauchas
simpliciella*
♀

♂

Southern Longhorn
Moth
(*Adela caeruleella*)

LEAFCUTTER MOTHS

Phylloporia bistrigella

Maple Leafcutter Moth
(*Paraclemensia acerifoliella*)

BAGWORMS

Bagworm Moth
(*Cryptothelea* sp.)

Bagworm
case

Superfamily Tineoidea cont.

DANCING MOTHS Family Dryadaulidae*

Spatula-shaped tip of labial palp. HW venation reduced; M3 absent. FW slender, relatively long and narrow with subacute apices. Male with long, narrow retinaculum and M1 and M2 fused in HW. WS 5–13 mm; World 2,200, NA 2.

MEESSIID MOTHS Family Meessiidae*

FW slender with slightly pointed tip. HW M3 absent and venation reduced. Male genitalia symmetrical, female reproductive organ long, narrow extension 3 times longer than broad basal structure. WS 5–13 mm; World 71, NA 3.

Superfamily Gracillarioidea

Small with lanceolate wings. FW usually lacking accessory cell. Labial palps lack bristles. Caterpillars are usually leaf miners. Pupa lacks abdominal spines.

RIBBED COCOON-MAKER MOTHS Family Bucculatricidae

Small moths with top of **head covered in bristly scales.** Scales partially cover eyes. Narrow, lanceolate wings. HW Rs extending through center of wing. WS 5–16 mm; World 297, NA >100.

LEAF BLOTCH MINERS Family Gracillariidae

At rest, moths stand with head elevated above abdomen with anterior part of wings touching substrate. HW longer than wide. **Antennae usually longer than FW.** FW cell $1R_3$ small or absent. Hind tibia has hairs. WS 2–20 mm; World 297, NA 427.

Superfamily Yponomeutoidea

Males have an expanded eighth pleuron that covers the genitalia and a naked proboscis that separates them from Gelechioidea.

ERMINE MOTHS Family Yponomeutidae

Small, usually brightly patterned. R_5 extends to edge of wing. HW with M_1 widely separated from Rs at origin and M_1 and M_2 not stalked. Larvae feed on apple and cherry trees and create silk webs, feeding on leaves. WS 7–30 mm; World 365, NA 32.

YPSOLOPHID MOTHS Family Ypsolophidae

Small with broad wings, **hooked at tip.** Rs and M_1 stalked in HW. A couple of species are considered pests of winter wheat and rye in Europe and may become a similar pest in NA. WS 10–13 mm; World 160, NA 54.

DIAMONDBACK MOTHS Family Plutellidae

Adults smaller and more slender than Ypsolophidae. **Hold antennae forward at rest.** HW has M_1 and M_2 stalked. Can be a major pest of cruciferous plants. Larvae eat holes in leaves and make silk cocoon attached to leaves. WS 10–16 mm; World 200, NA 22.

BUTTERFLIES & MOTHS

DANCING MOTHS

Hawaiian Dancing Moth
(*Dryadaula terpsichorella*)

MEESSIID MOTHS

Eudarcia eunitariaeella

LEAF BLOTCH MINERS

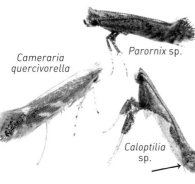

Cameraria quercivorella

Parornix sp.

Caloptilia sp.

RIBBED COCOON-MAKER MOTHS

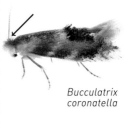

Bucculatrix coronatella

ERMINE MOTHS

American Ermine Moth
(*Yponomeuta multipunctella*)

Zelleria retiniella

YPSOLOPHID MOTHS

Honeysuckle Moth
(*Ypsolopha dentella*)

Euceratia securella

DIAMONDBACK MOTHS

Rhigognostis interrupta

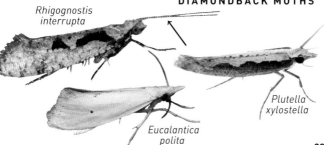

Eucalantica polita

Plutella xylostella

393

YPONOMEUTOIDEA

Superfamily Yponomeutoidea cont.

SEDGE MOTHS Family Glyphipterigidae
Small, diurnal moths with relatively large ocelli. FW pointed; HW usually longer than wide. Costa vein strongly arched. Proboscis naked. Larvae feed on sedges, rushes, and other monocots. WS 5–13 mm; World 500, NA 36.

SUN MOTHS Family Heliodinidae
Small, diurnal, with narrow, broadly fringed HW. Many individuals hold **hind legs elevated over wings at rest.** Many with metallic-colored scales over body and wings. Base of proboscis naked. Head scales closely appressed. FW has only 3–5 simple veins. WS 5–36 mm; World 72, NA 20.

NARROW-WINGED MOTHS Family Bedelliidae*
Small, tan-colored moths with narrow wings and lacking ocelli. R_{3-5} and M stalked in FW. Larvae are leaf miners of Morning Glory. WS 2–10 mm; World 4, NA 2.

LYONET MOTHS Family Lyonetiidae*
Similar to Narrow-winged Moths but **base of antennae modified to form eyecap.** Many species white with metallic scales distally on FW. Larvae are leaf and twig miners. WS 2–13 mm; World 260, NA 22.

SHINY HEAD-STANDING MOTHS Family Argyresthiidae
At rest, stand with abdomen elevated, performing a headstand. Labial palp 3-segmented and curved up. Maxillary palp 1-segmented. Wings have shiny pearl-like reflections. Larvae are miners of mainly conifers. WS 5–16 mm; World 160, NA 52.

TROPICAL ERMINE MOTHS Family Attevidae
Distinctive species in NA, *Atteva aurea*; widespread east of the Rockies. Orange FW with 4 rows of white dots with black borders. WS 23–34 mm; World 48, NA 2.

FALSE ERMINE MOTHS Family Praydidae*
Diverse, rarely encountered group. Labial palp 3-segmented and held straight out from head. Setal sensory structure present on head. WS 5–40 mm; World 26, NA >3.

Superfamily Incertae sedis (undefined)

DOUGLAS MOTHS Family Douglasiidae* Not illus.
FW with tufts of metallic scales. Rs in HW connects M_1 and margin in middle of wing. *Tinagma gaedikei* is diurnal and found on its host plant, Miami mist (*Phacelia purshii*), and has a distinctive cream-colored stripe midway on dark FW. WS 5–13 mm; World 25, NA 7.

SEDGE MOTHS

Carrionflower Moth
(*Acrolepiopsis incertella*)

SUN MOTHS

Lithariapteryx jubarella

Aetole unipunctella

Yellow Nutsedge Moth
(*Diploschizia impigritella*)

LYONET MOTHS

Proleucoptera smilaciella

NARROW-WINGED MOTHS

Morning-glory Leafminer Moth
(*Bedellia somnulentella*)

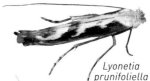

Lyonetia prunifoliella

SHINY HEAD-STANDING MOTHS

Bronze Alder Moth
(*Argyresthia goedartella*)

Cherry Fruit Moth
(*Argyresthia oreasella*)

TROPICAL ERMINE MOTHS

Ailanthus Webworm Moth
(*Atteva aurea*)

FALSE ERMINE MOTHS

Hop-tree Borer Moth
(*Prays atomocella*)

TORTRICOIDEA • GALACTICOIDEA • CHOREUTOIDEA • URODOIDEA • SCHRECKENSTEINIOIDEA • EPERMENIOIDEA

Superfamily Tortricoidea

LEAFROLLER MOTHS Family Tortricidae
Small, usually gray, tan, or brown. **FW usually with square tips and held roof-like over body and appear to have "shoulders" at rest with parallel sides.** FW and HW CuP weakly formed. Larvae roll leaves, feeding on plants or burrowing into them. Many are important pests of fruit trees to evergreens. Family also includes the Mexican Jumping Bean (*Cydia deshaisiana*).
WS 2–18 mm; World 10,000, NA 1,393.

Superfamily Galacticoidea

GALACTICID MOTHS Family Galacticidae*   Not illus.
FW shiny, gray with black spots; HW solid dark gray. Top of head with tuft of hairs. Ocelli present. M_1 and M_2 in HW stalked. Single species in NA; Mimosa Webworm (*Homadaula anisocentra*) was introduced in 1942 and is a major pest of mimosa and Honey Locust. WS 10–15 mm; World 33, NA 1.

Superfamily Choreutoidea

METALMARK MOTHS Family Choreutidae
Day-flying moths similar to leafrollers in appearance, with **broad and square-tipped FW.** Proboscis scaled. Large ocelli. These small moths often mimic jumping spiders. Most have metallic patches on wings. WS <10 mm; World 415, NA 40.

Superfamily Urodoidea

FALSE BURNET MOTHS Family Urodidae*
Dark brown or gray, with wings sometimes appearing greasy; HW translucent. R_4 and R_5 in FW not stalked; M_1 and M_2 in HW stalked. Larvae feed on *Persea*, *Bumelia*, and *Hibiscus*.
WS 12–18 mm; World 80, NA 3.

Superfamily Schreckensteinioidea

BRISTLE-LEGGED MOTHS Family Schreckensteiniidae
FW slender with **pointed triangular tip.** Hind legs held over wings at rest. Hind leg tarsal segments with whorls of bristles. HW with fringe 2–3 times length of membranous part of wing. Larvae feed on sumac and *Rubus*. WS 7–12 mm; World 4, NA 3.

Superfamily Epermenioidea

FRINGE-TUFTED MOTHS Family Epermeniidae*
FW with triangular-shaped group of scales at tip. HW margin with projecting tufts of scales. Hind tibia stiffly bristled with tufts alongside spurs. Larvae of most species miners on parsley (Apiaceae). WS 5–20 mm; World 100, NA 11.

LEAFROLLER MOTHS

Pseudexentera sp.

Garden Tortrix
(*Clepsis peritana*)

Pitch Pine Tip Moth
(*Rhyacionia rigidana*)

Oblique-banded
Leafroller
(*Choristoneura
rosaceana*)

Jack Pine
Budworm
(*Choristoneura
pinus*)

Oak Leafroller
(*Argyrotaenia
quercifoliana*)

Filigreed Chimoptesis
(*Chimoptesis
pennsylvaniana*)

Paralobesia cyclopiana

Robinson's Pelochrista Moth
(*Pelochrista robinsonana*)

METALMARK MOTHS

*Tebenna
carduiella*

Everlasting
Tebenna Moth
(*Tebenna gnaphaliella*)

Ofatulena duodecemstriata

FRINGE-TUFTED MOTH

Epermenia albapunctella

BRISTLE-
LEGGED
MOTHS

Blackberry
Skeletonizer
(*Schreckensteinia
festaliella*)

FALSE BURNET MOTHS

Bumelia Webworm Moth
(*Urodus parvula*)

Superfamily Alucitoidea

MANY-PLUME MOTHS Family Alucitidae

Six feather-like plumes in each wing displayed when at rest with wings spread. Larvae are pink and borers of honeysuckle (*Lonicera* spp.) and snowberry (*Symphoricarpos* spp.).
WS 7–16 mm; World 146, NA 3.

Superfamily Pterophoroidea

PLUME MOTHS Family Pterophoridae

Wings slender, held at right angles when at rest, forming a T shape appearance. FW split into 2 feather-like divisions; HW with **3 feather-like divisions.** Legs with prominent spines. Species are nocturnal and regularly seen at lights. Larvae feed on various plants. WS 12–40 mm; World 1,000, NA 147.

Superfamily Copromorphoidea

Broad-winged, medium-sized moths; usually camouflaged. Proboscis naked. Enlarged spiracles on prothorax and S8.

TROPICAL FRUITWORM MOTHS Family Copromorphidae

Tropical group occurring only from BC to Mexico and east to TX in NA. Gray or brown with lanceolate **FW coming to point.** All 3 branches of M are present in HW. WS 10–38 mm; World 55, NA 5.

FRUITWORM MOTHS Family Carposinidae

Gray or brown with lanceolate **FW coming to point.** Raised scale tufts on FW, sometimes only in females. No M_2 vein and usually no M_1 vein. Larvae bore into fruit trees.
WS 7–20 mm; World 275, NA 11.

Superfamily Hyblaeoidea

TEAK MOTHS Family Hyblaeidae*   Not illus.

Tropical group with single species, *Hyblaea puera*, in NA (AZ, FL, TX). Gray FW; HW with 2 lateral reddish bands. Similar to Noctuidae but with 2 distinct anal veins in FW. Males with hair pencil on hind leg. Larvae feed on *Crescentia cujete, Spatodea campanulata, Tabebuia heterophylla*, and *Petitia domingensis*.
WS 22–30 mm; World 18, NA 1.

Superfamily Thyridoidea

WINDOW-WINGED MOTHS Family Thyrididae*

Small, dark-colored day-flying moths; most with **clear spots in wings.** Color pattern on both wings similar. All branches of R present, arising from discal cell. Some larvae burrow into stems and can induce galls. WS 15–45 mm; World >1,000, NA 11.

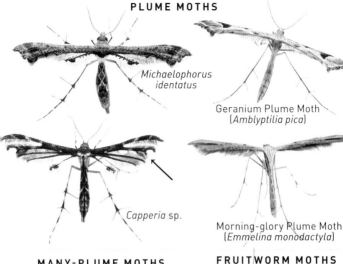

PLUME MOTHS

Michaelophorus identatus

Geranium Plume Moth
(*Amblyptilia pica*)

Capperia sp.

Morning-glory Plume Moth
(*Emmelina monodactyla*)

MANY-PLUME MOTHS

Montana Six-plume Moth
Alucita montana

FRUITWORM MOTHS

Crescent-marked Bondia
(*Bondia crescentella*)

TROPICAL FRUITWORM MOTHS

Lotisma trigonana

WINDOW-WINGED MOTH

Meskea dyspteraria

Spotted Thyris Moth
(*Thyris maculata*)

Superfamily Gelechioidea

Many with large, curved labial palps reaching over head. Proboscis densely scaled. Maxillary palps scaled and folded over proboscis.

GRASS MINER MOTHS Family Elachistidae

Usually gray or black with white markings. Antennal scape forming eyecap. HW without median vein in discal cell. Pupae have knob between segments 5 and 6 and 6 and 7. Larvae are leaf miners of grasses and sedges. WS 5–23 mm; World 545, NA 337.

TWIRLER MOTHS Family Gelechiidae

FW narrowly rounded or pointed at end. HW in the shape of a trapezoid. **Labial palps with segment 2 bearing tufted scales and segment 3 tapered over head.** FW with R_4 and R_5 stalked; R_5 reaches margin of wing. Hind tibia covered with long hair-like scales. WS 2.5–25 mm; World >4,600, NA 886.

AUTOSTICHID MOTHS Family Autostichidae*

Abdominal tergites with transverse band of spines. Female retinaculum consisting of a group of anteriorly directed scales between Sc and R. Larvae feed on rotting plant tissue. WS 7–20 mm; World >300, NA 25.

BATRACHEDRID MOTHS Family Batrachedridae*   Not illus.

Slender wings with long fringes wrapped tightly around body when at rest. HW broadly fringed. Group contains the palm leaf skeletonizers; others feed on poplar and willow. WS 17–30 mm; World 136, NA 24.

SCAVENGER MOTHS Family Blastobasidae

Small, slender, and dull colored. HW narrower than FW. FW slightly thickened along anterior margin. **Head scales project down over face.** Some larvae live inside acorns that have been hollowed out by acorn weevils. Others are associated with various detritus left by other insects. WS 7–18 mm; World 430, NA 17.

CASEBEARER MOTHS Family Coleophoridae

Wings lanceolate; usually solidly colored with streaks of other colors. **Base of antennae thickly scaled.** FW Cu_1 and Cu_2 very short. Hind tibia with rough hair-like scales. Larvae live in cases made out of frass and leaf material and carry the case around on their back. Many are associated with fruit trees. WS 2–28 mm; World 1,400, NA 300.

COSMET MOTHS Family Cosmopterigidae

Labial palps long and curved upward. FW narrow, HW more so. Wings often brightly marked. FW R_4 and R_5 stalked; R_5 running to edge of wing. Generalist feeders, including nuts, seeds, and galls. WS 5–28 mm; World 1,628, NA 181.

GRASS MINER MOTHS

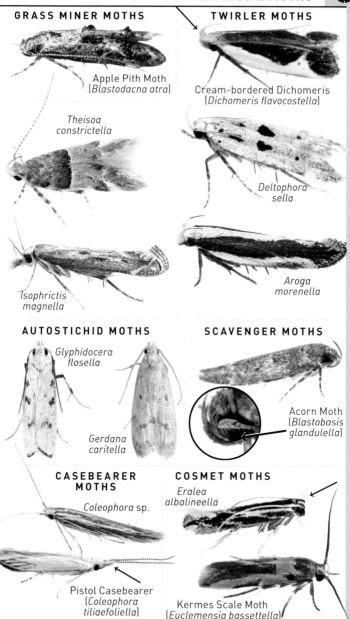

Apple Pith Moth
(*Blastodacna atra*)

Theisoa constrictella

Isophrictis magnella

TWIRLER MOTHS

Cream-bordered Dichomeris
(*Dichomeris flavocostella*)

Deltophora sella

Aroga morenella

AUTOSTICHID MOTHS

Glyphidocera flosella

Gerdana caritella

SCAVENGER MOTHS

Acorn Moth
(*Blastobasis glandulella*)

CASEBEARER MOTHS

Coleophora sp.

Pistol Casebearer
(*Coleophora tiliaefoliella*)

COSMET MOTHS

Eralea albalineella

Kermes Scale Moth
(*Euclemensia bassettella*)

401

Superfamily Gelechioidea cont.

DEPRESSARIID MOTHS Family Depressariidae
Small and flattened with broad, **apically rounded wings.** Labial palps upturned. CuP present in FW; R_4 and R_5 stalked or coalesced for their length. WS 10–32 mm; World 2,300, NA 111.

MOMPHID MOTHS Family Momphidae
Small moths, with some species feeding on evening primroses and other relatives. **Wings long and narrow, sharply pointed.** WS 7–18 mm; World 60, NA 30.

CONCEALER MOTHS Family Oecophoridae
Some brightly colored. FWs, oval and come to a point with complete venation. **Smooth scales on head**. Larvae feed on dead plants and fungus concealed in a leaf or webbing. WS 2–30 mm; World 3,150, NA 37.

LANCE-WING MOTHS Family Pterolonchidae*
Maxillary palps absent; labial palps 3-segmented extending forward. Proboscis reduced. Wings lanceolate. FW with vein in discal cell. Larvae are root borers. WS 22–28 mm; World 30, NA 4.

FLOWER MOTHS Family Scythrididae
Brownish diurnal moths attracted to composites (Asteraceae). Head with smooth scales. Maxillary palps are 4-segmented; labial palps 3-segmented and curved upward. FW lanceolate; R_1 stalked beyond middle of discal cell. WS 2–22 mm; World 669, NA 43.

STATHMOPODID MOTHS Family Stathmopodidae*   Not illus.
Can have colorful FW. Hind tibia with rows of long scales; held off to the side of body or in the air at rest. WS 5–30 mm; World 100, NA 2.

LECITHOCERID MOTHS Family Lecithoceridae*   Not illus.
Stout antennae with lower edge serrate. Labial palpi long, curved upward. Maxillary palpi small. Reduced genital plate on the nineth abdominal tergum. Recently split from Oecophoridae. WS 2–30 mm; World 900, NA 2.

Superfamily Zygaenoidea
Adults usually woolly with stout bodies and bipectinate antennae. Late-instar larva have retractable head.

PLANTHOPPER PARASITES Family Epipyropidae*
Fulgoraecia exigua larvae are parasites of planthoppers (Fulgoroidea). Larva feeds on dorsal side of abdomen, attaching itself to underside of wings. Adults are dark brown/black to purple. WS 2–15 mm; World 32, NA 1.

DEPRESSARIID MOTHS

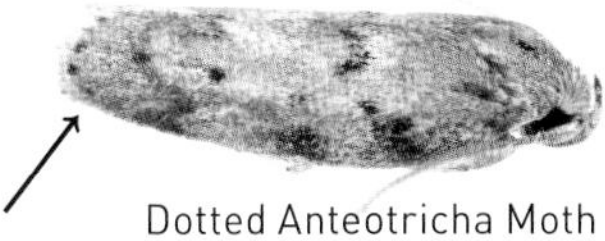

Dotted Anteotricha Moth
(*Antaeotricha humilis*)

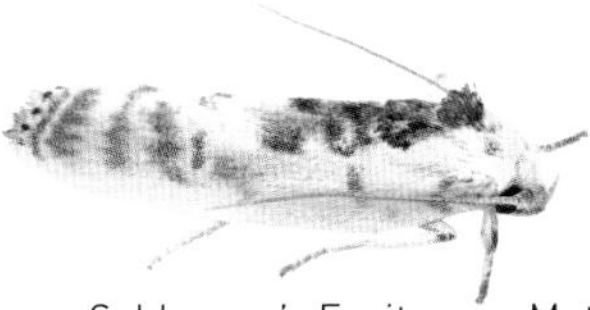

Schlaeger's Fruitworm Moth
(*Antaeotricha schlaegeri*)

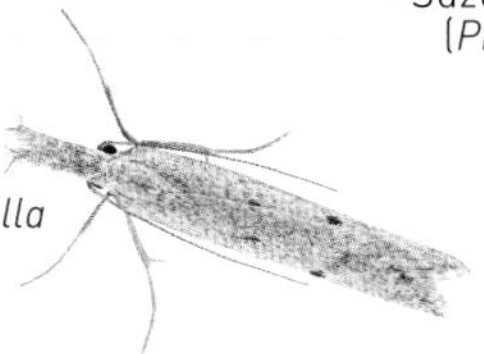

Pleurota albastrigulella

MOMPHID MOTHS

Mompha sp.

CONCEALER MOTHS

Orange-headed Epicallima
(*Epicallima argenticinctella*)

Suzuki's Promalactis Moth
(*Promalactis suzukiella*)

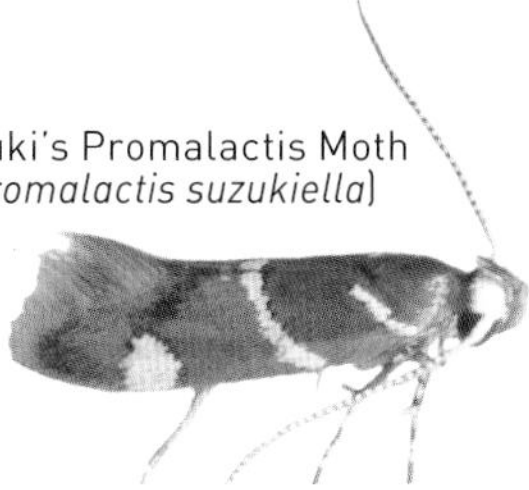

LANCE-WING MOTHS

Exclamation Moth
(*Homaledra heptathalama*)

FLOWER MOTHS

Scythris sp.

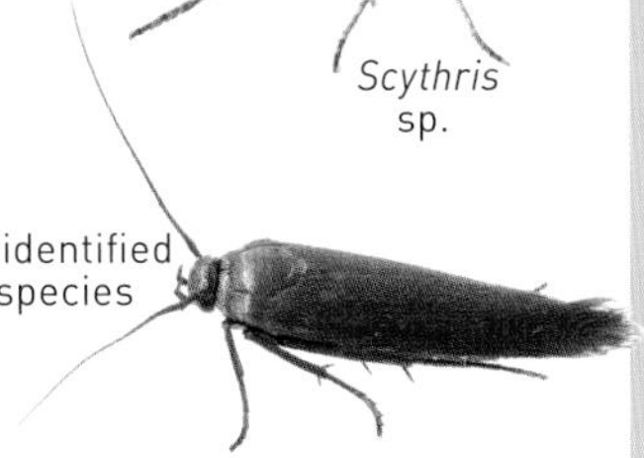

unidentified species

PLANTHOPPER PARASITES

Infected Planthopper

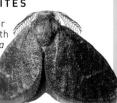

Planthopper Parasite Moth
(*Fulgoraecia exigua*)

TWIRLER MOTHS & RELATIVES • PLANTHOPPER PARASITES

403

ZYGAENOIDEA • SESIOIDEA • COSSOIDEA

Superfamily Zygaenoidea cont.

FLANNEL MOTHS Family Megalopygidae
Adults usually yellow to brown with broad wings and no markings.
Dense scales mixed with **fine hairs appearing woolly.** Larvae with
dense hairs and stinging spines; also with 7 pairs of prolegs.
WS 22–40 mm; World 263, NA 11.

SLUG CATERPILLAR MOTHS Family Limacodidae
Adults stout, typically brownish with irregular spots of color. **Wings
rounded.** Head small and retracted, with bipectinate antennae
in males. CuP is complete in both FW and HW. Abdomen often
elevated above wings at rest. Larvae are short and have short
thoracic legs and no prolegs so they move and look like a slug.
WS 12–44 mm; World 1,000, NA 50.

DALCERID MOTHS Family Dalceridae*   Not illus.
Only 1 species in NA (*Dalcerides ingenita*) known from TX to AZ.
Woolly moth with orange coloring, similar to flannel moths. FW
triangular, twice as long wide. WS 17–25 mm; World 80, NA 1.

TROPICAL BURNET MOTHS Family Lacturidae*
Southern species with white FW and black spots. HW, legs, and
head markings are distinctly pink, sometimes fading to orange
in older specimens. Larval food plant Saffron Plum (*Sideroxylon
celastrinum*) in FL. WS 15–25 mm; World 138, NA 6.

BURNET MOTHS Family Zygaenidae
Day-flying moths. Usually with black or dark gray wings and often
metallic scales. Many with orange or brightly marked prothorax.
Wings are 3 or more times as long as wide.
WS 7–28 mm; World 1,000, NA 25.

Superfamily Sesioidea

CLEARWING MOTHS Family Sesiidae
Often resemble bees and wasps, but usually lacking restriction
of abdomen and have different antennae. **One or both wings lack
scales.** FW are 4 times as long as wide, with anal veins reduced.
HW broad with well-developed anal veins. Larvae bore into roots
and trunks. WS 12–49 mm; World 1,452, NA 25.

Superfamily Cossoidea

CARPENTER AND LEOPARD MOTHS Family Cossidae
Often look like broken twigs or branches. Superficially resemble
sphinx moths, but **wings held tent-like over abdomen, twice as
long as wide and usually spotted or mottled.** Heavy bodied. Males
with bipectinate antennae. Larvae bore into wood and can cause
damage. WS 17–72 mm; World 700, NA 50.

FLANNEL MOTHS

Black-waved
Flannel Moth
(*Megalopyge
crispata*)

larva

TROPICAL BURNET MOTHS

Speckled Lactura
(*Lactura subfervens*)

CLEARWING MOTHS

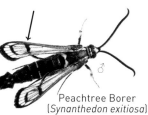

♂

Peachtree Borer
(*Synanthedon exitiosa*)

Maple Callus Borer
(*Synanthedon acerni*)

SLUG CATERPILLAR MOTHS

Spiny Oak-
Slug Moth
(*Euclea
delphinii*)

Warm-chevroned Moth
(*Tortricidia testacea*)

BURNET MOTHS

Orange-patched
Moth
(*Pyromorpha
dimidiata*)

Grapeleaf
Skeletonizer
(*Harrisina americana*)

CARPENTER AND LEOPARD MOTHS

Pecan Carpenterworm Moth
(*Cossula magnifica*)

Carpenter-
worm Moth
(*Prionoxystus
robiniae*)

SLUG CATERPILLAR MOTHS • CLEARWING MOTHS • CARPENTER MOTHS

405

Superfamily Papilionoidea

Antennae capitate or clavate. HW humeral lobe lacking a frenulum.
Diverse group, but typically, diurnal flying in open sunny fields.

SKIPPERS Family Hesperiidae

Small to medium with large eyes, **hooked antennae,** and stout
bodies. Fast fliers, usually landing with wings over back. R in FW is
5-branched, not stalked, and arising from discal cell. Larvae have
large heads compared to other caterpillars.
WS 19–71 mm; World 3,675, NA 305.

SWALLOWTAILS and PARNASSIANS Family Papilionidae

Swallowtails are large butterflies, HW of most **with tail(s).**
Parnassians are smaller and lack tails. 2A in FW extends up to wing
margin, not linking with 1A. Forelegs well developed; tarsal claws
not bifid. Larvae have a forked, eversible scent gland called the
osmeterium on the prothorax. WS 62–170 mm; World 560, NA 42.

SWALLOWTAILS

Black Swallowtail
(*Papilio polyxenes*)

Anise Swallowtail
(*Papilio zelicaon*)

Zebra Swallowtail
(*Eurytides marcellus*)

Rocky Mountain Parnassian
(*Parnassius smintheus*)

SKIPPERS

Silver-spotted Skipper
(*Epargyreus clarus*)

Horace's Duskywing
(*Erynnis horatius*)

Dun Skipper
(*Euphyes vestris*)

Common Checkered-skipper
(*Pyrgus communis*)

Fiery Skipper
(*Hylephila phyleus*)

Yucca Giant-Skipper
(*Megathymus yuccae*)

SWALLOWTAILS

Long-tailed Skipper (*Urbanus proteus*)

Giant Swallowtail
(*Papilio cresphontes*)

Eastern Tiger Swallowtail
(*Papilio glaucus*)

Pipevine Swallowtail
(*Battus philenor*)

407

Superfamily Papilionoidea cont.

WHITES, SULPHURS, & YELLOWS Family Pieridae

Medium sized with white, orange, or yellow wings, often with black markings. Many species sexually dimorphic. FW R with 3 or 4 branches. Forelegs well developed in both sexes; tarsal claws bifid. Chrysalis hangs at angle by silk girdle like Papilionidae.
WS 17–82 mm; World 1,100, NA 82.

BLUES, COPPERS, HAIRSTREAKS, & HARVESTERS

Family Lycaenidae

Small, delicate, and often brightly colored with blues, grays and oranges. **Antennae usually banded.** Eyes indented near antennae. Forelegs of males reduced. Many species with **eyespot distally on HW, and some bearing thin tails that mimic antennae** when they move. WS 12–15 mm; World 5,075, NA 178.

PAPILIONOIDEA

BLUES, COPPERS, HAIRSTREAKS, & HARVESTERS

Red-banded Hairstreak
(*Calycopis cecrops*)

Banded Hairstreak
(*Satyrium calanus*)

Gray Hairstreak
(*Strymon melinus*)

Eastern Tailed-blue
(*Cupido comyntas*)

Eastern Pine Elfin
(*Callophrys niphon*)

Northern Oak Hairstreak
(*Satyrium favonius*)

Striped Hairstreak
(*Satyrium liparops*)

Western Pygmy-Blue
(*Brephidium exilis*)

WHITES, SULPHURS, & YELLOWS

Pink-edged Sulphur
(*Colias interior*)

Dainty Sulphur
(*Nathalis iole*)

Little Yellow
(*Pyrisitia lisa*)

Sleepy Orange in flight
(*Abaeis nicippe*)

Southern Dogface
(*Zerene cesonia*)

Orange-barred Sulphur
(*Phoebis philea*)

Orange Sulphur
(*Colias eurytheme*)
open and closed

Cabbage White
(*Pieris rapae*)

Checkered White
(*Pontia protodice*)

409

Superfamily Papilionoidea cont.

BRUSH-FOOTED BUTTERFLIES Family Nymphalidae
Large group of butterflies containing some of the most commonly
encountered species. Brush-footed refers to the **front pair of legs
that are greatly reduced,** making it appear as though there are only
4 legs instead of 6. Antennae always have 2 grooves on underside.
WS 22–180 mm; World 6,000, NA 230.

METALMARKS Family Riodinidae
Tropical family with a small number of species found in NA. Named
for small metallic patches on the wings of some species. Most
are brown, orange, or gray. Costa of HW thickened out to humeral
angle. Larvae feed on ragwort, thistle, and related plants.
WS 17–25 mm; World 1,400, NA 29.

BRUSH-FOOTED BUTTERFLIES

Red Admiral (*Vanessa atalanta*)

Common Buckeye (*Junonia coenia*)

Common Wood Nymph (*Cercyonis pegala*)

Carolina Satyr (*Hermeuptychia sosybius*)

American Lady (*Vanessa virginiensis*)

METALMARKS

Fatal Metalmark (*Calephelis nemesis*)

Mormon Metalmark (*Apodemia mormo*)

Blue Metalmark (*Lasaia sula*)

Red-Bordered Pixie (*Melanis pixe*)

PAPILIONOIDEA

410

Monarch
(*Danaus plexippus*)

Viceroy
(*Limenitis archippus*)

Gulf Fritillary
(*Agraulis vanillae*)

Queen Butterfly
(*Danaus gilippus*)

Red-spotted Purple
(*Limenitis arthemis astyanax*)

Question Mark
(*Polygonia interogations*)

Tawny Emperor
(*Asterocampa clyton*)

Mourning Cloak
(*Nymphalis antiopa*)

American Snout
(*Libytheana carinenta*)

BRUSH-FOOTED BUTTERFLIES • METALMARKS

411

Superfamily Pyraloidea

Scales on proboscis near base. Long upturned labial palps. Maxillary palpi generally present. At rest, antennae held on top of wings along body. Tympanum located on S2.

PYRALID SNOUT MOTHS Family Pyralidae

Variable in shape and color. **Proboscis scaled. Labial palpi 3-segmented and upturned.** Small maxillary palpi 3- to 4-segmented or reduced. FW at least twice as long as wide; R_5 stalked or fused with R_{3+4}. Tympanal case closed medially. WS 7–38 mm; World 16,000, NA 681.

CRAMBID SNOUT MOTHS Family Crambidae

Proboscis scaled at base. Labial palpi 3-segmented and upturned. Small maxillary palpi 3- to 4-segmented or reduced. FW R_5 free; HW M_1 basally approximated to Rs. Tympanal case open medially. Feed on grasses, trees, shrubs, and herbs. WS 7–35 mm; World 11,630, NA 860.

Superfamily Drepanoidea

HW with veins Sc and R parallel for length of discal cell or forming basal areole as long as discal cell. Tympanal organs on S1 or S2.

HOOKTIP & FALSE OWLET MOTHS Family Drepanidae*

Small-headed moths. Some with **FW forming hooks at tip.** Labial palpi 3-segmented, curved upward; maxillary palpi reduced, 1-segment Cu in FW 4-branched. Larvae feed on various trees and shrubs, including birch and viburnum. WS 15–45 mm; World 790, NA 21.

DOID MOTHS Family Doidae*   Not Illus.

White with black spots. Ocelli present. HW M_3 and Cu_1 usually stalked. Last abdominal segment with 2 rounded elevations the length of the segment. Southwestern US and Mexico. WS 25–40 mm; World 6, NA 2.

Superfamily Geometroidea

Naked proboscis. Tympanal organs on abdomen. Ocelli absent. Spur present on tibia.

SWALLOWTAIL & SCOOPWING MOTHS Family Uraniidae*

Swallowtail moths resemble butterflies, including swallowtails. Scoopwing moths **crease their HW, leaving a gap between wings.** Discal cell open distally. WS 17–88 mm; World 700, NA 11.

GEOMETER MOTHS Family Geometridae

Small, delicate wings often marked with wavy lines. Slender body with long legs, usually hairy or scaly. Wings often broad and angulate. HW Sc creates triangular-shaped cell. Larvae, called inchworms, lack middle prolegs. WS 12–50 mm; World 35,000, NA 1,441.

SEMATURID MOTHS Family Sematuridae*   Not Illus.

Tailed day flying or night flying. Only family that lacks abdominal tympana. One species, *Anurapteryx crenulata*, found in west TX, NM, and AZ. WS 40–100 mm; World 41, NA 1.

PYRALID SNOUT MOTHS

American Plum Borer
(*Euzophera semifuneralis*)

Dimorphic Macalla
(*Epipaschia superatalis*)

Maple Webworm Moth
(*Pococera asperatella*)

Black-patched Glaphyria
(*Glaphyria fulminalis*)

CRAMBID SNOUT MOTHS

Grape Leaffolder
(*Desmia funeralis*)

Distinguished Colomychus
(*Colomychus talis*)

HOOKTIP & FALSE OWLET MOTHS

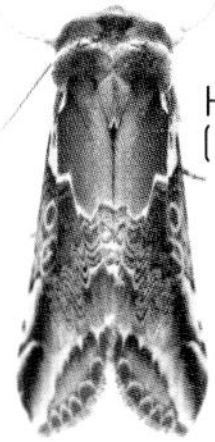

Glorious Habrosyne
(*Habrosyne gloriosa*)

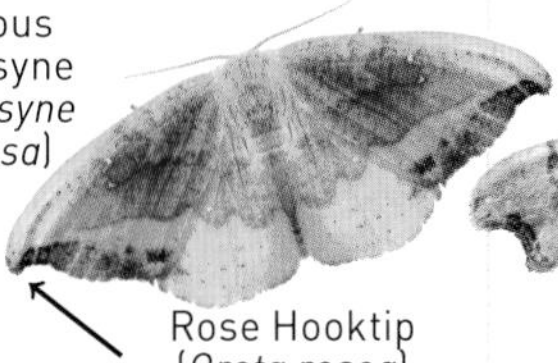

SCOOPWING MOTHS

Rose Hooktip
(*Oreta rosea*)

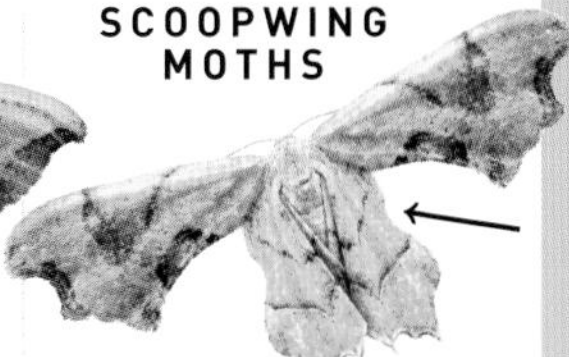

Brown Scoopwing
(*Calledapteryx dryopterata*)

GEOMETER MOTHS

Red-fringed Emerald
(*Nemoria bistriaria*)

Tulip-tree Beauty
(*Epimecis hortaria*)

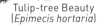

Chickweed Geometer
(*Haematopis grataria*)

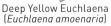

Ceratonyx satanaria

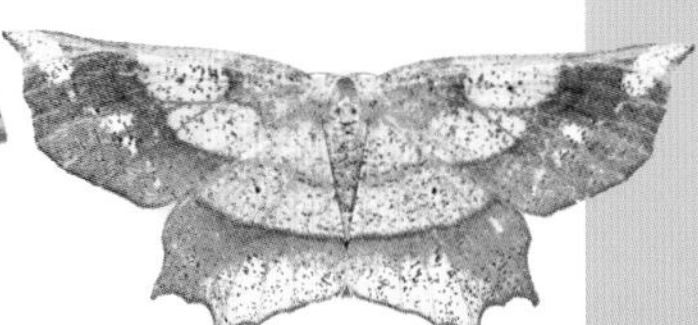

Deep Yellow Euchlaena
(*Euchlaena amoenaria*)

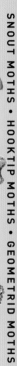

SNOUT MOTHS • HOOKTIP MOTHS • GEOMETRID MOTHS

Superfamily Noctuoidea
This superfamily contains some of the largest and most diverse families in the order.

PROMINENT MOTHS Family Notodontidae
Brownish-yellow medium-sized moths similar to Noctuidae. Hold wings roof-like over abdomen or rolled at rest. Tufts of scales on inner margins of wings near thorax; often erect when wings are folded. Sc+R_1 run closely parallel along discal cell near base of wing. Rs and M_1 stalked shortly after discal cell.
WS 20–120 mm; World 3,500, NA 140.

OWLET MOTHS Family Noctuidae
Diverse family with variable morphology. FW with Cu stem appearing to have 4 veins originating along it. HW Sc+R forming small basal areole. Tympanal organ present on metathorax. Larvae occupy wide array of niches.
WS 7–152 mm; World 35,000, NA 3,040.

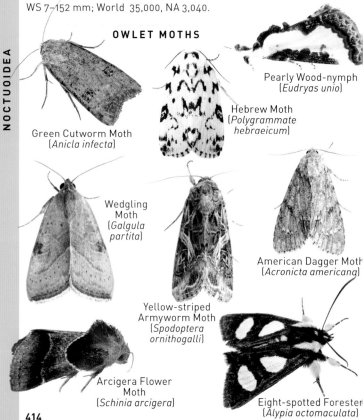

OWLET MOTHS

Green Cutworm Moth
(*Anicla infecta*)

Hebrew Moth
(*Polygrammate hebraeicum*)

Pearly Wood-nymph
(*Eudryas unio*)

Wedgling Moth
(*Galgula partita*)

Yellow-striped Armyworm Moth
(*Spodoptera ornithogalli*)

American Dagger Moth
(*Acronicta americana*)

Arcigera Flower Moth
(*Schinia arcigera*)

Eight-spotted Forester
(*Alypia octomaculata*)

NOCTUOIDEA

PROMINENT MOTHS

Datana sp.

Red-washed Prominent
(*Oedemasia semirufescens*)

White-dotted Prominent
(*Nadata gibbosa*)

Unicorn Caterpillar Moth
(*Coelodasys unicornis*)

White-blotched Heterocampa
(*Heterocampa pulverea*)

Oblique
Heterocampa
(*Heterocampa
obliqua*)

White-streaked
Prominent
(*Ianassa lignicolor*)

Mottled Prominent
(*Macrurocampa
marthesia*)

Angulose Prominent
(*Peridea angulosa*)

Black-etched Prominent
(*Cerura scitiscripta*)

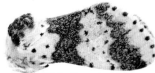

White Furcula
(*Furcula borealis*)

Superfamily Noctuoidea cont.

EREBID MOTHS Family Erebidae

Large diverse family united by wing characteristics and molecular evidence. FW Cu runs outward from base of wing to outer margin, typically splitting into 4 veins.
WS 7–152 mm; World 11,000, NA 956.

EUTELIID MOTHS Family Euteliidae

When at rest, **FW crumpled and partly rolled around HW with abdomen curled upward.** Eyes naked; ocelli present. HW with Cu stem appearing to have 4 veins originating along it. Tibia lacking spines. WS 25–36 mm; World 250, NA 18.

NOLID MOTHS Family Nolidae

Males with bar-shaped retinaculum. FW with Rs veins arising from a single stalk and tufts of raised scales or FW with accessory cell and frons glossy on lower half. Larvae feed on woody plants, forbs, or grasses. WS 10–16 mm; World 1,400, NA 40.

NOLID MOTHS

Sharp-blotched
Nola Moth
(*Nola pustulata*)

Eyed Baileya
(*Baileya ophthalmica*)

Coastal Plain
Meganola Moth
(*Meganola phylla*)

Three-spotted Nola
(*Nola triquestrana*)

EUTELIID MOTHS

Dark Marathyssa
(*Marathyssa inficita*)

Eyed Paectes
(*Paectes oculatrix*)

Large Paectes Moth
(*Paectes abrostoloides*)

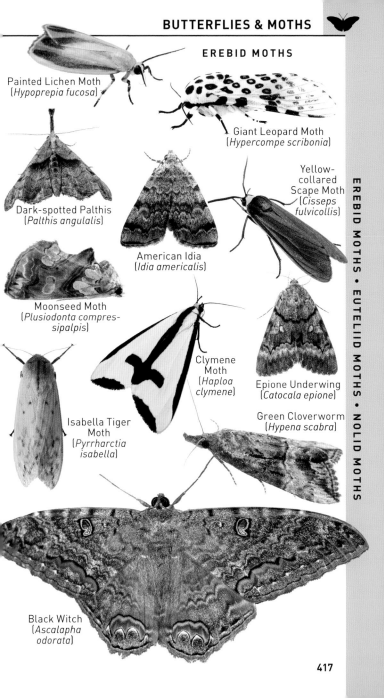

EREBID MOTHS

Painted Lichen Moth
(*Hypoprepia fucosa*)

Giant Leopard Moth
(*Hypercompe scribonia*)

Dark-spotted Palthis
(*Palthis angulalis*)

Yellow-collared Scape Moth
(*Cisseps fulvicollis*)

American Idia
(*Idia americalis*)

Moonseed Moth
(*Plusiodonta compressipalpis*)

Clymene Moth
(*Haploa clymene*)

Epione Underwing
(*Catocala epione*)

Isabella Tiger Moth
(*Pyrrharctia isabella*)

Green Cloverworm
(*Hypena scabra*)

Black Witch
(*Ascalapha odorata*)

EREBID MOTHS • EUTELIID MOTHS • NOLID MOTHS

LEPIDOPTERA

LONOIDEA • LASIOCAMPOIDEA • BOMBYCOIDEA

Superfamily Mimallonoidea

SACK-BEARERS Family Mimallonidae

Larvae make sacks out of leaves, carrying them around and using them to overwinter. Some adults have **scalloped FW** and are brownish-gray; others are reddish-gray with mottled wings. HW has 2 anal veins. WS 17–50 mm; World 255, NA 4.

Superfamily Lasiocampoidea

TENT CATERPILLARS Family Lasiocampidae

Brownish-gray moths. Body, legs, and eyes hairy. Proboscis vestigial or absent. **Antennae bipectinate** in both sexes. HW lacks frenulum, replaced with an enlarged humeral angle. Caterpillars create tent-like webbing, live communally, and can cause damage to trees. WS 22–58 mm; World 2,050, NA 35.

Superfamily Bombycoidea

FW with R_2 and R_3 closely paralleled, almost touching; fused with stem of R_4+R_5. Prothoracic coxae of last-instar larva fused anteriorly.

AMERICAN SILKWORM MOTHS Family Apatelodidae

Some have window near FW apex. M_2 arises halfway between M_1 and M_3. WS 30–50 mm; World 60, NA 7.

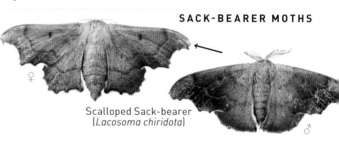

SACK-BEARER MOTHS

Scalloped Sack-bearer (*Lacosoma chiridota*)

AMERICAN SILKWORM MOTHS

Spotted Apatelodes (*Apatelodes torrefacta*)

The Seraph (*Olceclostera seraphica*)

Small Tolype
(*Tolype notialis*)

Eastern Tent Caterpillar &
moth (*Malacosoma americana*)

Forest Tent Caterpillar & moth
(*Malacosoma disstria*)

Dot-lined
White
(*Artace
cribrarius*)

Western Tent
Caterpillar moth & egg
mass
(*Malacosoma
californicum*)

BOMBYCOIDEA

Superfamily Bombycoidea cont.

SPHINX & HUMMINGBIRD MOTHS Family Sphingidae
Large-bodied moths with tapered abdomen. **Distinctive triangular
shape with narrow FW** and large tapering abdomen. **Antennae
thickened and slightly bulging at end.** No ocelli or tympanal
organs. Proboscis usually well developed and noticeably
elongated. Often fly at dusk; some fly during the day and resemble
a hummingbird. WS 27–175 mm; World 1,450, NA 124.

SPHINX MOTHS

caterpillar

White-lined Sphinx
(*Hyles lineata*)

Banded Sphinx
(*Eumorpha
fasciatus*)

pupa

Carolina Sphinx
(*Manduca sexta*)

Bald Cypress
Sphinx
(*Isoparce cupressi*)

caterpillar

Achemon
Sphinx
(*Eumorpha
achemon*)

Pandora Sphinx
(*Eumorpha
pandorus*)

Virginia Creeper Sphinx
(*Darapsa myron*)

SPHINX MOTHS

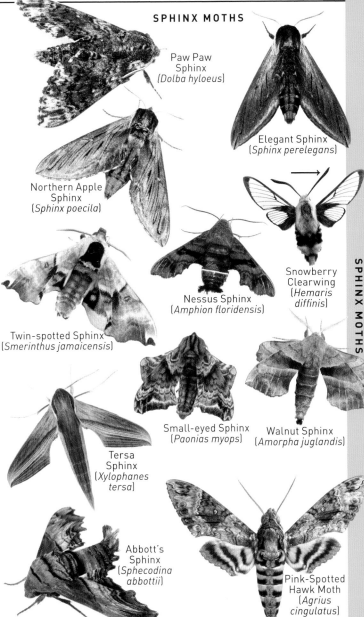

Paw Paw Sphinx
(*Dolba hyloeus*)

Elegant Sphinx
(*Sphinx perelegans*)

Northern Apple Sphinx
(*Sphinx poecila*)

Snowberry Clearwing
(*Hemaris diffinis*)

Twin-spotted Sphinx
(*Smerinthus jamaicensis*)

Nessus Sphinx
(*Amphion floridensis*)

Small-eyed Sphinx
(*Paonias myops*)

Walnut Sphinx
(*Amorpha juglandis*)

Tersa Sphinx
(*Xylophanes tersa*)

Abbott's Sphinx
(*Sphecodina abbottii*)

Pink-Spotted Hawk Moth
(*Agrius cingulatus*)

Superfamily Bombycoidea cont.

GIANT SILKWORM & ROYAL MOTHS Family Saturniidae

Some of the largest moths in NA. Robust hairy bodies and small heads with vestigial mouthparts. **Antennae bipectinate or quadripectinate.** Some with **wings showing eye-spots.** Frenulum reduced or absent. HW with only 1 anal vein. FW M_1 not stalked with R. Larvae create large silk cocoons.
WS 20–150 mm; World 1,480, NA 72.

GIANT SILKWORM MOTHS

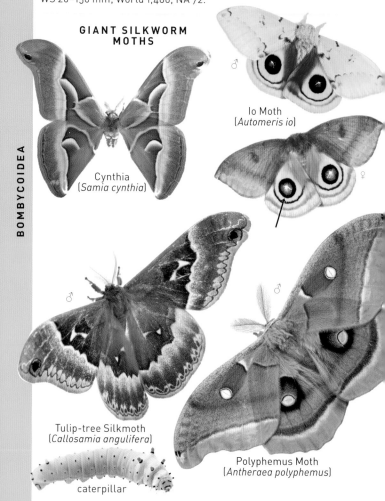

Cynthia
(*Samia cynthia*)

♂
Io Moth
(*Automeris io*)

♀

♂

♂

Tulip-tree Silkmoth
(*Callosamia angulifera*)

caterpillar

Polyphemus Moth
(*Antheraea polyphemus*)

BOMBYCOIDEA

GIANT SILKWORM MOTHS

Rosy Maple Moth
(*Dryocampa rubicunda*)

Cecropia
Moth
(*Hyalophora
cecropia*)

Calleta Silkmoth
(*Eupackardia calleta*)

♀

Luna Moth &
caterpillar
(*Actias luna*)

Southern Pink-
striped
Oakworm Moth
(*Anisota virginiensis*)

ROYAL
MOTHS

Pine Devil Moth
(*Citheronia
sepulcralis*)

Regal Moth
(*Citheronia regalis*)

Imperial Moth
(*Eacles
imperialis*)

Cat Flea
(*Ctenocephalides felis*)
Pulicidae

Cat Flea
(*Ctenocephalides felis*)
larva

FLEAS
ORDER SIPHONAPTERA

Derivation of Name
- Siphonaptera: *siphon*, tube; *aptera*, lacking wings. Referring to the tube-like, sucking mouthparts used to feed on the blood of other species. This order is also secondarily wingless.

Identification
- 1–10 mm TL.
- Brown to black and flattened laterally, enabling navigation through hairs of host.
- Head fused to small thorax and flattened oval abdomen.
- Hard body covered in small, posteriorly projecting hairs and usually spinelets, making it difficult for the host to dislodge them.
- Antennae are short and lay flat against the body in grooves.
- Mouthparts are modified into a short tube for piercing the host and sucking their blood.
- No evidence of wings or wing pads.
- Legs are modified for jumping and are covered in spines and setae.
- All are parasites that require a blood meal from the host to complete their life cycle.

Classification
- Evolved more recently than Mecopterans, their closest relative.
- Evolution of the group has presumably been due to their close association with mammalian and avian hosts.
- Eight families in NA; all specialize on specific groups of mammals and birds.
- Family-level revision is needed for the order.
- World: 2,500 species; North America: 325 species.

Range
- Found worldwide; most diverse in temperate areas.

Similar orders
- Psocodea (p. 220): not flattened laterally; legs not modified to jump.
- Mecoptera—Boreidae (p. 430): have reduced wing pads and are spherical in shape; mouthparts form a beak.

Food
- Adults feed on blood from their mammal or bird host.
- Larvae feed on host nest debris, including dried blood and feces from adults.

Behavior
- When adults emerge from the pupa, they immediately seek out a host for feeding.
- Unlike mites, fleas usually jump off their host when the host leaves its nest and jump back on when the host returns.
- Larvae lack eyes but do have light receptors and will move away from light.
- Feed throughout their second instar.
- Some groups come out at dusk or early morning to look for food and females, while other groups mate at night.

Life Cycle
- Holometabolous.
- Eggs typically laid while adult is feeding and subsequently drop off the host onto the ground.
- Incubation may take a few days to a few months, depending on temperature.
- Larvae are soft bodied with only a sclerotized head capsule (the width of which is useful for distinguishing species). Lack thoracic legs and prolegs.
- Three larval instars.
- Larvae spin a silken cocoon incorporating debris from the surroundings.
- Transformation from larvae to cocoon takes a few days.
- Adult uses an environmental cue like an increase in CO_2 to trigger eclosion from pupa.
- Some individuals take up to 450 days to complete their life cycle.
- Females require a blood meal for egg formation.

Importance to Humans
- Vector of several diseases; capable of wiping out colonies of mammals.
- Brown Rat, *Rattus rattus,* carries plague that is transmitted to humans via fleas even today.
- *Pulex irritans* (Pulicidae) carried the black death (bacterial disease caused by *Yersenia pestis*) that killed 25 million people across Europe in the 14th century.

Collecting and Preserving
- Hand collect from host using a comb or the host's environment.
- Preserve adults and larvae in 70% ethanol or mounted on a slide.

Resources
- General Information—www.zin.ru/Animalia/Siphonaptera
- *Functional and Evolutionary Ecology of Fleas* by Krasnov, 2008

Flea Anatomy

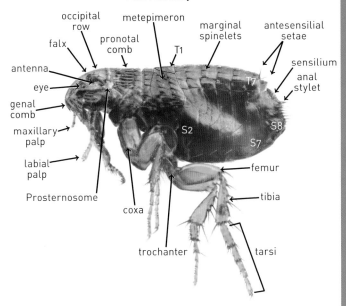

Labels: occipital row, metepimeron, marginal spinelets, antesensilial setae, falx, pronotal comb, T1, sensilium, antenna, anal stylet, eye, T7, genal comb, maxillary palp, S2, S8, labial palp, S7, Prosternosome, femur, coxa, tibia, trochanter, tarsi

Families of Fleas and Their Hosts

Bird and Rodent Fleas	pikas, domestic poultry, birds, small carnivores, small rodents (ground squirrels, tree squirrels, flying squirrels, groundhog, deer mice)
Bat Fleas	bats
Scaled Fleas	foxes, small rodents (house mouse), mountain beaver, woodpeckers, rabbits and hares, pikas
Small Mammal Fleas	small rodents
Rodent Fleas	small rodents (insectivores, ground squirrels, voles, wood rats, kangaroo rats, deer mice, flying squirrels, mountain beaver, short-tailed shrew)
Club Fleas	small rodents
Common Fleas	birds, rodents, livestock, humans, cats, dogs, domestic poultry, hedgehogs, large carnivores, peccaries, lagomorphs
Carnivore Fleas	carnivores, but uncommon

427

BIRD AND RODENT FLEAS Family Ceratophyllidae
Mesonotum with marginal spinelets. Dorsal surface of sensilium flat. Genal comb never present. Eyes present, but can be reduced. Parasites of animals other than bats, usually rodents and birds. Largest family of fleas in NA. TL ~2 mm; World 397, NA 125.

BAT FLEAS Family Ischnopsyllidae*   Not illus.
Pronotal comb present, but reduced. Genal comb consisting of 2 broad, blunt spines. Eyes are absent or vestigial. Mesonotum with marginal spinelets. Sensilium not convex. Parasites of bats. TL 2–4 mm; World 122, NA 11.

SCALED FLEAS Family Leptopsyllidae*   Not illus.
Mesonotum with marginal and submarginal spinelets. Sensilium convex. Genal and pronotal combs present. Eyes with wavy margins. Most found in Palearctic areas as parasites of small rodents and beavers. TL 2–6 mm; World 235, NA 18.

SMALL MAMMAL FLEAS Family Ctenophthalmidae
Club of male antenna does not extend onto prosternosome. **Sensilium not strongly convex.** Parasites of small rodents and insectivores. TL 2–3 mm; World 790, NA 114.

RODENT FLEAS Family Hystrichopsyllidae
Club of male antennae extends onto prosternosome. Lack eyes, mesonotum marginal spinelets, and genal and pronotal combs. Largest flea in the world, *Hystrichopsylla schefferi*, is in this family; found on the Mountain Beaver (*Aplodontia rufa*) in northern CA into BC. Parasites of small rodents and insectivores in Nearctic areas. TL 7–10 mm; World 570, NA 7.

CLUB FLEAS Family Rhopalopsyllidae
Hind tibia usually with apical tooth. Genal and pronotal combs absent. **Fifth tarsal segment with 4 pairs of bristles.** Parasites of armadillos, opossums, and various small mammals. TL 1.5–3 mm; World 122, NA 4.

COMMON FLEAS Family Pulicidae
Middle coxa without outer internal ridge. **Hind tibia without apical tooth. Sensilium with 8 or 14 pits per side.** Dog and cat fleas belong to this family; also parasitize rodents, birds, and humans. TL 1–2.5 mm; World 181, NA 16.

CARNIVORE FLEAS Family Vermipsyllidae*   Not illus.
Hind tibia usually with apical tooth. Lack combs, marginal spinelets, antesensilial setae, bristles on inner side of hind coxae, and anal stylets in females. Parasites of carnivores in colder Nearctic and Palearctic regions. TL 1.5–5 mm; World 46, NA 6.

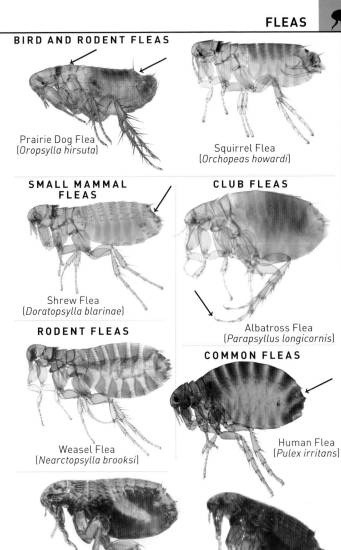

BIRD AND RODENT FLEAS

Prairie Dog Flea
(*Oropsylla hirsuta*)

Squirrel Flea
(*Orchopeas howardi*)

SMALL MAMMAL FLEAS

Shrew Flea
(*Doratopsylla blarinae*)

CLUB FLEAS

Albatross Flea
(*Parapsyllus longicornis*)

RODENT FLEAS

Weasel Flea
(*Nearctopsylla brooksi*)

COMMON FLEAS

Human Flea
(*Pulex irritans*)

Cat Flea
(*Ctenocephalides felis*)

Dog Flea
(*Ctenocephalides canis*)

Wingless Scorpionfly
mating with nuptial gift
(*Apterobittacus apterus*)
Bittacidae

Panorpid larva

HANGINGFLIES & SCORPIONFLIES
ORDER MECOPTERA

Derivation of Name
- Mecoptera: *meco*, long; *ptera*, wings. North American species have long, oval wings or have lost them completely.

Identification
- 2–25 mm TL.
- Usually elongated and brown to red with yellowish brown wings.
- Head forming an elongated beak with well-developed mandibles.
- Tarsi 5-segmented with either 1 or 2 claws.
- Males often have enlarged genitals forming terminal claspers that resemble earwigs and are held over the abdomen like a scorpion.
- Found on low vegetation within forests or along forest edges.

Classification
- One of the oldest insect groups with complete metamorphosis; fossils have been found as early as the Permian, 250 million years ago. Wing venation has remained relatively unmodified since then.
- Closely related to fleas, as is evident in some snow scorpionflies.
- Five families in NA separated by leg and wing characters.
- World: 600 species; North America: 85 species.

Range
- Can be found worldwide except Antarctica. Most diverse in southeast Asia.

Similar orders
- Diptera (p. 436): 1 pair of wings.
- Neuroptera (p. 280): have numerous crossveins in wings and lack long face.
- Hymenoptera (p. 236): lack long face, HW smaller than FW.

Common Scorpionfly
(*Panorpa nuptialis*)
Panorpidae
feeding on grasshopper

Food
- Larvae are scavengers, feeding on soft-bodied dead insects and vegetation.
- Adults feed mostly on other insects but will also eat vegetation, nectar, fruit, and pollen.

Behavior
- Some groups have complicated mating rituals involving nuptial gifts.
- Males of some species will mimic females in order to receive free food from other males.
- Mating happens in the understory of woods or near dense vegetation.
- Some groups come out at dusk or early morning to look for food and females, while other groups mate at night.

Life Cycle
- Holometabolous.
- Eggs laid in moist areas.
- Larvae go through 4 instars and can be found in soil near vegetation. Some larvae have compound eyes; unique among complete metamorphosis insects.
- Pupation is normally underground and lasts no more than 2 weeks.
- Different species will mate and lays eggs in spring, summer, or fall.
- Thought to be univoltine, but little is known about their life cycles.

Importance to Humans
- Predators on other insects and plants.
- Not known to be pests.

Collecting and Preserving
- Hand collect adults with net, then pin.
- Preserve larvae in 70% ethanol.

Resources
- *Scorpionflies, Hangingflies, and other Mecoptera* by Byers, 2002
- World Checklist of Extant Mecoptera Species— tinyurl.com/MecopteraWorldList

Common Scorpionfly
(*Panorpa nuptialis*)
Panorpidae

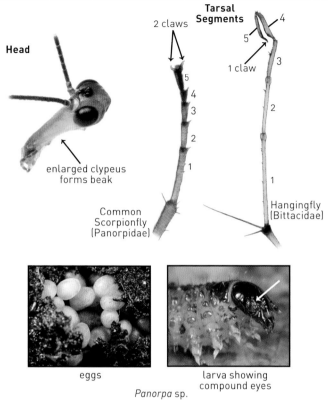

Head

enlarged clypeus
forms beak

**Tarsal
Segments**

2 claws

1 claw

Common
Scorpionfly
(Panorpidae)

Hangingfly
(Bittacidae)

eggs

larva showing
compound eyes

Panorpa sp.

Female Mid-winter Boreus on moss (*Boreus brumalis*) Boreidae.

433

COMMON SCORPIONFLIES Family Panorpidae
Wings are mottled or banded with dark and light coloring. Abdomen can be dirty yellow to a dark reddish-brown or almost black. **Tip of male abdomen enlarged** and held over abdomen resembling a scorpion. **Females have tapered abdomen.** Both larvae and adults feed on insects. Found on low shrubs in dense forests, often near water. TL 10–25 mm; World 360, NA 55.

HANGINGFLIES Family Bittacidae
Yellowish to reddish-brown abdomen. Long, narrow abdomen. **Fifth tarsal segment folds back against the fourth**, which is used to hang from vegetation and hold onto prey. Tarsi raptorial, with single claw. Resembles crane flies, especially in flight. One species in California (*Apterobittacus apterus*) is wingless.
TL 12–23 mm; World 170, NA 10.

SHORT-FACED SCORPIONFLIES Family Panorpodidae
Yellow-brown scorpionflies with a **short beak**. Males have enlarged abdomen like in Panorpidae, but do not hold tip over abdomen. Wings are well developed in males, but shortened in females to varying degrees. TL 7–10 mm; World 10, NA 5.

EARWIGFLIES Family Meropeidae*
Only 1 species in NA, *Merope tuber*, found from ON to GA and west to MN. It is brown with well-developed wings that are off-white to brown. **Large forcep-like claspers at tip of abdomen resembling earwigs.** Ocelli absent.
TL 7–16 mm; World 3, NA 1.

SNOW SCORPIONFLIES Family Boreidae*
Wings greatly reduced and females lack them; males have them modified to grasp the female, who rides on his back while mating. Become adults in winter, where dark color contrasts against snow. Can also be found on moss, which the larvae feed on.
TL 2.5–8 mm; World 30, NA 15.

Wingless Scorpionfly
(*Apterobittacus apterus*)
Bittacidae

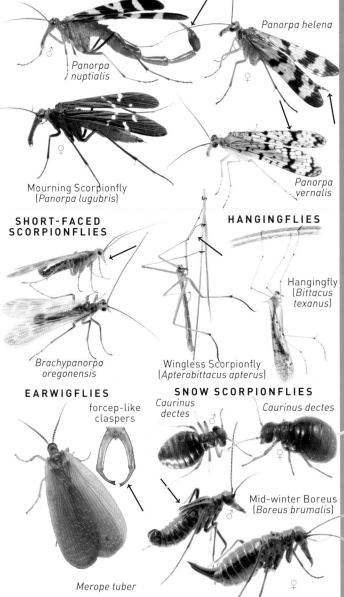

COMMON SCORPIONFLIES

Panorpa nuptialis ♂

Panorpa helena ♀

Mourning Scorpionfly (*Panorpa lugubris*) ♀

Panorpa vernalis

SHORT-FACED SCORPIONFLIES

Brachypanorpa oregonensis

HANGINGFLIES

Hangingfly (*Bittacus texanus*)

Wingless Scorpionfly (*Apterobittacus apterus*)

EARWIGFLIES

forcep-like claspers

Merope tuber

SNOW SCORPIONFLIES

Caurinus dectes ♂

Caurinus dectes ♀

Mid-winter Boreus (*Boreus brumalis*) ♂

♀

435

Longlegged Fly
(*Condylostylus* sp.)
Dolichopodidae

aggregation of larvae
Calliphoridae

FLIES
ORDER DIPTERA

Derivation of Name
- Diptera: *di*, meaning 2; *ptera*, winged. Referring to the identifying character that there are only 2 wings or 1 pair.

Identification
- 0.5–40 mm TL
- Usually small and elongate, with one pair of prominent wings (FW); HW reduced to knob-like structures called halteres.
- Mouthparts sucking, but may be modified for piercing, sponging, or lapping; antennae are variable but often short and 3-segmented; compound eyes large, often meeting on top of head; often adorned with spines, setae ,and/or sometimes scales.

Classification
- Large, diverse order with several non-monophyletic groupings (e.g., Nematocera, Aschiza, and Acalyptratae). No overarching consensus for the taxonomy of the order has been reached to date.
- Traditionally, 2 suborders recognized: Nematocera and Brachycera.
- Numerous infraorders are often recognized.
- World: 153,000 species; North America: 17,000 species.

Range
- Found on every continent, including Antarctica (*Belgica antarctica*: Chironomidae).

Similar orders
- Strepsiptera (p. 288): FW are reduced, HW are membranous.
- Hymenoptera (p. 236): 2 pairs of wings that are joined together via hamuli; FW larger than HW; chewing mouthparts.
- Neuroptera (p. 280): have 2 pairs of wings that are similar in size; chewing mouthparts.

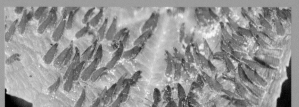

Grape Tube Gallmaker (*Schizomyia viticola*) Cecidomyiidae

Food
- Adults feed on flowers, plants, dung, blood, carrion, fungi, detritus, other fluids, or are predators.
- Larvae feed on food similar to adults, but can also be predators on other insects; some are parasitic.

Behavior
- Many species are aquatic in the larval stage; adults can usually be found around the larval habitat.
- Many species found in large numbers, particular around aquatic habitats, including those that are freshwater, brackish, and alkaline.
- Many flies are mimics of wasps or bees.
- Many species engage in elaborate courtship behaviors.

Life Cycle
- Holometabolous.
- Eggs often laid near food source.
- Most species are oviparous, laying eggs, but some are ovoviviparous or viviparous, especially when dependent on short-lived food sources.
- Larvae, or maggots, lack segmented legs, but some have prolegs; head is well developed in some (Nematocera) and reduced, absent, and/or capable of being retracted into thorax (most Brachycera). Eyes and antennae reduced in Brachycera.
- Pupae may have appendages visible and adhering to the body (Nematocera); others form a protective puparium (Cyclorrhapha).
- Pupae of aquatic species are often capable of swimming movements.
- Adults usually short-lived.

Importance to Humans
- Considerable economic importance, with many species as pests on agricultural crops and others attacking livestock.
- Species of mosquito, black fly, and sand fly are vectors of major tropical diseases, such as malaria, river blindness, Zika virus, dengue fever, and more, causing great impact and even death to humans.
- Important food for many larger insects and other animals.
- Many are parasitoids on other groups of insects.
- Can be a nuisance because of their large numbers.
- *Drosophila melanogaster* is a model organism for scientific research.

Collecting and Preserving
Immatures
- Collect live immatures, parboil them, and preserve in 70% ethanol.
- Whole body or mouthpart slide mounts is needed for some groups.
Adults
- Easily collected by hand, at lights and malaise traps.
- Parasitic species may have to be reared from hosts.
- Pin or mount on a point. Critical point dry small specimens.

Resources
- *Flies: The Natural History and Diversity* of Diptera by Marshall, 2012
- *Field Guide to the Flower Flies of NE NA* by Skevington et al., 2019
- *Manual to Nearctic Diptera*, vols. 1–3 (1981–1989)
- The (new) Diptera Site—diptera.myspecies.info

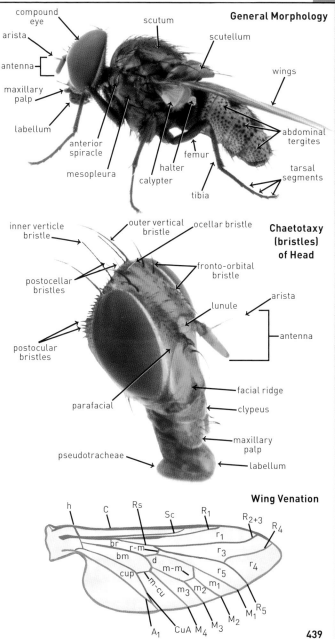

General Morphology

compound eye
arista
antenna
maxillary palp
labellum
anterior spiracle
mesopleura
calypter
halter
femur
tibia
scutum
scutellum
wings
abdominal tergites
tarsal segments

Chaetotaxy (bristles) of Head

inner verticle bristle
outer vertical bristle
ocellar bristle
postocellar bristles
fronto-orbital bristle
arista
lunule
antenna
postocular bristles
parafacial
facial ridge
clypeus
maxillary palp
pseudotracheae
labellum

Wing Venation

h
C
Rs
Sc
R_1
R_{2+3}
R_4
r_1
br
r-m
r_3
bm
d
r_5
r_4
cup
m-m
m_1
m_2
m_3
m-cu
A_1
CuA
M_4
M_3
M_2
M_1
R_5

439

SILHOUETTES OF MAJOR FLY GROUPS

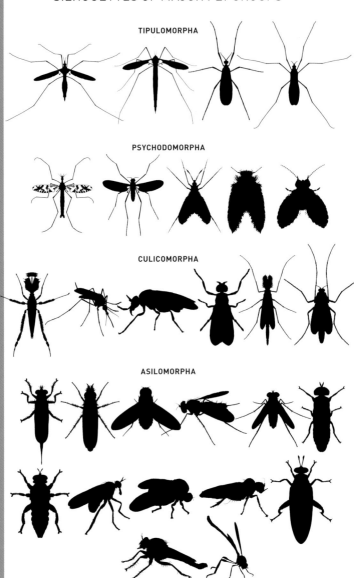

TIPULOMORPHA

PSYCHODOMORPHA

CULICOMORPHA

ASILOMORPHA

BIBIONOMORPHA

STRATIOMYOMORPHA

TABANOMORPHA

ASCHIZA

"ACALYPTRATAE FLIES"

CALYPTRATAE FLIES

441

Bee Killer
(*Mallophora fautrix*)
Asilidae

SUBGROUPS OF DIPTERA

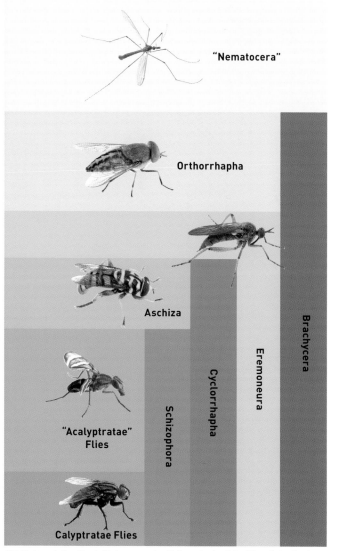

"Nematocera"

Orthorrhapha

Aschiza

"Acalyptratae" Flies

Calyptratae Flies

Schizophora

Cyclorrhapha

Eremoneura

Brachycera

The higher relationships in Diptera are not yet fully understood. Traditionally 2 suborders, Nematocera and Brachycera, were recognized. Studies now show Nematocera as being non-monophyletic and the Brachycera within the Nematocera. The above figure shows groupings that are convienient for our use.

DIPTERA

KEY 1 TO COMMON FLY FAMILIES

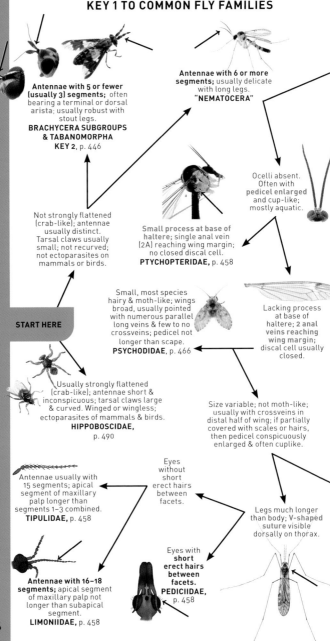

Antennae with 5 or fewer (usually 3) segments; often bearing a terminal or dorsal arista; usually robust with stout legs.
BRACHYCERA SUBGROUPS & TABANOMORPHA
KEY 2, p. 446

Antennae with 6 or more segments; usually delicate with long legs.
"NEMATOCERA"

Ocelli absent. Often with **pedicel enlarged and cup-like;** mostly aquatic.

Not strongly flattened (crab-like); antennae usually distinct. Tarsal claws usually small; not recurved; not ectoparasites on mammals or birds.

Small process at base of haltere; single anal vein (2A) reaching wing margin; no closed discal cell.
PTYCHOPTERIDAE, p. 458

Small, most species hairy & moth-like; wings broad, usually pointed with numerous parallel long veins & few to no crossveins; pedicel not longer than scape.
PSYCHODIDAE, p. 466

Lacking process at base of haltere; **2 anal veins reaching wing margin;** discal cell usually closed.

START HERE

Usually strongly flattened (crab-like); antennae short & inconspicuous; tarsal claws large & curved. Winged or wingless; ectoparasites of mammals & birds.
HIPPOBOSCIDAE, p. 490

Size variable; not moth-like; usually with crossveins in distal half of wing; if partially covered with scales or hairs, then pedicel conspicuously enlarged & often cuplike.

Antennae usually with 15 segments; apical segment of maxillary palp longer than segments 1–3 combined.
TIPULIDAE, p. 458

Eyes without short erect hairs between facets.

Legs much longer than body; **V-shaped suture visible dorsally on thorax.**

Antennae with 16–18 segments; apical segment of maxillary palp not longer than subapical segment.
LIMONIIDAE, p. 458

Eyes with **short erect hairs between facets.**
PEDICIIDAE, p. 458

FLIES

Ocelli present; first 2 antennal segments subequal in length; mostly terrestrial.
**Terrestrial Nematocera
KEY 7,** p. 456

Postscutellum lacking longitudinal groove; radial veins usually short, ending well before wing tip.
CERATOPOGONIDAE,
p. 462

Postscutellum usually with prominent longitudinal groove; radial veins usually extending close to wing tip.
CHIRONOMIDAE,
p. 464

Proboscis elongate and covered with scales; body, legs and **wings covered in scales.**
CULICIDAE, p. 464

Eight or fewer longitudinal wing veins reaching wing margin; wings usually lacking scales or hairs.

Nine or more longitudinal wing veins reaching wing margin; wings usually bearing conspicuous scales or hairs.

Proboscis short, not covered in scales; scales restricted to hind margin of wings.
CHAOBORIDAE, p. 464

Antennae usually longer than head with enlarged, cup-like pedicel; typically slender.

Antennae not longer than head with pedicel not enlarged; stout bodied, usually hump-backed.
SIMULIIDAE, p. 464

First tarsomere as long or longer than second; tarsi 5-segmented; 6 or more veins reaching wing margin; size variable.

Legs variable, but thorax never with V-shaped suture visible dorsally.

First tarsomere reduced or absent, so tarsi appear 4-segmented; venation reduced with few veins; tiny <3 mm.
CECIDOMYIIDAE (in part), p. 460

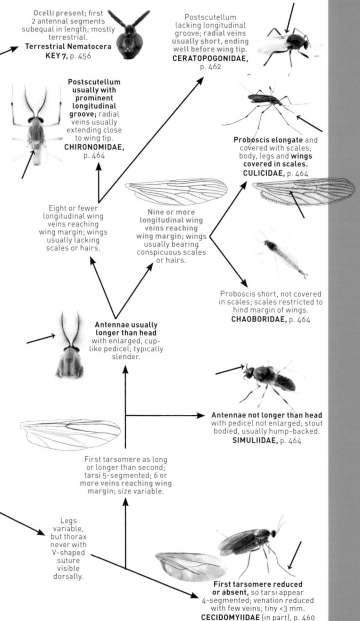

445

KEY 2 TO COMMON FLY FAMILIES

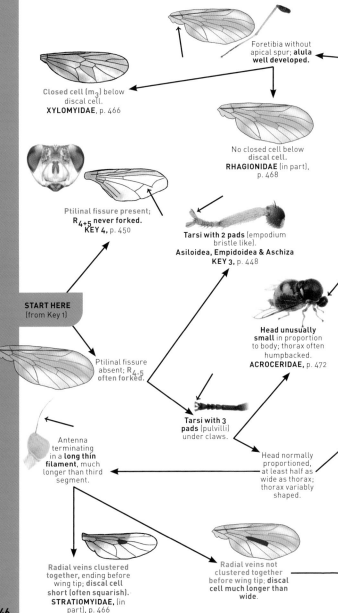

Foretibia without apical spur; **alula well developed.**

Closed cell (m₃) below discal cell.
XYLOMYIDAE, p. 466

No closed cell below discal cell.
RHAGIONIDAE (in part), p. 468

Ptilinal fissure present; R₄₊₅ never forked.
KEY 4, p. 450

Tarsi with 2 pads (empodium bristle like).
Asiloidea, Empidoidea & Aschiza
KEY 3, p. 448

START HERE
(from Key 1)

Ptilinal fissure absent; R₄₊₅ often forked.

Head unusually **small** in proportion to body; thorax often humpbacked.
ACROCERIDAE, p. 472

Antenna terminating in a **long thin filament,** much longer than third segment.

Tarsi with 3 pads (pulvilli) under claws.

Head normally proportioned, at least half as wide as thorax; thorax variably shaped.

Radial veins clustered together, ending before wing tip; discal cell short (often squarish).
STRATIOMYIDAE, (in part), p. 466

Radial veins not clustered together before wing tip; discal cell much longer than wide.

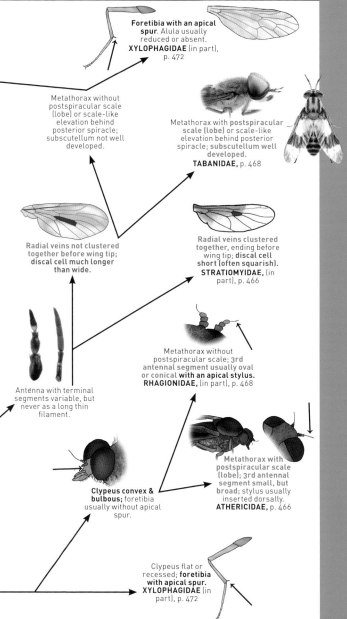

Foretibia with an apical spur. Alula usually reduced or absent.
XYLOPHAGIDAE (in part), p. 472

Metathorax without postspiracular scale (lobe) or scale-like elevation behind posterior spiracle; subscutellum not well developed.

Metathorax with **postspiracular scale (lobe)** or scale-like elevation behind posterior spiracle; **subscutellum well developed**.
TABANIDAE, p. 468

Radial veins not clustered together before wing tip; **discal cell much longer than wide**.

Radial veins clustered together, ending before wing tip; **discal cell short (often squarish)**.
STRATIOMYIDAE, (in part), p. 466

Metathorax without postspiracular scale; 3rd antennal segment usually oval or conical **with an apical stylus**.
RHAGIONIDAE, (in part), p. 468

Antenna with terminal segments variable, but never as a long thin filament.

Clypeus convex & bulbous; foretibia usually without apical spur.

Metathorax with postspiracular scale (lobe); 3rd antennal segment small, but broad; stylus usually inserted dorsally.
ATHERICIDAE, p. 466

Clypeus flat or recessed; **foretibia with apical spur**.
XYLOPHAGIDAE (in part), p. 472

447

KEY 3 TO COMMON FLY FAMILIES

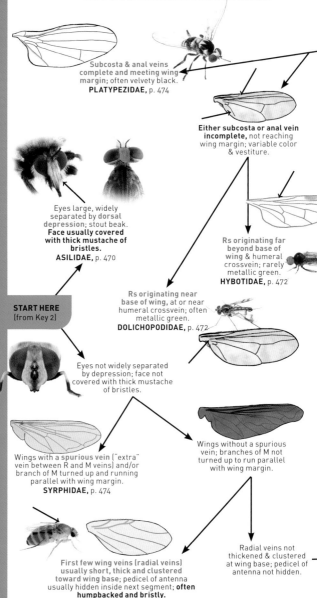

Subcosta & anal veins complete and meeting wing margin; often velvety black.
PLATYPEZIDAE, p. 474

Either subcosta or anal vein **incomplete,** not reaching wing margin; variable color & vestiture.

Eyes large, widely separated by **dorsal depression; stout beak. Face usually covered with thick mustache of bristles.**
ASILIDAE, p. 470

Rs originating far beyond base of wing & humeral crossvein; rarely metallic green.
HYBOTIDAE, p. 472

Rs originating near base of wing, at or near humeral crossvein; often metallic green.
DOLICHOPODIDAE, p. 472

START HERE (from Key 2)

Eyes not widely separated by depression; face not covered with thick mustache of bristles.

Wings with a **spurious vein** ("extra" vein between R and M veins) and/or branch of M turned up and running parallel with wing margin.
SYRPHIDAE, p. 474

Wings without a spurious vein; branches of M not turned up to run parallel with wing margin.

Radial veins not thickened & clustered at wing base; pedicel of antenna not hidden.

First few wing veins (radial veins) usually short, thick and clustered toward wing base; pedicel of antenna usually hidden inside next segment; **often humpbacked and bristly.**
PHORIDAE, p. 474

448

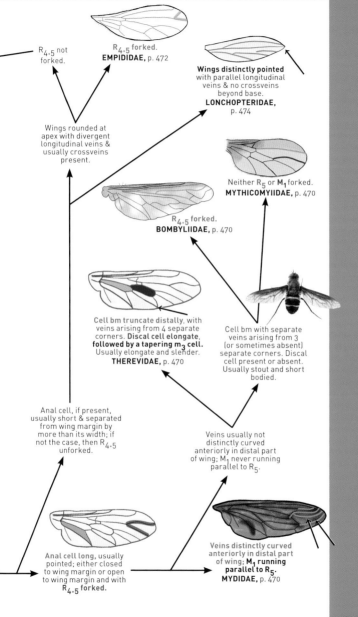

R$_{4.5}$ not forked.

R$_{4.5}$ forked.
EMPIDIDAE, p. 472

Wings distinctly pointed with parallel longitudinal veins & no crossveins beyond base.
LONCHOPTERIDAE, p. 474

Wings rounded at apex with divergent longitudinal veins & usually crossveins present.

Neither R$_5$ or **M$_1$** forked.
MYTHICOMYIIDAE, p. 470

R$_{4.5}$ forked.
BOMBYLIIDAE, p. 470

Cell bm truncate distally, with veins arising from 4 separate corners. **Discal cell elongate, followed by a tapering m$_3$ cell.** Usually elongate and slender.
THEREVIDAE, p. 470

Cell bm with separate veins arising from 3 (or sometimes absent) separate corners. Discal cell present or absent. Usually stout and short bodied.

Anal cell, if present, usually short & separated from wing margin by more than its width; if not the case, then R$_{4.5}$ unforked.

Veins usually not distinctly curved anteriorly in distal part of wing; M$_1$ never running parallel to R$_5$.

Anal cell long, usually pointed; either closed to wing margin or open to wing margin and with R$_{4.5}$ forked.

Veins distinctly curved anteriorly in distal part of wing; **M$_1$ running parallel to R$_5$.**
MYDIDAE, p. 470

449

KEY 4 TO COMMON FLY FAMILIES

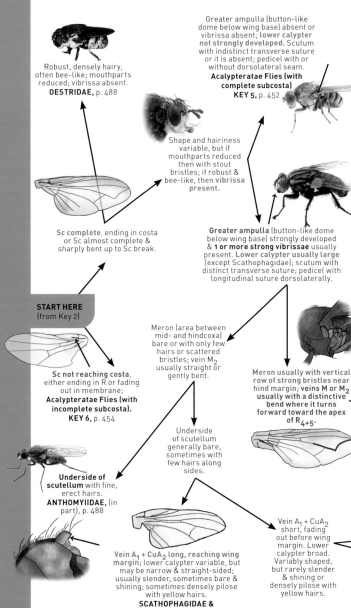

Robust, densely hairy, often bee-like; mouthparts reduced; vibrissa absent.
OESTRIDAE, p. 488

Greater ampulla (button-like dome below wing base) absent or vibrissa absent; **lower calypter not strongly developed.** Scutum with indistinct transverse suture or it is absent; pedicel with or without dorsolateral seam.
Acalypteratae Flies (with complete subcosta)
KEY 5, p. 452

Shape and hairiness variable, but if mouthparts reduced then with stout bristles; if robust & bee-like, then **vibrissa present.**

Sc complete, ending in costa or Sc almost complete & sharply bent up to Sc break.

Greater ampulla (button-like dome below wing base) strongly developed & **1 or more strong vibrissae** usually present. **Lower calypter usually large** (except Scathophagidae); scutum with distinct transverse suture; pedicel with longitudinal suture dorsolaterally.

START HERE (from Key 2)

Meron (area between mid- and hindcoxa) bare or with only few hairs or scattered bristles; vein M_2 usually straight or gently bent.

Meron usually with vertical row of strong bristles near hind margin; veins M or M_2 usually with a distinctive bend where it turns forward toward the apex of R_{4+5}.

Sc not reaching costa, either ending in R or fading out in membrane;
Acalypteratae Flies (with incomplete subcosta).
KEY 6, p. 454

Underside of scutellum generally bare, sometimes with few hairs along sides.

Underside of **scutellum** with fine, erect hairs.
ANTHOMYIIDAE, (in part), p. 488

Vein $A_1 + CuA_2$ long, reaching wing margin; lower calypter variable, but may be narrow & straight-sided; usually slender, sometimes bare & shining; sometimes densely pilose with yellow hairs.
SCATHOPHAGIDAE & ANTHOMYIIDAE, p. 488

Vein $A_1 + CuA_2$ short, fading out before wing margin. Lower calypter broad. Variably shaped, but rarely slender & shining or densely pilose with yellow hairs.

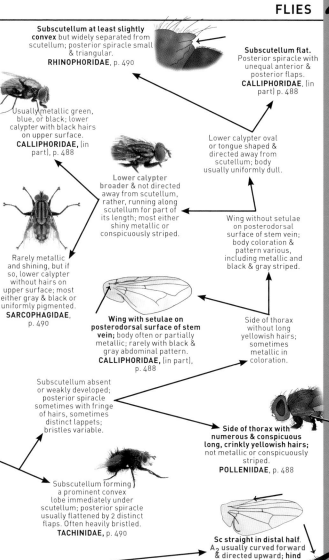

Subscutellum at least slightly convex but widely separated from scutellum; posterior spiracle small & triangular.
RHINOPHORIDAE, p. 490

Subscutellum flat. Posterior spiracle with unequal anterior & posterior flaps.
CALLIPHORIDAE, (in part) p. 488

Usually metallic green, blue, or black; lower calypter with black hairs on upper surface.
CALLIPHORIDAE, (in part), p. 488

Lower calypter **broader** & not directed away from scutellum, rather, running along scutellum for part of its length; most either shiny metallic or conspicuously striped.

Lower calypter oval or tongue shaped & directed away from scutellum; body usually uniformly dull.

Wing without setulae on posterodorsal surface of stem vein; body coloration & pattern various, including metallic and black & gray striped.

Rarely metallic and shining, but if so, lower calypter without hairs on upper surface; most either gray & black or uniformly pigmented.
SARCOPHAGIDAE, p. 490

Wing with setulae on posterodorsal surface of stem vein; body often or partially metallic; rarely with black & gray abdominal pattern.
CALLIPHORIDAE, (in part), p. 488

Side of thorax without long yellowish hairs; sometimes metallic in coloration.

Subscutellum absent or weakly developed; posterior spiracle sometimes with fringe of hairs, sometimes distinct lappets; bristles variable.

Side of thorax with numerous & conspicuous long, crinkly yellowish hairs; not metallic or conspicuously striped.
POLLENIIDAE, p. 488

Subscutellum forming a prominent convex lobe immediately under scutellum; posterior spiracle usually flattened by 2 distinct flaps. Often heavily bristled.
TACHINIDAE, p. 490

Sc straight in distal half. A$_2$ usually curved forward & directed upward; hind tibia with dorsal bristle near middle.
FANNIIDAE, p. 488

Sc curved toward costa in distal half. A$_2$ not curved forward & upward; hind tibia often with an **anterodorsal bristle near middle** (not truly dorsal).
MUSCIDAE, p. 488

DIPTERA

KEY 5 TO COMMON FLY FAMILIES

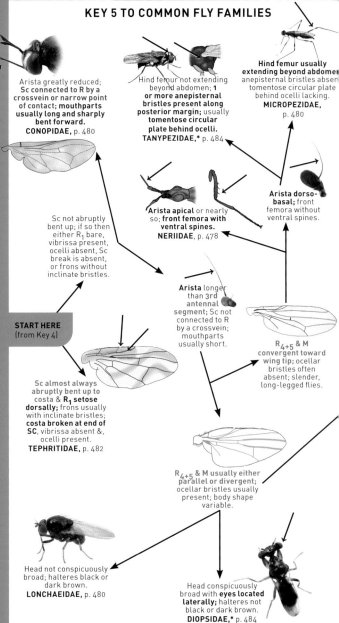

Arista greatly reduced; **Sc connected to R by a crossvein or narrow point of contact; mouthparts** usually long and sharply bent forward. **CONOPIDAE,** p. 480

Hind femur not extending beyond abdomen; **1 or more anepisternal bristles present along posterior margin;** usually **tomentose circular plate behind ocelli. TANYPEZIDAE,*** p. 484

Hind femur usually **extending beyond abdomen** anepisternal bristles absent tomentose circular plate behind ocelli lacking. **MICROPEZIDAE,** p. 480

Arista dorso-basal; front femora without ventral spines.

Sc not abruptly bent up; if so then either R₁ bare, vibrissa present, ocelli absent, Sc break is absent, or frons without inclinate bristles.

Arista apical or nearly so; **front femora with ventral spines. NERIIDAE,** p. 478

Arista longer than **3rd antennal segment;** Sc not connected to R by a crossvein; mouthparts usually short.

START HERE (from Key 4)

R₄₊₅ & M convergent toward wing tip; ocellar bristles often absent; slender, long-legged flies.

Sc almost always abruptly bent up to costa & **R₁ setose dorsally;** frons usually with inclinate bristles; **costa broken at end of SC,** vibrissa absent &, ocelli present. **TEPHRITIDAE,** p. 482

R₄₊₅ & M usually either **parallel or divergent;** ocellar bristles usually present; body shape variable.

Head not conspicuously broad; halteres black or dark brown. **LONCHAEIDAE,** p. 480

Head conspicuously broad with **eyes located laterally;** halteres not black or dark brown. **DIOPSIDAE,*** p. 484

452

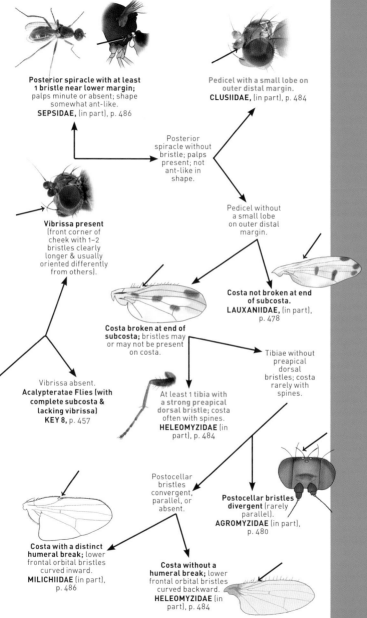

Posterior spiracle with at least 1 bristle near lower margin; palps minute or absent; shape somewhat ant-like.
SEPSIDAE, (in part), p. 486

Pedicel with a small lobe on outer distal margin.
CLUSIIDAE, (in part), p. 484

Posterior spiracle without bristle; palps present; not ant-like in shape.

Pedicel without a small lobe on outer distal margin.

Vibrissa present (front corner of cheek with 1–2 bristles clearly longer & usually oriented differently from others).

Costa not broken at end of subcosta.
LAUXANIIDAE, (in part), p. 478

Costa broken at end of subcosta; bristles may or may not be present on costa.

Tibiae without preapical dorsal bristles; costa rarely with spines.

Vibrissa absent.
Acalypteratae Flies (with complete subcosta & lacking vibrissa)
KEY 8, p. 457

At least 1 tibia with a strong preapical dorsal bristle; costa often with spines.
HELEOMYZIDAE (in part), p. 484

Postocellar bristles convergent, parallel, or absent.

Postocellar bristles divergent (rarely parallel).
AGROMYZIDAE (in part), p. 480

Costa with a distinct humeral break; lower frontal orbital bristles curved inward.
MILICHIIDAE (in part), p. 486

Costa without a humeral break; lower frontal orbital bristles curved backward.
HELEOMYZIDAE (in part), p. 484

453

KEY 6 TO COMMON FLY FAMILIES

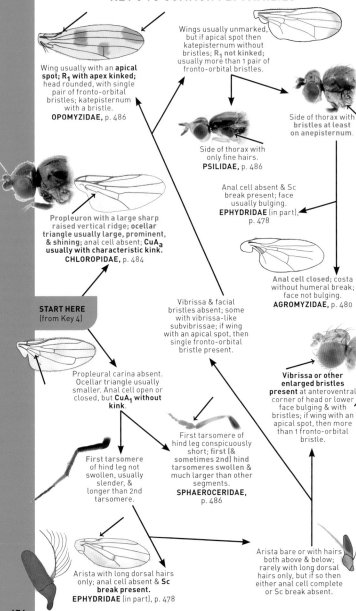

Wings usually unmarked, but if apical spot then katepisternum without bristles; R$_1$ not kinked; usually more than 1 pair of fronto-orbital bristles.

Wing usually with an **apical spot**; R$_1$ with apex kinked; head rounded, with single pair of fronto-orbital bristles; katepisternum with a bristle.
OPOMYZIDAE, p. 486

Side of thorax with **bristles at least on anepisternum**.

Side of thorax with only fine hairs.
PSILIDAE, p. 486

Anal cell absent & Sc break present; face usually bulging.
EPHYDRIDAE (in part), p. 478

Propleuron with a large sharp raised vertical ridge; ocellar triangle usually large, prominent, & shining; anal cell absent; CuA$_a$ usually with characteristic kink.
CHLOROPIDAE, p. 484

Anal cell closed; costa without humeral break; face not bulging.
AGROMYZIDAE, p. 480

START HERE (from Key 4)

Vibrissa & facial bristles absent; some with vibrissa-like subvibrissae; if wing with an apical spot, then single fronto-orbital bristle present.

Propleural carina absent. Ocellar triangle usually smaller. Anal cell open or closed, but **CuA$_1$ without kink**.

Vibrissa or other enlarged bristles present at anteroventral corner of head or lower face bulging & with bristles; if wing with an apical spot, then more than 1 fronto-orbital bristle.

First tarsomere of hind leg not swollen, usually slender, & longer than 2nd tarsomere.

First tarsomere of hind leg conspicuously short; first (& sometimes 2nd) hind tarsomeres swollen & much larger than other segments.
SPHAEROCERIDAE, p. 486

Arista with long dorsal hairs only; anal cell absent & **Sc break present**.
EPHYDRIDAE (in part), p. 478

Arista bare or with hairs both above & below; rarely with long dorsal hairs only, but if so then either anal cell complete or Sc break absent.

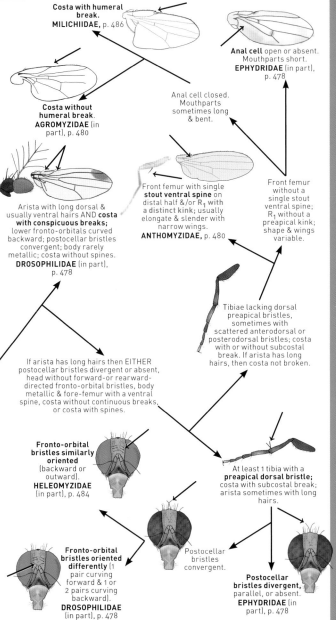

Costa with humeral break.
MILICHIIDAE, p. 486

Anal cell open or absent.
Mouthparts short.
EPHYDRIDAE (in part),
p. 478

Costa without humeral break.
AGROMYZIDAE (in part), p. 480

Anal cell closed.
Mouthparts sometimes long & bent.

Front femur with single **stout ventral spine** on distal half &/or R_1 with a **distinct kink**; usually elongate & slender with narrow wings.
ANTHOMYZIDAE, p. 480

Front femur without a single stout ventral spine; R_1 without a preapical kink; shape & wings variable.

Arista with long dorsal & usually ventral hairs AND **costa with conspicuous breaks**; lower fronto-orbitals curved backward; postocellar bristles convergent; body rarely metallic; costa without spines.
DROSOPHILIDAE (in part), p. 478

Tibiae lacking dorsal preapical bristles, sometimes with scattered anterodorsal or posterodorsal bristles; costa with or without subcostal break. If arista has long hairs, then costa not broken.

If arista has long hairs then EITHER postocellar bristles divergent or absent, head without forward- or rearward-directed fronto-orbital bristles, body metallic & fore-femur with a ventral spine, costa without continuous breaks, or costa with spines.

Fronto-orbital bristles similarly oriented (backward or outward).
HELEOMYZIDAE (in part), p. 484

At least 1 tibia with a **preapical dorsal bristle**; costa with subcostal break; arista sometimes with long hairs.

Postocellar bristles convergent.

Fronto-orbital bristles oriented differently (1 pair curving forward & 1 or 2 pairs curving backward).
DROSOPHILIDAE (in part), p. 478

Postocellar bristles divergent, parallel, or absent.
EPHYDRIDAE (in part), p. 478

DIPTERA

KEY 7 TO COMMON FLY FAMILIES

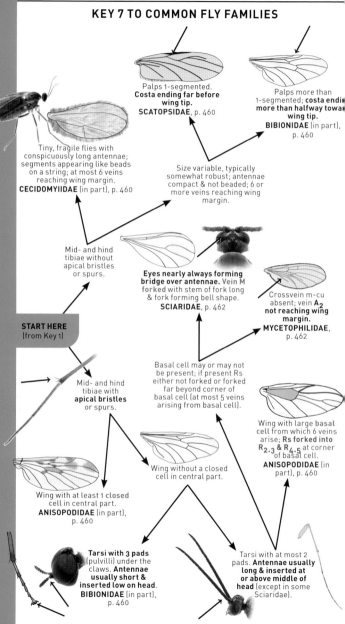

Palps 1-segmented.
Costa ending far before wing tip.
SCATOPSIDAE, p. 460

Palps more than 1-segmented; **costa ending more than halfway toward wing tip.**
BIBIONIDAE (in part), p. 460

Tiny, fragile flies with conspicuously long antennae; segments appearing like beads on a string; at most 6 veins reaching wing margin.
CECIDOMYIIDAE (in part), p. 460

Size variable, typically somewhat robust; antennae compact & not beaded; 6 or more veins reaching wing margin.

Mid- and hind tibiae without apical bristles or spurs.

Eyes nearly always forming bridge over antennae. Vein M forked with stem of fork long & fork forming bell shape.
SCIARIDAE, p. 462

Crossvein m-cu absent; vein **A₂ not reaching wing margin.**
MYCETOPHILIDAE, p. 462

START HERE (from Key 1)

Basal cell may or may not be present; if present Rs either not forked or forked far beyond corner of basal cell (at most 5 veins arising from basal cell).

Mid- and hind tibiae with **apical bristles or spurs.**

Wing with large **basal cell** from which 6 veins arise; **Rs forked into R₂.₃ & R₄.₅ at corner of basal cell.**
ANISOPODIDAE (in part), p. 460

Wing without a closed cell in central part.

Wing with at least 1 closed cell in central part.
ANISOPODIDAE (in part), p. 460

Tarsi with 3 pads (pulvilli) under the claws. **Antennae usually short & inserted low on head.**
BIBIONIDAE (in part), p. 460

Tarsi with at most 2 pads. **Antennae usually long & inserted at or above middle of head** (except in some Sciaridae).

456

KEY 8 TO COMMON FLY FAMILIES

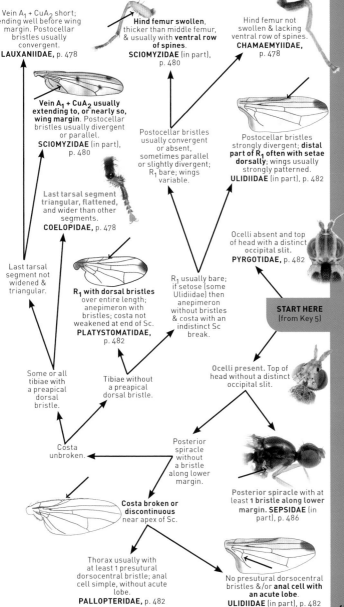

Vein A$_1$ + CuA$_2$ short; ending well before wing margin. Postocellar bristles usually convergent. **LAUXANIIDAE**, p. 478

Hind femur swollen, thicker than middle femur, & usually with **ventral row of spines**. **SCIOMYZIDAE** (in part), p. 480

Hind femur not swollen & lacking ventral row of spines. **CHAMAEMYIIDAE**, p. 478

Vein A$_1$ + CuA$_2$ usually **extending to, or nearly so, wing margin**. Postocellar bristles usually divergent or parallel. **SCIOMYZIDAE** (in part), p. 480

Postocellar bristles usually convergent or absent, sometimes parallel or slightly divergent; R$_1$ bare; wings variable.

Postocellar bristles strongly divergent; **distal part of R$_1$ often with setae dorsally**; wings usually strongly patterned. **ULIDIIDAE** (in part), p. 482

Last tarsal segment triangular, flattened, and wider than other segments. COELOPIDAE, p. 478

Last tarsal segment not widened & triangular.

R$_1$ with dorsal bristles over entire length; anepimeron with bristles; costa not weakened at end of Sc. **PLATYSTOMATIDAE**, p. 482

R$_1$ usually bare; if setose (some Ulidiidae) then anepimeron without bristles & costa with an indistinct Sc break.

Ocelli absent and top of head with a distinct occipital slit. **PYRGOTIDAE**, p. 482

START HERE (from Key 5)

Ocelli present. Top of head without a distinct occipital slit.

Some or all tibiae with a preapical dorsal bristle.

Tibiae without a preapical dorsal bristle.

Costa unbroken.

Posterior spiracle without a bristle along lower margin.

Posterior spiracle with at least **1 bristle along lower margin. SEPSIDAE** (in part), p. 486

Costa broken or discontinuous near apex of Sc.

Thorax usually with at least 1 presutural dorsocentral bristle; anal cell simple, without acute lobe. **PALLOPTERIDAE**, p. 482

No presutural dorsocentral bristles &/or **anal cell with an acute lobe. ULIDIIDAE** (in part), p. 482

DIPTERA

Suborder "Nematocera"
Antennae with 6 or more antennomeres. Larvae (except Cecidomyiidae & Tipulidae) with well-developed head.

MOUNTAIN MIDGES Family Deuterophlebiidae*   Not Illus

Delicate, midge-like, dark brown to black bodies with small head and arched thorax. Males with antennae 4 times as long as body; short, < 0.5 mm in female. Wings broad in male (similar to mayflies); smaller in female. Larvae found in swift-flowing streams. TL 2–4 mm; World 14, NA 6

NYMPHOMYIIDS Family Nymphomyiidae*   Not illus

Elongated, pale, slender, sub-cylindrical, and largely membranous. Vestigial wings long, thin, fringed, and feather-like. Antennae clubbed. Single genus, *Nymphomyia*. TL 1.5–2.5 mm; World 8, NA 2.

PHANTOM CRANE FLIES Family Ptychopteridae

Slender, often with **long legs, banded black and white** in some species. Lack ocelli. Antennae long with many antennomeres. One anal vein reaching wing margin; lack closed discal cell. **Wings banded in some.** Larvae in marshy and swampy ponds. TL 6–14 mm; World 160, NA 16.

Infraorder Tipulomorpha

CYLINDROTOMID CRANE FLIES Family Cylindrotomidae* Not illus

Large, pale brown crane flies with long legs and abdomen. Antennae long, slender with 16 antennomeres. Terminal segment of maxillary palpus short. Tip of $R_{1,2}$ usually atrophied; R_1 gradually converging toward R_3. TL 11–16 mm; World 115, NA 8.

LIMONIID CRANE FLIES Family Limoniidae

Antennae generally 14 to 16 antennomeres. Apical segment of maxillary palpus short or subequal to subapical segment. V-shaped suture between wings. R_2 fused with R_3. Ocelli absent. TL 10–16 mm; World 10,800, NA 900.

HAIRY-EYED CRANE FLIES Family Pediciidae

Eyes pubescent with short erect hairs between facets. Antennae long and thin with 12–17 antennomeres. V-shaped suture between wing bases. Wing with 2 anal veins; apical crossveins and M-Cu form oblique line. Ocelli absent. TL 5–35 mm; World 500, NA 150.

LARGE CRANE FLIES Family Tipulidae

Elongated abdomen, long thin legs, and short rostrum. Terminal segment of maxillary palpus longer than 1–3 combined. Antennae thin, usually with 13 antennomeres. V-shaped suture present. Ocelli absent. Wings sometimes maculated. TL 10–40 mm; World 4,400, NA 560.

WINTER CRANE FLIES Family Trichoceridae*

Antennae elongate, somewhat setaceous with 18 antennomeres. Three ocelli present. V-shaped suture between wing bases. Some found active or cool days, including on snow. TL 6–16 mm; World 180, NA 28.

NEMATOCERA • TIPULOMORPHA

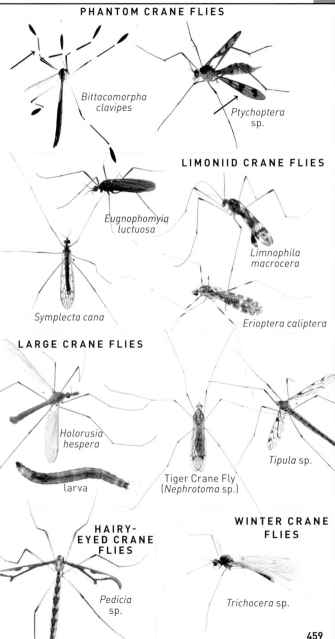

PHANTOM CRANE FLIES

Bittacomorpha clavipes

Ptychoptera sp.

LIMONIID CRANE FLIES

Eugnophomyia luctuosa

Limnophila macrocera

Symplecta cana

Erioptera caliptera

LARGE CRANE FLIES

Holorusia hespera

larva

Tiger Crane Fly (*Nephrotoma* sp.)

Tipula sp.

HAIRY-EYED CRANE FLIES

Pedicia sp.

WINTER CRANE FLIES

Trichocera sp.

459

DIPTERA

NEMATOCERA • BIBIONOMORPHA

Infraorder Bibionomorpha

WOOD GNATS Family Anisopodidae
Body slender, elongate; legs long and slender; head with 3 ocelli; wings often with faint spots. **Antennae as long as head and thorax combined;** 16 antennomeres with uniform flagellomeres. Larvae feed on decaying matter and sap. TL 4–10 mm; World 200, NA 9.

AXYMYIID FLIES Family Axymyiidae*
Stout bodied with **large eyes divided longitudinally** and relatively short antennae. Eyes holoptic in male; divided in female. Three ocelli on prominent convex tubercle. Antennae with 16 antennomeres; longer in males. **Thorax with prominent posterior ridge.** Found near slow water, larvae in saturated rotten wood. TL 7–9 mm; World 8, NA 2.

MARCH FLIES Family Bibionidae
Robust, generally dark, and often with contrasting colored thorax. Antennae usually short, inserted low on face. Eyes holoptic in male; divided in female. **Tibia with prominent apical spurs.** Wing with long basal radial and basal medial cells; Rs, when present, arising at or beyond bm-cu. Adults can sometimes be abundant and a pest. TL 5–12 mm; World 1,000, NA 60.

CANTHYLOSCELID FLIES Family Canthyloscelidae*   Not illus.
Small, slender, dark, and shiny with bead-like antennae. Maxillary palp 4-segmented. Eyes holoptic. Antennae with 12–16 antennomeres. TL 2–3.5 mm; World 16, NA 2.

PACHYNEURID FLIES Family Pachyneuridae*   Not illus.
Superficially similar to crane flies. Antennae as long as head and thorax; 15–17 antennomeres. Crossvein between branches of Rs. R and M with 3 branches reaching wing margin; closed discal cell. Single species, *Cramptonomyia spenceri*, in NA restricted to the northwest. TL 11–13 mm; World 8, NA 1.

MINUTE BLACK SCAVENGER FLIES Family Scatopsidae
Robust, dark, sometimes shiny. Antennae with 7–12 antennomeres. Maxillary palp 1-segmented. Wings with clear veins along margin, others faint. Larvae feed in decaying organic material and excrement. TL 0.6–5 mm; World 400, NA 72.

BOLITOPHILID FLIES Family Bolitophilidae*   Not illus.
Similar to Mycetophilidae. Crossvein r-m distinct; crossvein bm-cu or point of contact of M and CuA$_1$ well before level of base of Rs. Single genus, *Bolitophila*, in NA. TL 3–13 mm; World 60, NA 20.

GALL MIDGES & WOOD MIDGES Family Cecidomyiidae
Most minute and delicate with long legs and antennae. Wing venation reduced; costa usually continuous around wing. Tibial spurs absent. Galls often more recognizable than adults.
TL 1–8 mm (usually < 5 mm); World 1,000, NA 6,300.

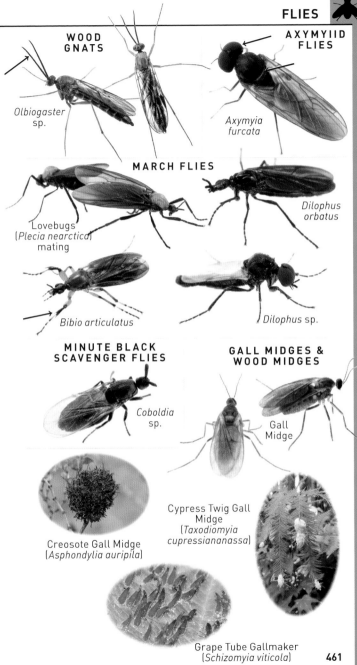

WOOD GNATS

Olbiogaster sp.

AXYMYIID FLIES

Axymyia furcata

MARCH FLIES

Lovebugs (*Plecia nearctica*) mating

Dilophus orbatus

Bibio articulatus

Dilophus sp.

MINUTE BLACK SCAVENGER FLIES

Coboldia sp.

GALL MIDGES & WOOD MIDGES

Gall Midge

Creosote Gall Midge (*Asphondylia auripila*)

Cypress Twig Gall Midge (*Taxodiomyia cupressiananassa*)

Grape Tube Gallmaker (*Schizomyia viticola*)

461

NEMATOCERA • BIBIONOMORPHA • CULICOMORPHA

Infraorder Bibionomorpha cont.

DIADOCIDIID FLIES Family Diadocidiidae*   Not illus.

Elongated, pale with long legs, and antennae longer than head
and thorax combined. Segments 2–4 of foretarsus swollen below
in female. Previously considered Mycetophilidae; single genus,
Diadocidia. TL 4–11 mm; World 40, NA 3.

DITOMYIID FLIES Family Ditomyiidae*   Not illus.

Elongated, with long legs and antennae as long as head and thorax
combined. Thorax strongly arched. M-Cu crossvein present; R_4
at least half as long as R_5; Sc short and ending free. Previously
considered Mycetophilidae. Found on rotting wood or fungi.
TL 4–7 mm; World 100, NA 6.

PREDATORY FUNGUS GNATS Family Keroplatidae*

Elongated, with long legs and antennae of variable length. Wing with
M and Cu sectors connected by m-cu; A_1 sometimes reaching wing
margin. Previously considered Mycetophilidae. Larvae often found
under bracket fungi in damp, forested environments; some species
found in caves. TL 8–13 mm; World 1,000, NA 80.

LONG-BEAKED FUNGUS GNATS
Family Lygistorrhinidae*   Not illus.

Elongated, with long legs, distinct rostrum, and antennae not
reaching to posterior portion of thorax. Male with distinctly banded
abdomen; female with uniformly dark, shorter abdomen. R_{2+3} absent.
Previously considered Mycetophilidae. Single species, *Lygistorrhina
sanctaecatharinae*, in eastern NA. TL 4–11 mm; World 30, NA 1.

FUNGUS GNATS Family Mycetophilidae

Elongate, slender, generally dull yellow, brown, or black, with long
legs and often pronounced thoracic and tibial bristles. **Tibia often
with pronounced apical spurs.** Wings often marked with maculation.
Usually found in moist areas feeding on decaying vegetation and fungi.
TL 2–14 mm; World 4,500, NA 620.

DARK-WINGED FUNGUS GNATS Family Sciaridae

Dark bodied, with long antennae with **eyes meeting above antennal
bases.** Crossvein r-m in line with and appearing as basal extension of
Rs. Typically found in wet, shady habitats where larvae feed on fungi.
TL 1–11 mm (usually <5 mm); World 2,500, NA 170.

Infraorder Culicomorpha

BITING MIDGES, NO-SEE-UMS Family Ceratopogonidae

Minute, ocelli absent, generally robust. Female with heavily setose
filiform antennae; male plumose. Wings held flat over abdomen when
at rest and sometimes patterned. Females bite vertebrates (important
disease vectors) and other insects. TL 1–4 mm; World 6,000, NA 600.

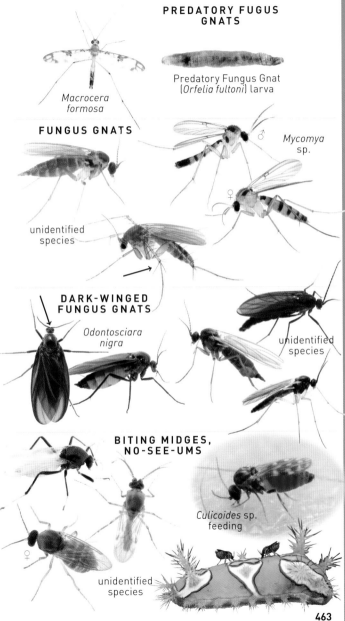

PREDATORY FUGUS GNATS

Macrocera formosa

Predatory Fungus Gnat (*Orfelia fultoni*) larva

FUNGUS GNATS

Mycomya sp.

♂

♀

unidentified species

DARK-WINGED FUNGUS GNATS

Odontosciara nigra

unidentified species

BITING MIDGES, NO-SEE-UMS

♀

unidentified species

Culicoides sp. feeding

NEMATOCERA • CULICOMORPHA

Infraorder Culicomorpha cont.

PHANTOM MIDGES Family Chaoboridae
Similar to Culicidae, usually pale yellow, gray, or brown with short proboscis and fewer scales on wings. Ocelli absent. Antennae plumose in male with enlarged pedicel. Adults don't bite. Larvae aquatic with raptorial antennae. TL 1.5–10 mm; World 90, NA 19.

MIDGES Family Chironomidae
Delicate, often elongated with long narrow wings lacking scales. **Front legs longer than others and often held out in front and above head when at rest; front tarsi very long in most.** Males with plumose antennae. Adults often found in large aggregations. Most larvae aquatic; others in decaying matter, soil, and under bark. TL 1–13 mm (usually <10 mm); World 7,300, NA 1,050.

FROG-BITING MIDGES Family Corethrellidae*   Not illus.
Similar to Chaoboaridae, but has swollen middle femur, clypeus with few setae and R_1 terminates closer to Sc than to R_2. Females attracted to calls of mating frogs, which are external parasites. Single genus, *Corethrella*, in NA. TL 1.5–2.5 mm; World 110, NA 4.

MOSQUITOES Family Culicidae
Slender, delicate bodies with long thin legs and **proboscis longer than head. Wing veins and often most of body covered with scales**. Antennae plumose, with whorls of setae longer and more dense in males. Labrum, mandibles, laciniae, and hypopharynx all greatly elongated. Larvae found in a variety of aquatic habitats. Females in many species are obligatory blood feeders and act as vectors of several important diseases. TL 3–15 mm; World 3,700, NA 174.

MENISCUS MIDGES Family Dixidae*   Not illus.
Similar to Culicidae, but with short proboscis. Body usually yellowish to dark brown. Body and wings lacking scales. R_{2+3} moderately to strongly arched. Adults do not bite. Larvae are aquatic, feeding at surface. TL 4–8 mm; World 200, NA 42.

BLACK FLIES Family Simuliidae
Small, robust, strongly **humpbacked,** short legs appearing black, gray, or yellow, small head with prominent round eyes. Ocelli lacking. Wings clear, broad, and lacking scales or hairs; heaviest veins along anterior wing margin. Females are bloodsucking and can be a considerable pest to both humans and livestock. Larvae live in streams and rivers. TL 1–5.5 mm; World 2,100, NA 255.

TRICKLE MIDGES Family Thaumaleidae*   Not illus.
Stocky, head arising low on thorax, shiny brown to yellow. Antennae short (no longer than head), slender, and arista-like. Eyes holoptic in both sexes; ocelli absent. Wings with 7 veins reaching margin; C extends around entire margin. TL 2–4.5 mm; World 200, NA 26.

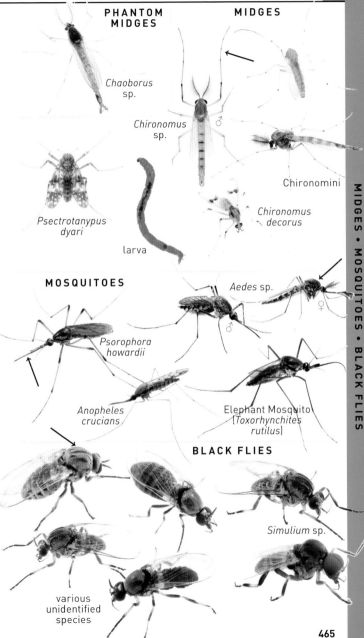

PHANTOM MIDGES

MIDGES

Chaoborus sp.

Chironomus sp. ♂

Chironomini

Psectrotanypus dyari

Chironomus decorus

larva

MOSQUITOES

Aedes sp. ♀

Psorophora howardii

Anopheles crucians

Elephant Mosquito
(*Toxorhynchites rutilus*)

BLACK FLIES

Simulium sp.

various unidentified species

465

DIPTERA

NET-WINGED MIDGES Family Blephariceridae*   Not illus.

Similar to crane flies with long legs, but lack V-shaped suture between wings. Wings with network of fine lines and folds between veins; anal angle well developed; base of M3 lacking. Larvae found in swift water clinging to rocks. TL 3–13 mm; World 330, NA 40.

MOTH FLIES & SAND FLIES Family Psychodidae

Very hairy with broad wings, appearing moth-like; many species hold wings roof-like over abdomen. **Antennae long with 12–16 bulbous antennomeres encircled by numerous hairs.** Larvae aquatic; living in decaying organic matter. TL 1.5–4 mm; World 3,000, NA 113.

PRIMITIVE CRANE FLIES Family Tanyderidae*

Similar to Tipulidae, but **dark with mottled or banded wings**. Ocelli absent. Antennae usually with 16 antennomeres. Wings with 3A present and 5 R veins. Larvae found along streams in sandy soil; adults hang from riparian vegetation. TL 10–22 mm; World 55, NA 4.

Suborder Brachycera
Antennae 10 or fewer antennomeres; maxillary palp with 2 or fewer segments; larvae without distinct head capsule.

Infraorder Stratiomyomorpha

SOLDIER FLIES Family Stratiomyidae

Variable in color and appearance, but many colorful and wasp-like; wings held flat over abdomen. Antennae in many, with **terminal flagellomeres typically directed laterally;** others are aristate. Wing with branches of R strong and crowded together toward costal margin; discal cell is small. TL 2–18 mm; World 2,700, NA 250.

WOOD SOLDIER FLIES Family Xylomyidae

Similar to Stratiomyidae, some slender and ichnuemonid-like.
Antennae with 8 flagellomeres tapering and diverging laterally. Spurs present on mid- and hind tibiae. Found in wooded areas, where larvae are often predaceous under bark. TL 5–15 mm; World 140, NA 11.

Infraorder Tabanomorpha

WATER SNIPE FLIES Family Athericidae

Similar to Rhagionidae, but lack spurs on front tibia, wings with R_1 cell closed at wing margin and dark brown to black banded abdomen. Antennae aristate. Eyes contiguous in males; separated in females. Found on riparian vegetation. TL 7–8 mm; World 130, NA 5.

BOLBOMYIID FLIES Family Bolbomyiidae*   Not illus.

Similar to Rhagionidae, but with clypeus flat. Brown or black and shiny; wings overlapping and held back over abdomen and slightly darkened. Single genus, *Bolbomyia*, in NA. TL 2–6 mm; World 4, NA 4.

MOTH FLIES & SAND FLIES

Psychoda sp.

Filter Fly (*Clogmia albipunctata*) larva & pupa

Filter Fly (*Clogmia albipunctata*)

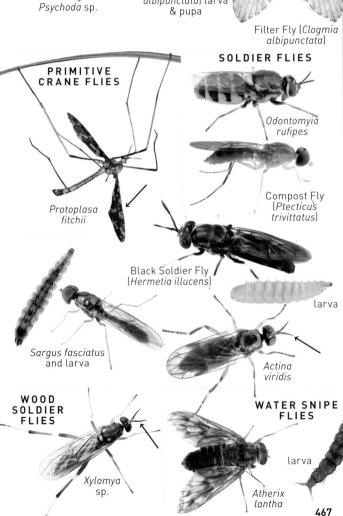

PRIMITIVE CRANE FLIES

Protoplasa fitchii

SOLDIER FLIES

Odontomyia rufipes

Compost Fly (*Ptecticus trivittatus*)

Black Soldier Fly (*Hermetia illucens*)

larva

Sargus fasciatus and larva

Actina viridis

WOOD SOLDIER FLIES

Xylomya sp.

WATER SNIPE FLIES

larva

Atherix lantha

467

DIPTERA

Infraorder Tabanomorpha cont.

OREOLEPTID FLIES Family Oreoleptidae*   Not illus.
Dull gray with stylate antennae; similar to Pelecorhynchidae. Wing
with cell r1 open. Single species, *Oreoleptis torrenticola*, found
in torrential streams of the Rocky Mountains. Larvae with long
crocheted prolegs on S2–7. TL 5–6 mm; World 1, NA 1.

PELECORHYNCHID FLIES Family Pelecorhynchidae* Not illus.
Dull brown to shiny black, clypeus and front of face swollen and
convex. Antennae projecting forward and laterally with stylate
flagellum. Wings clear or with some fusion of color. Larvae found in
soil of swamps and stream banks. TL 4–15 mm; World 50, NA 8.

SNIPE FLIES Family Rhagionidae
Generally brownish, gray, or black with elongated legs. Clypeus
bulbous. **Abdomen long and tapering posteriorly;** body either
bare or covered with short hairs. Antennae variable; 8 tapering
flagellomeres or with enlarged basal flagellomere with
unsegmented stylus or arista. Common in riparian wooded habitats.
TL 4–15 mm; World 750, NA 100.

HORSE FLIES & DEER FLIES Family Tabanidae
Robust, lacking bristles, but with large scales above halteres
and tarsi with 3 pads. **Antennae elongated and fused of several
antennomeres (stylate);** third segment bearing a prominent tooth
at base in some species. Wing veins R_4 and R_5 fork to form large
Y across wing tip. Females are blood feeders; males and some
females visit flowers. Many species with colorfully patterned eyes.
TL 8–25 mm; World 4,500, NA 350.

WORMLIONS Family Vermileonidae*   Not illus.
Slender, elongate flies with thin long legs and stylate antennae.
Similar to Rhagionidae but with wings elongate, decidedly
narrowed at base and apical spurs on front tibiae. Larvae, known as
wormlions, are predaceous, constructing pitfall traps in sand.
TL 4–5 mm; World 60, NA 2.

Infraorder Asilomorpha

APIOCERID FLIES Family Apioceridae*
Relatively large, elongate, gray to brown, bare or moderately setose.
Abdomen tapers posteriorly; male with enlarged terminalia and
female with stout spines on tergite 10. Similar to Asilidae, but without
notch between eyes. Single genus, *Apiocera*, found in arid west.
TL 7.5–35 mm (most < 20 mm); World 150, NA 65.

APSILOCEPHALID FLIES Family Apsilocephalidae*   Not illus.
Single species, *Apsilocephala longistyla*, in NA. Similar to Therevidae,
from which they can be distinguished only by genitalic characters.
TL 2–3 mm; World 3, NA 1.

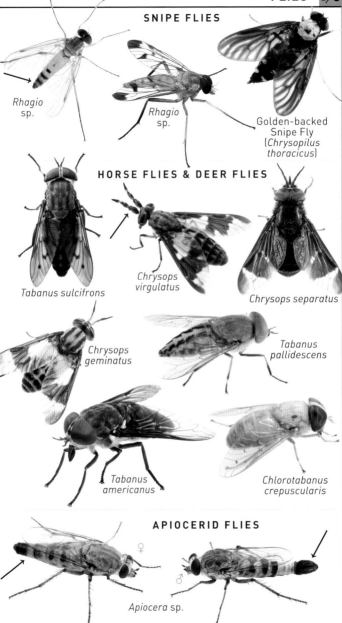

SNIPE FLIES

Rhagio sp.

Rhagio sp.

Golden-backed Snipe Fly (*Chrysopilus thoracicus*)

HORSE FLIES & DEER FLIES

Tabanus sulcifrons

Chrysops virgulatus

Chrysops separatus

Chrysops geminatus

Tabanus pallidescens

Tabanus americanus

Chlorotabanus crepuscularis

APIOCERID FLIES

♀

♂

Apiocera sp.

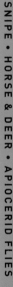

DIPTERA

Infraorder Asilomorpha cont.

ROBBER FLIES Family Asilidae

Variable in form, profusely hairy to nearly bare, typically elongate with tapered abdomen, **head with depression or notch between eyes** containing 3 ocelli and usually with very **hairy or "bearded" face.** Antennae with 3 antennomeres; third elongated and often with stout bristles. Mouthparts stout, injecting prey with saliva. Some species mimic bees and wasps. TL 5–30 mm; World 7,500, NA 1,040.

BEE FLIES Family Bombyliidae

Most are stout bodied, densely hairy with **long proboscis** and slender legs. Eyes large, nearly touching above (especially in males). **Wings variously maculated and held outstretched at rest;** m-cu absent. Larvae are largely ectoparasitoids of soil-inhabiting holometabolous larvae. TL 4–40 mm; World 5,000, NA 800.

MICRO BEE FLIES Family Mythicomyiidae

Most with strongly humpbacked thorax and not densely hairy. **Proboscis stout,** not as long as in Bombyliidae. Wings held together over abdomen at rest; R_{4+5} unbranched (branched in Bombyliidae). TL 0.5–5 mm; World 350, NA 170.

MYDAS FLIES Family Mydidae

Most with large, elongated bodies with slightly tapering abdomen, **long 4-segmented clubbed antennae,** and few bristles or setae except on legs. **Hind legs longer and more robust than others; femora swollen.** Wings long; veins ending at upper margin before apex. Many are mimics of wasps. Larvae are predaceous and live in decaying wood. TL 9–30 mm; World 500, NA 80.

WINDOW FLIES Family Scenopinidae*

Dark brown to black with very short proboscis, glabrous or few hairs, and antennae with 3 antennomeres. Eyes generally holoptic in males; divided in females; 3 ocelli. Wings with large discal cell, subapical cell, and long closed anal cell. One species, *Scenopinus fenestralis*, sometimes common on windows. TL 2–5 mm; World 420, NA 160.

STILETTO FLIES Family Therevidae

Moderately robust, body moderately hairy, **with abdomen tapering posteriorly.** Wings with R_4 elongate and sinuous; cell d elongate with 3 veins extending from its apex; m-cu present. Larvae predaceous, living in soil and decaying wood. TL 2.5–15 mm; World 1,600, NA 150.

HILARIMORPHID FLIES Family Hilarimorphidae*   Not illus.

Robust, dark colored, with wings clear to pale brown usually with pterostigma. Antennae with apical 2 antennomeres forming stylus. Short proboscis. Eyes holoptic in males, divided in females. Single genus, *Hilarimorpha*, in NA. TL 1.8–7.2 mm; World 36, NA 26.

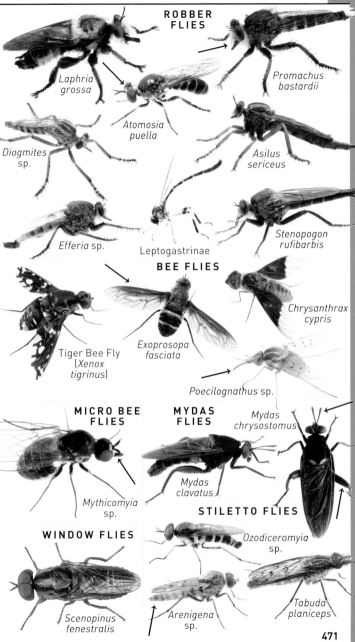

ROBBER FLIES

Laphria grossa

Atomosia puella

Promachus bastardii

Diogmites sp.

Asilus sericeus

Efferia sp.

Leptogastrinae

Stenopogon rufibarbis

BEE FLIES

Tiger Bee Fly (*Xenox tigrinus*)

Exoprosopa fasciata

Chrysanthrax cypris

Poecilognathus sp.

MICRO BEE FLIES

Mythicomyia sp.

MYDAS FLIES

Mydas clavatus

Mydas chrysostomus

STILETTO FLIES

Ozodiceromyia sp.

WINDOW FLIES

Scenopinus fenestralis

Arenigena sp.

Tabuda planiceps

Infraorder Asilomorpha cont.

SMALL-HEADED FLIES Family Acroceridae*
Robust, somewhat or strongly humpbacked, with relatively **small head consisting of large compound eyes;** some with proboscis extending length of body or longer. Dull or shiny, often metallic green, red, blue, or purple; others orange or yellow. Larvae are internal parasites of spiders. TL 2.5–21 mm; World 400, NA 60.

TANGLE-VEINED FLIES Family Nemestrinidae*   Not illus.
Robust, sometimes hairy, and bee-like. Wings clear or washed with brown; numerous cells in apical third of wing; radial and most medial veins run parallel to wing margin, terminating before apex. Parasitoids of other insects. TL 7–14 mm; World 300, NA 6.

XYLOPHAGID FLIES Family Xylophagidae*
Robust, similar to Rhagionidae; body generally not hairy and variously colored with abdomen tapering rearward. Wings spotted in many species; costa continues around wing margin; discal cell large, as close to posterior margin as anterior margin. TL 2–25 mm; World 150, NA 28.

APYSTOMYIID FLIES Family Apystomyiidae*   Not illus.
Black, shiny with clear wings; M_1 and M_2 joining beyond pointed apex of cell dm. Single species, *Apystomyia elinguis*, found in CA. Little is known of its biology. Taxonomic uncertainty has resulted in its placement in several families. TL <5 mm; World 1, NA 1.

ATELESTID FLIES Family Atelestidae*   Not illus.
Small, grayish; antennae with stylus. $M_{1,2}$ forked; Sc distinctly separated from R_1. Taxonomic placement is uncertain; placed in Empididae and Platypezidae at times. Single genus, *Meghyperus*, found in western NA. TL 2–3 mm; World 8, NA 2.

LONGLEGGED FLIES Family Dolichopodidae
Usually metallic green, blue, or copper with long legs. Lack frontal suture. Wings with r-m very short or absent, located in basal fourth; Rs often swollen fork. Many species engage in mating dance. TL 1–9 mm (usually <5 mm); World 7,400, NA 1,300.

DANCE FLIES Family Empididae
Generally dark with elongate body, **nearly spherical head with distinct neck**, humpbacked thorax, and tapered abdomen. **Legs long and relatively slender**. Adults can occur in large swarms, some with elaborate mating rituals. TL 1.5–12 mm; World 3,140, NA 460.

HYBOTID DANCE FLIES Family Hybotidae
Formerly considered subfamily of Empididae, but differ in thinner proboscis that points forward or diagonally, vein R_{4+5} never forked, and discal cell with 2 or fewer veins on the distal side.
TL 2–8 mm; World 2,000, NA 300.

SMALL-HEADED FLIES

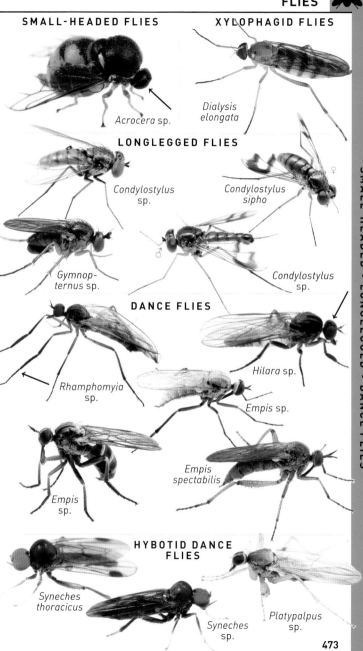

Acrocera sp.

XYLOPHAGID FLIES

Dialysis elongata

LONGLEGGED FLIES

Condylostylus sp.

Condylostylus sipho ♀

Gymnopternus sp.

♂ *Condylostylus* sp.

DANCE FLIES

Rhamphomyia sp.

Hilara sp.

Empis sp.

Empis sp.

Empis spectabilis

HYBOTID DANCE FLIES

Syneches thoracicus

Syneches sp.

Platypalpus sp.

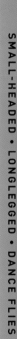

473

DIPTERA

BRACHYSTOMATID FLIES Family Brachystomatidae*   Not illus.
Nearly spherical head made up of large eyes, usually darkly colored, and humpbacked thorax and elongated abdomen. Wings held obliquely from body when at rest. Legs long and slender. Formerly placed in Empididae. TL 2–4 mm; World 150, NA 6.

OREOGETONID FLIES Family Oreogetonidae*   Not illus.
Single genus, *Oreogeton*, previously placed in Empididae. Dark or orange with tapering abdomen and long legs. Antennae with indistinct 4-segmented arista-like stylus. TL 4–6 mm; World 36, NA 3.

"Aschiza"
3-segmented antenna with arista; lack ptilinal suture on head.

BIG-HEADED FLIES Family Pipunculidae*
Head unusually large, comprised nearly entirely of eyes, body usually black, wings elongate and narrowed at base. Antennae generally with long arista arising from underside of antenna. Hover near vegetation in forest clearings. TL 1.5–10 mm; World 1,400, NA 130.

HOVER FLIES Family Syrphidae
Variable body shape, but many species resemble wasps and bees. Spurious vein between R and M in wings. Usually with scattered short hairs on thorax and abdomen, but never with dense long hairs. Many adults found on flowers; some larvae are predaceous on aphids and small insects, others in standing water, including tree holes (e.g., rat-tailed maggot). TL 4–35 mm (usually 10–20); World 6,000, NA 813.

POINTED-WINGED FLIES Family Lonchopteridae
Slender, yellowish or brown, with **wings coming to an acute point** and overlapping on abdomen when at rest. Sexes with different venation; M_3 cell closed in female, open in male. Found in moist, shady, and grassy habitats. Larvae found under leaves and decaying vegetation. Single genus, *Lonchoptera*, in NA. TL 2–5 mm; World 55, NA 7.

SCUTTLE FLIES Family Phoridae
Dark or yellowish bodies with humpbacked thorax, relatively small **head set low on thorax.** Antennae appear 1-segmented. Wing with costa extending only about halfway along anterior margin; 2 strong longitudinal veins anteriorly and 4–5 posteriorly, not connected by crossveins. Hind femur usually enlarged and flattened. Some larvae live in moist decaying organic material as scavengers or predators; others are parasitoids. TL 0.5–7 mm (usually <4); World 4,200, NA 376.

FLAT-FOOTED FLIES Family Platypezidae
Slender to robust with body usually black, brown, gray, yellow, or orangish. Wings clear or sometimes with brownish tint. **Hind tarsi usually flattened;** more so in females. Larvae live in fungus; adults run about on leaves and may swarm. TL 2–5 mm; World 280, NA 70.

BIG-HEADED FLIES

POINTED-WINGED FLIES

Chalarus sp.

Lonchoptera sp.

HOVER FLIES

Syritta pipiens

Dusky-winged Hover Fly (*Ocyptamus fuscipennis*) and larva

Yellowjacket Hover Fly (*Milesia virginiensis*)

Spilomyia longicornis

Toxomerus geminatus

Ornidia obesa

American Hover Fly (*Eupeodes* sp.)

Eastern Swiftwing (*Volucella evecta*)

SCUTTLE FLIES

FLAT-FOOTED FLIES

Pseudacteon tricuspis

Polyporivora polypori

Megaselia scalaris

475

"Acalyptratae Schizophora"
Antennae 3-segmented with arista; lack enlarged lower calypter.

BEE LICE Family Braulidae* Not illus.
Brown, wingless (also lacking halteres) species covered in long setae, found in honey bee hives. Eyes are reduced, located just above antennae, which are hidden in grooves. Legs short and robust. Single introduced species, *Braula coeca*, in NA. Adults roam around on honey bees feeding on mouth secretions. TL 1–2 mm; World 7, NA 1.

CAMILLID FLIES Family Camillidae* Not illus.
Similar to Drosophilidae, but are shiny metallic, lack sternopleural bristles, and wings with anal cell open apically. Little is known of the biology, but larvae are thought to feed on decaying plant material and animal feces. TL 1.5–3.5 mm; World 42, NA 5.

SCALE PARASITE FLIES Family Cryptochetidae* Not illus.
Superficially similar to Simuliidae; black, largely dull and humpbacked in appearance. Antennae with third antennomere enlarged, nearly reaching lower edge of head; lacks arista but bears a short spine at apex. Larvae are thought to be parasites of scale insects. Single intentionally introduced species, *Cryptochetum iceryae*, in NA. TL 1.5–2 mm; World 34, NA 1.

CURTONOTID FLIES Family Curtonotidae* Not illus.
Superficially similar to Drosophilidae; light yellow to dark brown and somewhat humpbacked. Head with post-vertical bristles well developed and crossing one another; 3 orbital bristles on either side of frons. Antennae with long plumose arista. Wings with posterior basal and discoidal cells fused. Found in tall grass in moist habitats. Single genus, *Curtonotum*, in NA. TL 6–8 mm; World 65, NA 3.

DIASTATID FLIES Family Diastatidae* Not illus.
Superficially similar to Drosophilidae; grayish to black, usually with patterned wings. Costa is weakly spinose. Found along margins of bogs, marshes, and woodlands. TL 2.5–4 mm; World 50, NA 8.

CYPSELOSOMATIDS Family Cypselosomatidae* Not illus.
Elongate, heavily bristled, and often with yellowish and black markings and banded legs. Wings slender. Closely related to Micropezidae. Some species thought to be associated with bat guano, but nothing is known about NA species. Single genus, *Latheticomyia*, known in southwestern NA. TL 2–3.5 mm; World 13, NA 2.

DRYOMYZID FLIES Family Dryomyzidae* Not illus.
Yellow-orange, brown, or gray, moderately to extensively hairy, similar to Sciomyzidae. Sc complete, well separated from costa. Larvae typically found in decaying organic matter, carrion, dung, or fungi. TL 4–12 mm; World 30, NA 9.

"Acalyptratae Schizophora" cont.

HELCOMYZID FLIES Family Helcomyzidae* Not illus.
Gray body with spotted wings. Antennae separated by more than diameter of antenna. Formerly placed in Dryomyzidae. Single species, *Helcomyza mirabilis*, found in Pacific northwest on ocean beaches. TL 11–12 mm; World 12, NA 1.

HETEROCHEILID FLIES Family Heterocheilidae* Not illus.
Dark gray with dark spots on pronotum, brown wings with reddish-brown legs. Front and top of head reddish-brown. Formerly placed in Dryomyzidae. Single species, *Heterocheila hannai*, found in Pacific northwest on ocean beaches. TL 8 mm; World 2, NA 1.

ACARTOPHTHALMID FLIES Family Acartophthalmidae*
Not Illus.
Poorly known group with gray bodies and pubescent arista. Adults associated with forests; larvae reared in dead wood. Single genus recognized, *Acartophthalmus*. TL 2–3 mm; World 5, NA 2.

ASTEIID FLIES Family Asteiidae* Not illus.
Elongated, shiny black thorax and pale abdomen. Wings clear with R_{2+3} usually shortened, ending in C at or just beyond R_1 (except in *Leiomyza*, where R_{2+3} ends beyond R_1 at about three-quarters wing length); last section of M_1 forming long anteriorly concave arc. Found at windows, bleeding tree wounds, and fungi. TL 1–2.5 mm; World 140, NA 19.

AULACIGASTRID FLIES Family Aulacigastridae* Not illus.
Dark brown to black and shiny or nearly all yellowish. Some species with face banded white, brown, or orange. Found at sap flows or in association with grasses. TL 2.5–5 mm; World 19, NA 5.

BEACH FLIES Family Canacidae* Not Illus.
Gray, black, brown, or yellowish with contrasting gray pruinesence ventrally. Similar to Ephydridae, but C has a single break, anal cell is present, and ocellar triangle is large. Restricted to intertidal habitats. TL 2–5 mm; World 320, NA 30.

CARNID FLIES Family Carnidae* Not illus.
Shiny, black, with 2 upper pairs of lateroclinate orbital bristles and 2 lower pairs of medioclinate frontal bristles. Wings with breaks in both C and Sc. One species is a blood-feeding ectoparasite of birds, others saprophagous, associated with carrion or excrement. TL 1–3 mm; World 93, NA 16.

STRONGYLOPHTHALMYIID FLIES
Family Strongylophthalmyiidae* Not Illus.
Slender, black with yellowish legs, antennae, and frons. Abdomen tapers posteriorly. Closely related to Tanypezidae, in which it is sometimes included. Ocellar tubercle displaced forward. Single genus, *Strongylophthalmyia*, in NA. TL 3–4 mm; World 80, NA 2.

477

DIPTERA

VINEGAR FLIES Family Drosophilidae
Typically with pale body color, **eyes often red,** and **costa with a break near junction of Sc.** Frons typically with 1 strong bristle directed posteriorly. Typically found around decaying vegetation and fruits; may become a pest in houses. Larvae feed on yeasts growing on rotting fruits. TL 2.5–4.5 mm; World 4,000, NA 180.

SHORE FLIES Family Ephydridae
Generally dull and dark-colored bodies, sometimes with patterned wings and facing bulging anteriorly. Pseudopostocellar bristles divergent or lacking. Antennae with bare, pubescent, or pectinate arista. Wings with Sc rudimentary; R_1 merging with costa before middle of wing. Often found along shores of marshes, ponds, and streams; larvae are aquatic, with many species living in brackish or alkaline waters, even crude petroleum (*Helaeomyia petrolei*).
TL 2.5–9 mm; World 2,000, NA 430.

CHAMAEMYIID FLIES Family Chamaemyiidae
Usually **gray with black spots,** sometimes shiny black; frons wide with at most 2 pairs of bristles. Proboscis and antennae short. Prothorax lacks bristles; 1 katepisternal bristle. **Front femora bear bristles.** Larvae are predatory and used in the biological control of aphids and scale insects. TL 1–5 mm; World 350, NA 150.

LAUXANIID FLIES Family Lauxaniidae
Yellowish-brown or black, relatively robust, generally with clear or yellow-tinted wings (sometimes patterned) and iridescent reddish or greenish eyes. Frons typically with 2 strong bristles above each eye; oral vibrissae lacking. Wings with Sc complete. Found in woodland and forest habitats. TL 3–6 mm; World 1,900, NA 136.

CACTUS FLIES Family Neriidae
Slender, brownish-gray to black bodies, sometimes variegated with yellowish-orange and lacking many bristles. **Antennae elongated, broadened apically** with bare or densely hairy apical arista. **Front femora with spines ventrally.** Wings brownish, yellowish, or clear. Adults found on decaying cacti, where larvae feed.
TL 10–12 mm; World 110, NA 3.

KELP FLIES Family Coelopidae
Brown, grayish, or black; robust; and densely **bristly or hairy, including legs.** Thorax slightly to strongly flattened dorsally. Eyes small, wings unmarked with costa entire, Sc complete BM-Cu present, and anal cell closed. Antennae bare or pubescent. Found on seashores, where larvae feed on rotting seaweed.
TL 3.7–9 mm (usually <7 mm); World 35, NA 4.

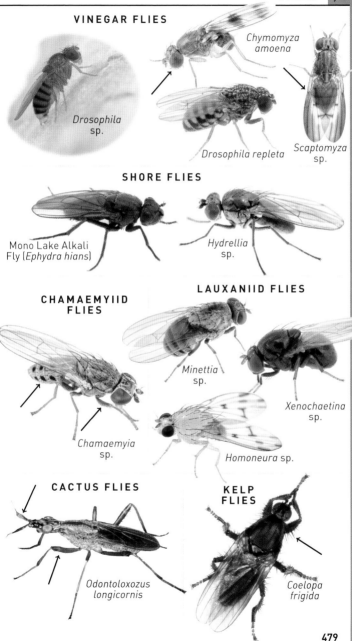

VINEGAR FLIES

Drosophila sp.

Chymomyza amoena

Drosophila repleta

Scaptomyza sp.

SHORE FLIES

Mono Lake Alkali Fly (*Ephydra hians*)

Hydrellia sp.

CHAMAEMYIID FLIES

LAUXANIID FLIES

Minettia sp.

Xenochaetina sp.

Chamaemyia sp.

Homoneura sp.

CACTUS FLIES

Odontoloxozus longicornis

KELP FLIES

Coelopa frigida

DIPTERA

STILT-LEGGED FLIES Family Micropezidae
Elongated, glabrous bodies, with very long, often **banded legs.**
Will walk around with front legs held out in front mimicking
Ichneumonidae with long antennae. Wings with R_5 narrowed
apically; anal cell often long and pointed. Found in marsh and wet
woodland habitats. TL 4–15 mm; World 600, NA 30.

THICK-HEADED FLIES Family Conopidae
Typically brown or black colored with red or yellow; abdomen
narrowed anteriorly and broadened posteriorly so as to appear like
thread-waisted wasps. Head with **long, slender proboscis** that may
be elbowed. **Antennae projecting forward, as long as or longer
than head.** Wings darkened, spotted, or unmarked; cell r_{4+5} closed
or nearly so. Some undergo significant color changes following
emergence. Often found on flowers; adults catch wasps in flight and
lay eggs. Larvae are internal parasites.
TL 4–18 mm; World 830, NA 67.

MARSH FLIES Family Sciomyzidae
Yellowish-brown or gray body, **elongated antennae extending
forward,** and **wings often patterned. Eyes red or multicolored
in many species.** Face often concave in profile. Middle femur with
bristle on anterior of face. Found in riparian habitats, where some
larvae feed on snails and slugs. TL 5–10 mm; World 620, NA 200.

LANCE FLIES Family Lonchaeidae
Shiny black with somewhat broad and flat abdomen that tapers
posteriorly. **Wings extend well beyond abdomen;** usually clear
or yellowish. Sc close to R_1, often with the space between them
darkened; wing margin bulges out before tip of Sc. Larvae generally
live under bark of dead trees. TL 3–6 mm; World 580, NA 120.

LEAF MINER FLIES Family Agromyzidae
Generally gray, black, or yellow; dull or shiny; with clear wings
and often reddish eyes. Most species more easily recognized by
leaf damage rather than individual fly. Postocellar bristles, when
present, divergent. TL 1–5 mm; World 3,000, NA 640.

ANTHOMYZID FLIES Family Anthomyzidae
Generally dull gray, brown, black, or yellow and similar in
appearance to Muscidae. Wings usually clear, sometimes brownish
or rarely spotted. Legs yellowish to black. Costa reaching to apex of
M_1; main veins, other than costa, lacking hairs. TL 2–5 mm; World
100, NA 14.

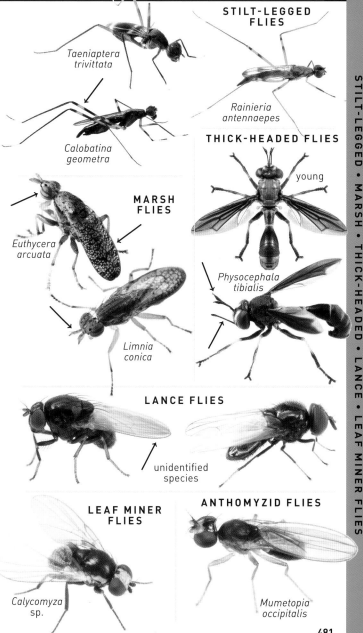

STILT-LEGGED FLIES

Taeniaptera trivittata

Rainieria antennaepes

Calobatina geometra

THICK-HEADED FLIES

young

MARSH FLIES

Euthycera arcuata

Physocephala tibialis

Limnia conica

LANCE FLIES

unidentified species

LEAF MINER FLIES

Calycomyza sp.

ANTHOMYZID FLIES

Mumetopia occipitalis

481

DIPTERA

"Acalyptratae Schizophora" cont.

FLUTTER FLIES Family Pallopteridae
Gray or yellowish bodies with **long wings marked with brown and held away from body.** Proboscis short and thick. Females with a non-retractable sheath over ovipositor. Found on flowers and low-hanging vegetation in shady riparian habitats.
TL 4–6 mm; World 70, NA 9.

CHEESE SKIPPERS AND KIN Family Piophilidae
Bodies typically shiny black or blue to dull yellowish with pronounced black bristles or sometimes densely hairy. Frons often conspicuously yellowish, at least in part. Wings unmarked or sometimes with brown markings. Typically found in shady, riparian habitats; some live in cheese and preserved meats.
TL 3–6 mm; World 80, NA 60.

SIGNAL FLIES Family Platystomatidae
Usually brightly and shiny colored with **strongly patterned wings.** Costa is uninterrupted; R_{4+5} with bristles; anal cell elongated, bordered on outer side by arcuate or straight vein. Found in moist habitats, where larvae feed on decaying or live vegetation.
TL 3–12 mm; World 1,200, NA 41.

PYRGOTID FLIES Family Pyrgotidae*
Elongate with **strongly patterned wings. Head rounded,** lacking ocelli. **Sponging mouthparts often enlarged and projected downward.** Antennae large, elongate, and projected anteriorly with bare subapical bristle on first flagellomere. Adults attracted to lights; larvae are internal parasitoids of scarab beetles in the genus *Phyllophaga*. TL 6–18 mm; World 350, NA 9.

RICHARDIID FLIES Family Richardiidae*   Not illus.
Body yellowish, brown, reddish-brown to black, with some areas shiny. Wings typically with brown bands or spots. Some species strongly sexually dimorphic. In NA, most diverse in the southwest at fruit-baited traps and on yucca. TL 3.5–15 mm; World 180, NA 10.

FRUIT FLIES Family Tephritidae
Wings usually **strongly patterned;** Sc bends apically forward at almost a right angle, then fades out; anal cell with acute distal projection posteriorly in most species. Adults found on flowers and vegetation. TL 2.5–10 mm; World 4,700, NA 360.

PICTURE-WINGED FLIES Family Ulidiidae
Many species brightly colored and/or shiny with **patterned wings.** Sc smoothly curving; some with acute distal projection posteriorly from anal cell. Often with **pronounced sponging mouthparts.** Some species pests on crops. TL 3–12 mm; World 680, NA 130.

FLUTTER FLIES

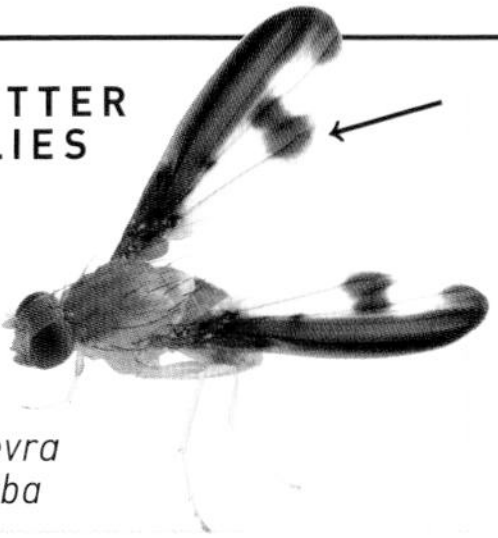

Toxonevra superba

CHEESE SKIPPERS AND KIN

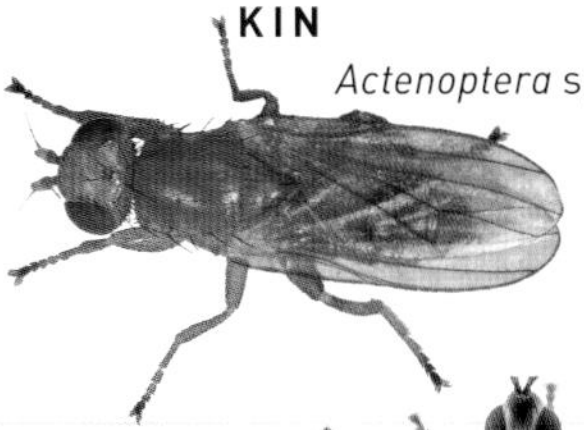

Actenoptera sp.

SIGNAL FLIES

Senopterina foxleei

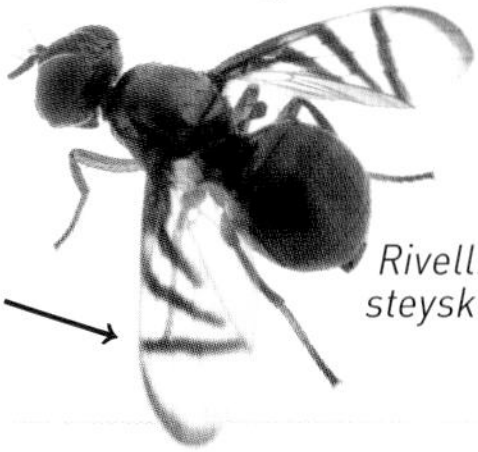

Rivellia steyskali

PYRGOTID FLIES

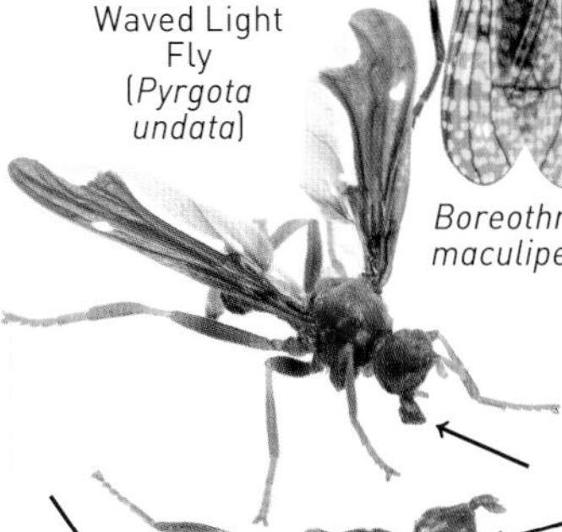

Waved Light Fly (*Pyrgota undata*)

Boreothrinax maculipennis

FRUIT FLIES

Trupanea sp.

Sphecomyiella valida

PICTURE-WINGED FLIES

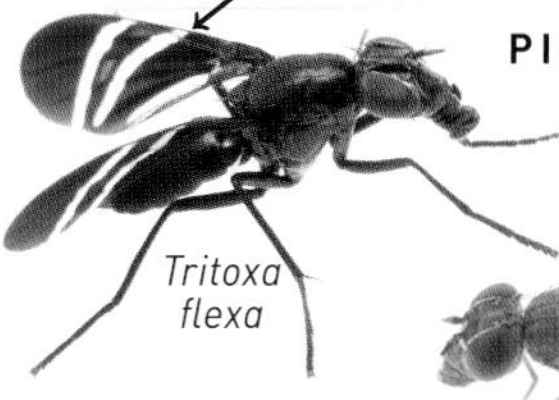

Tritoxa flexa

Delphinia picta

Tritoxa incurva

Chaetopsis massyla

Pseudotephritis vau

483

"Acalyptratae Schizophora" cont.

FRIT FLIES Family Chloropidae
Variable, black or gray, sometimes shiny to brightly colored with black markings. **Ocellar triangle well developed and shiny.** Wing venation somewhat reduced; costa with subcostal break; A_1, CuA_2 and cell cup absent. Can be common in grassy meadows, but found in a variety of habitats. TL 1.5–5 mm; World 3,000, NA 280.

CHYROMYID FLIES Family Chyromyidae*
Yellow bodies with yellow bristles, clear wings, and often **greenish eyes.** Superficially similar to Drosophilidae. Wings with well-developed anal lobe; costa extends to M_1 and broken by Sc; Sc complete, but weak in apical fifth. Found at windows, reared from bird nests and debris of hollow trees. TL 1–3 mm; World 140, NA 9.

CLUSIID FLIES Family Clusiidae
Slender, yellowish to black, dull to shiny, and variously ornamented. Antennae with extension on outer margin of pedicel and a dorsoapical arista. Wings often smoky or marked with brown; costa usually reaching M_1; Sc complete. Five or fewer fronto-orbital bristles. Larvae found in decaying wood. TL 2–6 mm; World 360, NA 43.

STALK-EYED FLIES Family Diopsidae*
Head somewhat triangular when viewed straight on with **eyes widely separated;** body brown to black. **Scutellum with pair of stout processes.** Front femora swollen, typically with short spines or bristles ventrally. Single genus, *Sphyracephala*, restricted to northeast in NA. TL 3.5–4.5 mm; World 200, NA 2.

HELEOMYZID FLIES Family Heleomyzidae
Brown, reddish-brown, yellow, to black with moderate bristles and often distinctly pruinose. Apparent third antennal segment (first flagellomere) is short, with a dorsal arista. Stout bristles on face below antennae. **Costa with prominent spines.** Typically found in shaded areas near rotting organic matter.
TL 2–7 mm; World 740, NA 113.

TANYPEZID FLIES Family Tanypezidae*
Dark with patches of **silvery tomentum on thorax and head.** Long legs and sometimes **first tarsomere yellowish.** Single genus, *Tanypeza*, in northeastern NA. Found in moist woods. TL 5–7 mm; World 70, NA 2.

PERISCELIDID FLIES Family Periscelididae*   Not illus.
Dull gray, black, or brown, sometimes with brownish-marked wings and appearing Drosophilidae like. Tibiae often banded. Wing with R_1 joining near middle of C; R_{4+5} and M convergent. Found at injured trees oozing sap. TL 3–4 mm; World 90, NA 3.

FRIT FLIES

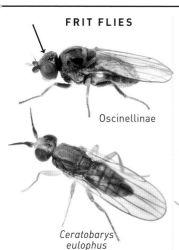

Oscinellinae

Ceratobarys eulophus

CHYROMYID FLIES

Gymnochiromyia sp.

CLUSIID FLIES

Clusia lateralis

HELEOMYZID FLIES

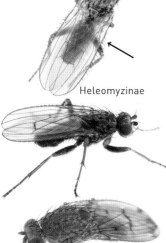

Heleomyzinae

Suillia sp.

STALK-EYED FLIES

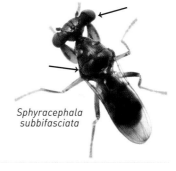

Sphyracephala subbifasciata

TANYPEZID FLIES

Tanypeza sp.

"Acalyptratae Schizophora" cont.

FREELOADER FLIES Family Milichiidae
Black, usually somewhat shiny appearing silvery. Short, slender proboscis in many species. Two or three pairs of orbital bristles (usually latero- or proclinate) and 2 pairs of medioclinate frontal bristles. Wings with humeral and Sc breaks in costa. Larvae feed on sap, dung, or other decaying organic matter; adults can be common in open areas. TL 1–5 mm; World 350, NA 36.

OPOMYZID FLIES Family Opomyzidae
Slender, usually shiny or pruinose yellow, brown, or black with wings having at least an **apical dark spot** and often more extensively marked. Found in grassy areas, where larvae feed on stems. Wings with anal lobe usually absent; costa broken only at apex of Sc and extending to end of M; cells in posterior half narrow. TL 2–5 mm; World 60, NA 13.

RUST FLIES Family Psilidae
Slender or moderately robust, yellow to red, brown, or black without many bristles. Antennae either short or relatively long with the first flagellomere elongated. Wings clear, yellowish, or brownish with darkened areas apically, along the costa, and sometimes on crossveins. Larvae live in roots or galls of plants. TL 3–8 mm; World 320, NA 30.

ROPALOMERID FLIES Family Ropalomeridae*   Not Illus.
Gray thorax with brown longitudinal stripes. Wings clear with darkened cells. Femora, especially hind femur, enlarged. Hind tibia laterally flattened. Wings with cell r_5 narrowed apically. Single species, *Rhytidops floridensis*, found in FL on fresh palm exudates. TL 6–8 mm; World 33, NA 1.

BLACK SCAVENGER FLIES Family Sepsidae
Black, shiny bodies, sometimes with reddish tinge, brown, or yellow; **spherical head;** and **abdomen distinctly narrowed anteriorly.** Some species with dark spot near apex of wing. Larvae feed in excrement and decaying organic matter. TL 2–6 mm; World 350, NA 34.

LESSER DUNG FLIES Family Sphaeroceridae
Black or brown, generally with clear to milky white wings, and **first tarsomere of hind leg short and thickened.** In many, longitudinal veins weak, not reaching wing margin. Can be common and numerous on excrement. TL 1–5 mm; World 1,600, NA 280.

ODINIID FLIES Family Odiniidae*   Not illus.
Usually dull gray with black or brown markings, compact, and strongly bristled. Legs robust, with banded tibiae. Wings with few to numerous brown spots; always darkened around Sc break. Often found on sap or fungi on dying trees. TL 2–5 mm; World 65, NA 11.

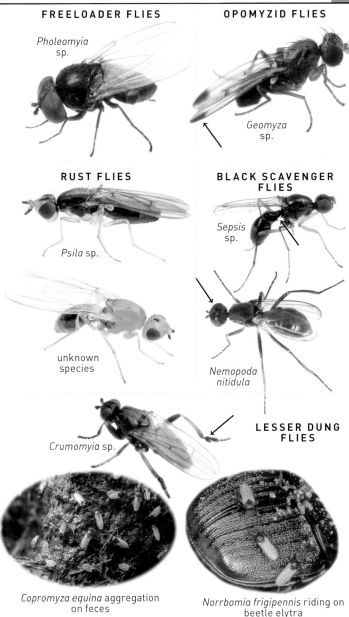

FREELOADER FLIES

Pholeomyia sp.

OPOMYZID FLIES

Geomyza sp.

RUST FLIES

Psila sp.

unknown species

BLACK SCAVENGER FLIES

Sepsis sp.

Nemopoda nitidula

Crumomyia sp.

LESSER DUNG FLIES

Copromyza equina aggregation on feces

Norrbomia frigipennis riding on beetle elytra

DIPTERA

"Calyptratae Schizophora"
Antennae 3-segmented with arista; enlarged lower calypter
usually present.

ROOT-MAGGOT FLIES Family Anthomyiidae
Dull yellowish, brown, gray, or black with well-developed calypter, resembling small Muscidae. Legs yellowish to black. Wings with r5 cell parallel sided; 2A reaching margin of wing. Most with fine hairs on underside of scutellum. Most larvae are plant feeders, some are pests on crops. TL 2–12 mm; World 2,000, NA 640.

HOUSE FLIES Family Muscidae
Slender to robust, with prominent bristles; typically dull black, gray, or yellow. Frontoclypeal suture present. Antennae with arista usually plumose its entire length. Wings usually unmarked; Rs with 2 branches; cell r5 either parallel sided or narrowed distally; 2A short, not reaching wing margin. Common around livestock, and some species enter houses. TL 2–14 mm; World 5,200, NA 700.

FANNIID FLIES Family Fanniidae
Body, legs gray or dark brown to black. Hind tibia with dorsal bristle at three-fifths its length. Sc and R_1 widely separated; Sc straight most of its length; axillary vein strongly curved toward wing tip. Larvae are scavengers. TL 2–12 mm; World 360, NA 110.

DUNG FLIES Family Scathophagidae
Slender, black, gray, brown, or yellow; strongly to weakly **armed with bristles;** some densely hairy. Wings most often clear, but some with spots or transverse bands. Abdomen in male often expanded apically. Underside of scutellum lacks fine hairs. Some larvae are leaf miners, others feed in dung or are aquatic predators. TL 3–11 mm; World 400, NA 150.

BLOW FLIES Family Calliphoridae
Most shiny blue, green, or black with antennae plumose throughout its length. Most with 2, rarely 3, notopleural bristles. Sexually dimorphic; males with frons narrower and lacking orbital and outer vertical setae. Females lay eggs on living or dead animal tissue; can be common. TL 4–16 mm; World 1,500, NA 84.

CLUSTER FLIES Family Polleniidae
Similar to house flies, but slightly larger and narrower, without distinct thoracic stripes, and wings overlap at tips when at rest. Thorax with numerous short yellowish hairs. Previously considered a subfamily of Calliphoridae. TL 6–11 mm; World 190, NA 21.

BOT FLIES Family Oestridae
Robust and often bee-like, with vestigial mouthparts and short antennae. Bodies hairy or densely pilose. Larvae are endoparasites of mammals. TL 9–25 mm; World 180, NA 40.

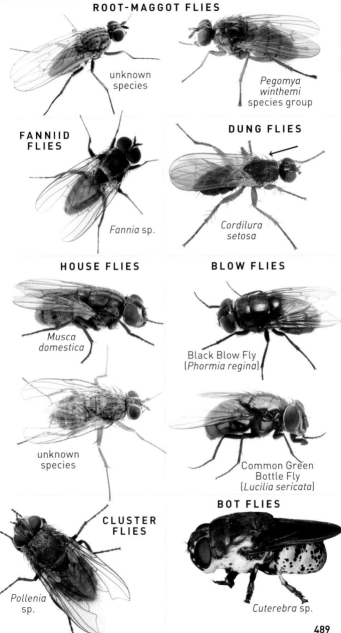

ROOT-MAGGOT FLIES

unknown species

Pegomya winthemi species group

FANNIID FLIES

Fannia sp.

DUNG FLIES

Cordilura setosa

HOUSE FLIES

Musca domestica

unknown species

BLOW FLIES

Black Blow Fly (*Phormia regina*)

Common Green Bottle Fly (*Lucilia sericata*)

CLUSTER FLIES

Pollenia sp.

BOT FLIES

Cuterebra sp.

"Calypteratae Schizophora" cont.

WOODLOUSE FLIES Family Rhinophoridae   Not illus.
Black or gray, similar to Tachinidae and Sarcophagidae, but have a weakly developed subscutellum and narrow calypteres. Body bristly. Wings with r5 cell closed. Either dark winged with M meeting R_{4+5} perpendicularly away from wing margin or clear winged with M meeting R_{4+5} obtusely near wing margin. Parasitic on woodlouse and other arthropods.
TL ~5 mm; World 170, NA 5.

FLESH FLIES Family Sarcophagidae
Generally dull gray or black, often with **3 gray thoracic dorsal stripes.** Abdomen often checkered, striped, or banded. Antennae plumose in basal half; bare in distal half. Many species sexually dimorphic; male with narrower frons and elongated tarsal claws and pulvilli. Most species are very difficult to identify. Larvae in many species feed on carrion, others parasitize arthropods.
TL 3–18 mm; World 3,100, NA 400.

PARASITIC FLIES Family Tachinidae
Large and variable family, but generally with **heavily bristled body** (including abdomen) and subscutellum well developed, as convex on upper half as on lower half when viewed in profile. Antennae usually with bare arista. Hypopleural and pteropleural bristles are prominent. Found in nearly all habitats; larvae parasitize other arthropods. TL 3–25 mm; World 10,000, NA 1,350.

LOUSE FLIES Family Hippoboscidae
Robust, somewhat dorsoventrally flattened, and dull to somewhat shiny light to dark brown. Antennae highly modified, lying in single or paired antennal sockets. **Mouthparts also highly specialized,** placed anteriorly giving head a somewhat triangular appearance. Wings present in most species, but sometimes reduced or absent. Adults feed on blood of birds and mammals.
TL 1.5–12 mm; World 210, NA 31.

NYCTERIBIID FLIES Family Nycteribiidae*   Not Illus.
Bristly, spider-like, lacking scutellum and wings, but with halteres present. Head folds back on thorax when at rest. Thorax dorsoventrally flattened; dorsally membranous, ventrally sclerotized. Adults are obligate blood feeders on bats and live on their hair. TL 1.5–5 mm; World 280, NA 6.

BAT FLIES Family Streblidae*   Not illus.
Either dorsoventrally flattened or laterally compressed, some densely setose. Usually with wings, but some reduced or absent. Head in some species with ventral comb. Adults are obligate blood feeders on bats, living on their hair. TL 2–3 mm; World 240, NA 8.

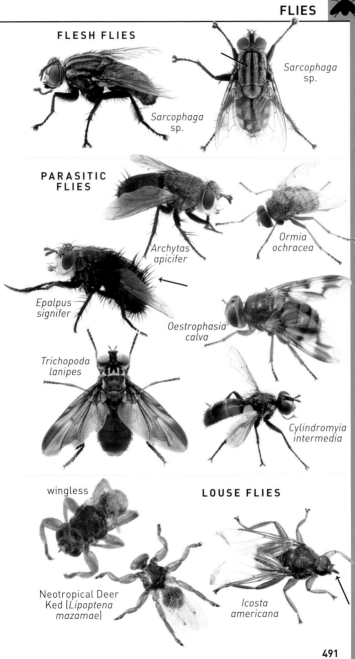

FLESH FLIES

Sarcophaga sp.

Sarcophaga sp.

PARASITIC FLIES

Archytas apicifer

Ormia ochracea

Epalpus signifer

Oestrophasia calva

Trichopoda lanipes

Cylindromyia intermedia

wingless

LOUSE FLIES

Neotropical Deer Ked (*Lipoptena mazamae*)

Icosta americana

Golden Silk Orbweaver
(*Trichonephila clavipes*)

NON-INSECT ARTHROPODS

There are a number of non-insect arthropod groups. While many are marine crustaceans, there are a number that you will no doubt encounter during your pursuit of insects. The major terrestrial groups that the average observer will encounter include the isopods, centipedes, millipedes, and arachnids. Many of these are closely associated with insects, either occurring in the same habitat as many insects or as predators of insects. The following pages are meant to represent just some of the more common or distinctive groups and species that you may encounter and provide a small guide to the taxonomic placement of others. The higher-level taxonomy for the non-insect arthropod groups that follow and their close relatives is provided below.

subphylum CHELICERATA
    class Arachnida
        superorder Acariformes—mites
        superorder Parasitiformes—mites & ticks
        order Amblypygi—tailless whipscorpions & whip spiders
        order Araneae—spiders
        order Opiliones—harvestmen
        order Palpigradi—micro-whipscorpions
        order Pseudoscorpiones—pseudoscorpions
        order Ricinulei—hooded tickspiders
        order Schizomida—short-tailed whipscorpions
        order Scorpiones—scorpions
        order Solifugae—camel spiders, & windscorpions
            order Thelyphonida (Uropygi)—whipscorpions & vinegaroons
    class Merostomata—horseshoe crabs
        order Xiphosura
    class Pycnogonida—sea spiders
        order Pantopoda
subphylum CRUSTACEA
    class Branchiopoda—water fleas & brine, fairy & tadpole shrimp
    class Remipedia
    class Cephalocarida
    class Maxillopoda—barnacles, copepods, & relatives
    class Thecostraca
    class Ostracoda—seed shrimp
    class Malacostraca (selected orders)
        order Decapoda—crabs, crayfishes, lobsters, & shrimp
        order Amphipoda—amphipods
        order Isopoda—isopods
subphylum MYRIAPODA
    class Chilopoda—centipedes
    class Diplopoda—millipedes
    class Symphyla—symphylans
    class Pauropoda—pauropods

NON-INSECT ARTHROPODS

SUBPHYLUM CHELICERATA

Body with 2 regions (prosoma and opisthosoma), though mites and harvestmen have lost a visible division; no antennae; possess chelicerate mouthparts, usually appearing as fangs or pincers; total of 6 pairs of appendages: chelicerae, pedipalps, 4 pairs of walking legs.

Class Arachnida

Nearly all members with 8 legs, first pair can be highly modified. Larval ticks and mites normally have 6 legs. Additional pair of appendages, known as pedipalps, anterior to legs. Nearly all terrestrial.

MITES & TICKS subclass Acari

Includes 2 large orders that may not be closely related. Have simple, unsegmented body (prosoma and opisthosoma are fused). Have a retractable feeding apparatus (including chelicerae, pedipalps, & oral cavity) called the *gnathosoma*, or capitulum. Some are parasitic and feed on a host, while others ingest solid food, including macroinvertebrates, fungi, and detritus. Most are less than 1 mm, but ticks are larger. TL <1 mm–3 mm; World 50,000, NA unknown.

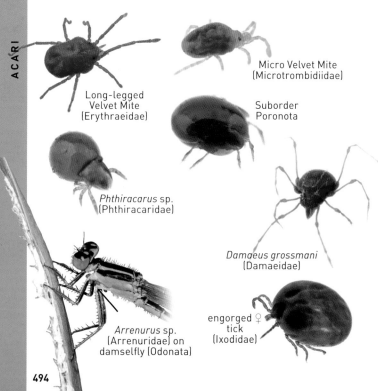

Long-legged Velvet Mite (Erythraeidae)

Micro Velvet Mite (Microtrombidiidae)

Suborder Poronota

Phthiracarus sp. (Phthiracaridae)

Damaeus grossmani (Damaeidae)

Arrenurus sp. (Arrenuridae) on damselfly (Odonata)

engorged ♀ tick (Ixodidae)

ACARI

494

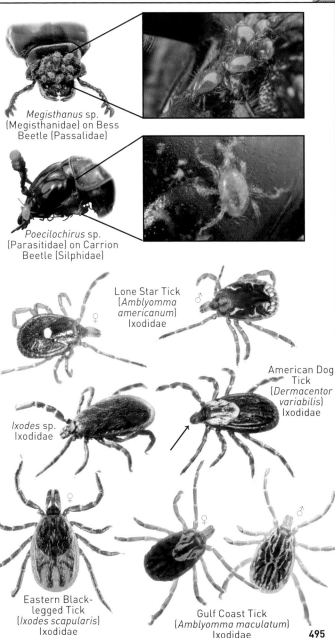

Megisthanus sp. (Megisthanidae) on Bess Beetle (Passalidae)

Poecilochirus sp. (Parasitidae) on Carrion Beetle (Silphidae)

Lone Star Tick (*Amblyomma americanum*) Ixodidae ♂

American Dog Tick (*Dermacentor variabilis*) Ixodidae

Ixodes sp. Ixodidae

Eastern Black-legged Tick (*Ixodes scapularis*) Ixodidae ♀

Gulf Coast Tick (*Amblyomma maculatum*) Ixodidae ♀ ♂

495

Class Arachnida cont.

TAILLESS WHIPSCORPIONS order Amblypygi

Broad and highly flattened bodies with segmented opisthosoma. Most with 4 pairs of eyes; 1 median pair and 2 clusters of 3 smaller eyes each laterally. **Pedipalps** modified for capturing and retaining prey. **First pair of legs long and slender, modified as sensory organs**. Lack venom. Restricted to southern states. TL 5–45 mm; World 155, NA 4.

PSEUDOSCORPIONES order Pseudoscorpions

Small, flattened , and usually somewhat pear shaped with **pedipalps** modified into pincers. Opisthosoma is 12-segmented. Harmless to humans, but have poison glands in their pincers for subduing prey. Found under bark and in leaf litter. Some phoretic, ride around on other insects. Make cocoons for mating using silk glands in their jaws. TL <3 mm; World 3,550, NA 450.

SCORPIONS order Scorpiones

Pedipalps large and modified as pincers. A pair of eyes medially on prosoma and 2–5 pairs anterolaterally. Segmented opisthosoma with long, 5-segmented tail modified with an **apical stinger**. A pair of pectines, or comb-like sensory structures, are found ventrally on sternite 2. While venomous, most are not seriously harmful to humans; 1 species in AZ, *Centruroides sculpturatus*, can be fatal. TL 18–100 mm; World 2,000, NA 50.

WINDSCORPIONS order Solifugae

Enlarged chelicerae, in many longer than prosoma; while not venomous, they can inflict a painful bite. **Pedipalps** are modified as sensory organs. A pair of large central eyes usually present, and lateral eyes usually lacking. Opisthosoma consists of 10 segments. Underside of coxae and trochanters on last pair of legs are fan-shaped sensory organs called *malleoli*. Move very quickly; voracious predators on insects and other arthropods. Known from southern and western US. TL ~25 mm; World >1,100, NA 200.

WHIPSCORPIONS order Thelyphonida (Uropygi)

Large, with **pedipalps** modified as pincers and **first pair of legs** modified to be long, slender, and antennae-like sensory structures. One pair of eyes at front of prosoma and three on each side of head. Lack venom, but can excrete a mixture of acetic and caprylic acid from glands at rear of opisthosoma, which bears **thread-like tail.** Single genus, *Mastigoproctus,* found in southern US (AZ, NM, TX & FL). TL ~85 mm; World >100, NA 3.

Cinteoti Vinegaroon
(*Mastigoproctus cinteoti*)
Thelyphonidae

ARACHNIDA

NON-INSECT ARTHROPODS

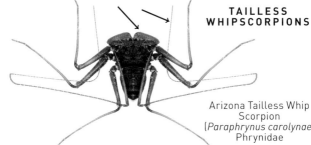

TAILLESS WHIPSCORPIONS

Arizona Tailless Whip Scorpion
(*Paraphrynus carolynae*)
Phrynidae

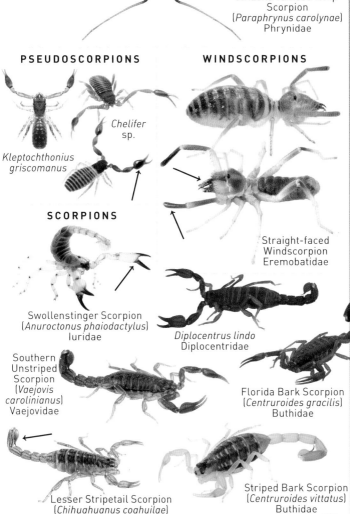

PSEUDOSCORPIONS

WINDSCORPIONS

Chelifer sp.

Kleptochthonius griscomanus

Straight-faced Windscorpion
Eremobatidae

SCORPIONS

Swollenstinger Scorpion
(*Anuroctonus phaiodactylus*)
Iuridae

Diplocentrus lindo
Diplocentridae

Southern Unstriped Scorpion
(*Vaejovis carolinianus*)
Vaejovidae

Florida Bark Scorpion
(*Centruroides gracilis*)
Buthidae

Lesser Stripetail Scorpion
(*Chihuahuanus coahuilae*)
Vaejovidae

Striped Bark Scorpion
(*Centruroides vittatus*)
Buthidae

SCORPIONS

497

NON-INSECT ARTHROPODS

Class Arachnida cont.

HARVESTMEN order Opiliones
Spider-like, but with the body regions fused, forming a singular segment. Many have very long legs. Lack venom and the ability to produce silk. Differ from most other arachnids in the ability to swallow solid food, not just liquids. Most have single pair of eyes in middle of head. TL (body) 1–7 mm; World >6,650, NA unknown.

SPIDERS order Araneae
Chelicerae usually well developed, terminating in fangs that are generally capable of injecting venom (except Uloboridae). Abdomen with **spinnerets** that extrude silk from glands. Usually with 4 pairs of eyes on anterior portion of prosoma. The pattern of eyes is diverse and variable; useful for family-level identification. They use silk in a variety of ways, including capturing prey, lining burrows, and locomotion. Most are not dangerous to humans, but a few like, the widow spiders and Brown Recluse, can be. TL 3–50 mm; World 44,000, NA 3,900.

OPILIONES • ARANEAE

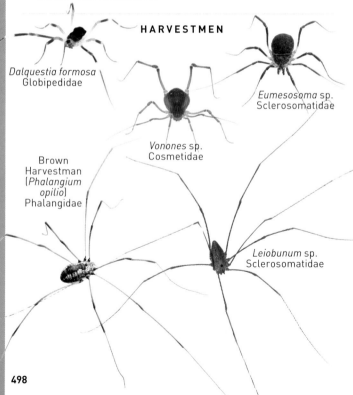

HARVESTMEN

Dalquestia formosa
Globipedidae

Eumesosoma sp.
Sclerosomatidae

Vonones sp.
Cosmetidae

Brown Harvestman (*Phalangium opilio*)
Phalangidae

Leiobunum sp.
Sclerosomatidae

SPIDERS

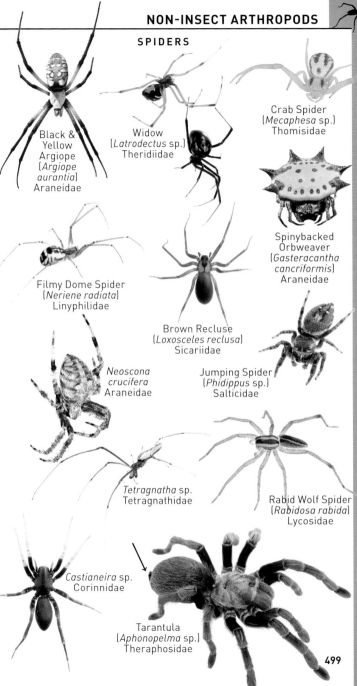

Black &
Yellow
Argiope
(*Argiope
aurantia*)
Araneidae

Widow
(*Latrodectus* sp.)
Theridiidae

Crab Spider
(*Mecaphesa* sp.)
Thomisidae

Filmy Dome Spider
(*Neriene radiata*)
Linyphilidae

Spinybacked
Orbweaver
(*Gasteracantha
cancriformis*)
Araneidae

Brown Recluse
(*Loxosceles reclusa*)
Sicariidae

*Neoscona
crucifera*
Araneidae

Jumping Spider
(*Phidippus* sp.)
Salticidae

Tetragnatha sp.
Tetragnathidae

Rabid Wolf Spider
(*Rabidosa rabida*)
Lycosidae

Castianeira sp.
Corinnidae

Tarantula
(*Aphonopelma* sp.)
Theraphosidae

HARVESTMEN • SPIDERS

499

SUBPHYLUM CRUSTACEA
Have 3 body regions: head, thorax, and abdomen; head and thorax often fused (cephalothorax) and covered by a large carapace. Two pairs of antennae. Most are marine, but a few are freshwater or terrestrial.

Class Malacostraca
Body contains 20 segments, including a 5-segmented head, 8-segmented thorax, 6-segmented abdomen, and a telson (tail). Most are marine.

CRAYFISH & SHRIMP order Decapoda
Head and thorax fused. Possess 10 legs; 1 pair each on the last 5 thoracic segments. In many, **first pair enlarged to form pincers** (chelae); legs behind called chelipeds. Five pairs of appendages on abdomen (pleopods) and 1 final pair (uropods) plus telson form tail fan. World 15,000, NA 500.

WOODLICE & RELATIVES order Isopoda
Can be strongly convex, capable of rolling into ball, to somewhat dorso-ventrally flattened. Head and thorax fused. Seven pairs of jointed appendages on the thorax and 5 pairs of branching appendages on abdomen. Nearly half are marine, the other half are found in moist areas on land, with relatively few species found in freshwater. Most of ours <12 mm in TL; World 10,000, NA 1,200.

AMPHIPODS order Amphipoda
Usually laterally compressed; body divided into 13 segments. Head and thorax fused. Thorax and abdomen usually distinct, with different kinds of legs. Gills present on thoracic segments. Most are marine, with some species found in freshwater environments and a few terrestrial. Most of ours <6 mm in TL; World 10,000, NA 300.

SUBPHYLUM MYRIAPODA
Single pair of antennae. Usually simple eyes. Mandibulate mouthparts. Numerous body segments not differentiated into thorax and abdomen, with at least 1 pair of appendages on most segments.

CENTIPEDES Class Chilopoda
Elongated, with 1 pair of appendages per segment. Most venomous, with pincer-like appendages; some can inflict a painful bite to humans. Predatory and quick moving. TL 10 mm–150 mm; World 3,000, NA 300.

MILLIPEDES Class Diplopoda
Elongated, with 2 pairs of appendages on most segments (result of every 2 segments fusing). Most cylindrical or flattened, with more than 20 segments. Most are detritivores. TL 3 mm–160 mm; World <12,000, NA 500.

SYMPHYLANS Class Symphyla
Pale, elongate, with 14 segments; first 12 with pairs of legs. Penultimate segment with cerci and pair of long sensory hairs; last segment fused to telson. Found in soil. TL 0.5 mm–8 mm; World 200, NA 30.

Scutigerella sp.
Scutigerellidae

CRAYFISH & SHRIMP

AMPHIPODS

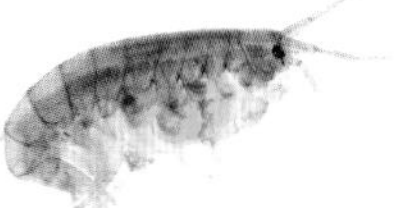

Scud
(*Hyalella* sp.)
Hyalellidae

Devil Crawfish
(*Cambarus diogenes*)
Cambaridae

Grass Shrimp
(*Palaemon* sp.)
Palaemonidae

WOODLICE & RELATIVES

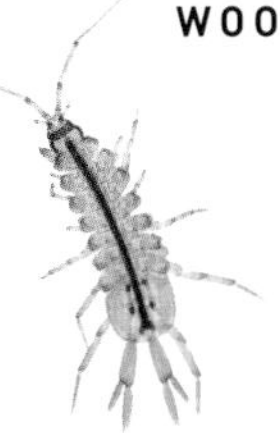

Rock Slater
(*Ligia* sp.)
Ligiidae

Common Pill Bug
(*Armadillidium vulgare*)
Armadillidiidae

Caecidotea sp.
Asellidae

CENTIPEDES

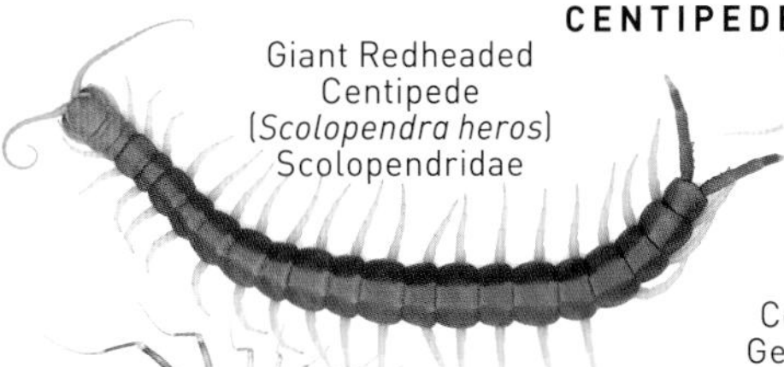

Giant Redheaded
Centipede
(*Scolopendra heros*)
Scolopendridae

Soil
Centipede
Geophilidae

House Centipede
(*Scutigera coleoptrata*)
Scutigeridae

MILLIPEDES

Bristly Millipede
(*Polyxenus lagurus*)
Polyxenidae

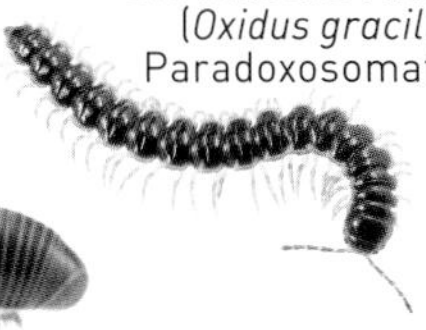

Greenhouse Millipede
(*Oxidus gracillis*)
Paradoxosomatidae

Narceus sp.
Spirobolidae

SHRIMP • ISOPODS • CENTIPEDES • MILLIPEDES

Miniature Trap-jaw Ant
(*Strumigenys hexamera*)
Formicidae

PHOTOGRAPHIC CREDITS

Most of the photos in this book were taken by John and Kendra Abbott. We are, however, eternally grateful to the following amazing photographers who allowed us to use their photos. Without the use of these images, this guide would have suffered.

Alice Abela—Anostostomatidae, p. 135; *Dracotettix monstrosus*, p. 139; *Cnemotettix bifasciatus*, p. 141; *Iris oratoria* male, p. 164; *Zootermopsis angusticollis* soldier, p. 173.

Gary Alpert—*Oropsylla hirsuta*, p. 425.

James Bailey—*Evalljapyx* sp., p. 73; Tanaoceridae, p. 134; *Tanaocerus koeblei*, p. 138; *Macrovelia hornii*, p. 203.

Thomas Barbin—*Vesicephalus occidentalis*, p. 69.

Bob Behrstock/Naturewide Images—Eumastacidae, p 134; *Eumorsea balli*, p. 139.

Matt Bertone—*Styletoentomon* sp., p. 59; *Icerya purchasi*, p. 189.

Dara Blumfield—*Fulgoraecia exigua* larva in planthopper, p. 403.

Ashley Bradford—*Trigoniophthalmus alternatus*, p. 77; *Tenuirostritermes cinereus*, p. 173.

Margarethe Brummermann—*Marginitermes hubbardi*, p. 173.

Nicholas Caffarilla—wing venation on end pages (CC-BY-SA; https://commons.wikimedia.org/wiki/File:Venation_of_insect_wing.svg, modified).

Ken Childs—*Acrolepiopsis incertella*, p. 395; *Diploschizia impigritella*, p. 395; *Eralea albalineella*, p. 403.

Ryan Cooke—*Gonatista grisea*, p. 163.

John Davis—Schreckensteinioidea, p. 387; *Euceratia securella*, p. 393; *Schreckensteinia festaliella*, p. 397.

Scott Ditzel—*Halictoxenos* sp., p. 293.

Charley Eiseman—*Pedetontus submutans*, p. 77; *Ectoedemia rubifoliella*, p. 389; *Etainia sericopeza*, p. 389; *Astrotischeria helianthi*, p. 389; *Coptotriche citrinipennella*, p. 389; *Antispila ampelopsifoliella*, p. 391; *Paraclemensia acerifoliella*, p. 391; *Argyresthia oreasella*, p. 395; *Aetole unipunctella*, p. 395; *Lithariapteryx jubarella*, p. 395; *Proleucoptera smilaciella*, p. 395; *Fulgoraecia exigua*, p. 403.

Lynette Elliott—*Zootermopsis angusticollis* worker, p. 173.

Jennifer Forman Orth—*Eriocrania semipurpurella*, p. 389.

Jeremy Gatten—*Cauchas simpliciella*, p. 391.

Eric Gofreed—*Parabacillus* sp., p. 157.

Elizabeth Golden—*Haplopus mayeri*, p. 157.

Nicolas Gompel—*Lutrochus laticeps*, 331.

Gary Goss—*Homaledra hepthalama*, p. 403.

Donald Griffiths (Spencer Entomological Collection, Beaty Biodiversity Museum, UBC)—*Doratopsylla blarinae*, p. 429; *Nearctopsylla brooksi*, p. 429; *Parasyllus longicornis*, p. 429.

Joyce Gross—Campodeidae, p. 71; *Trachypachus gibbsii*, p. 319; *Semijulistus flavipes*, p. 349.

Jeff Gruber—*Halictophagus* sp., p. 293; *Epimetopus* sp., p. 321.

Jan Hamrsky—Sisyridae larva, p. 287.

Guy Hanley—Amphizoidae, p. 301; *Amphizoa lecontei*, p. 319.

Maury Heiman—*Meskea dyspteraria*, p. 399.

Simon Hinkley & Ken Walker—*Thaneroclerus buquet*, p. 349.

Peter Hollinger—*Podura aquatica* (left image), p. 65.

Illinois Natural History Survey—antennal sensory organs (redrawn from Stannard with permission), p. 217.

Jena Johnson—*Perilampis* sp., p. 257; Mantispidae larvae, p. 286.

Jim Johnson—*Palaemnema domina*, p. 103; Alucitidae, p. 385; *Eucalantica polita*, p. 393; *Rhigognostis interrupta*, p. 393; *Alucita montana*, p. 399; *Lotisma trigonana*, p. 399.

Barrett Klein—Zyoptera (illustration), p. 100; *Hetaerina americana* (illustration), p. 103; *Hetaerina titia* (illustration), p. 103; *Argia plana* (illustration), p. 103; *Argia fumipennis* (illustration), p. 103; *Enallagma civile* (illustration), p. 103; *Telebasis salva* (illustration), p. 103; *Nehalennia gracilis* (illustration), p. 103.

Paul Lenhart—*Nicoletia wheeleri*, p. 81.

Robert Lord Zimlich—*Xiphydria tibialis*, p. 247.

David R. Madison—*Hydroscapha natans* (CC-BY; https://commons.wikimedia.org/wiki/File:Hydroscapha_natans01.jpg modified by removing background), p. 315.

Sean McCann—*Doru taeniatum*, p. 117.

Gary McClellan—*Neotermes castaneus*, p. 173.

Charles Melton—*Tanaostigmodes albiclavus* (both images), p. 257; *Helicopsyche* sp., p. 375.

Graham Montgomery—*Omoglymmius hamatus*, p. 315; *Peltis pippingskoeldi*, p. 349; *Eronyxa pallida*, p. 349.

Jim Moore—*Negha* sp. (all 3 images), p. 275; *Polystoechotes punctata*, p. 287; *Lichnanthe rathvoni*, p. 329.

Marcia Morris—*Pseudopostega cretea*, p. 385; *Yponomeuta multipunctella*, p. 393.

Andy Murray—*Acerentomon* sp., pp. 56, 57, 58 & 59; *Sminthurides malmgreni*, p. 62; *Podura aquatica* (right image), p. 65; *Cyphoderus similis*, p. 67; *Deuterosminthurus* sp., p. 69.

Tom Murray—*Janusius sylvestris*, p. 69; *Baetisca* sp., p. 89; *Livia saltatrix*, p. 191; *Hebrus* sp., p. 203; *Acordulecera dorsalis*, p. 247; *Xiphydria mellipes*, p. 247; *Torymus* sp., p. 257; *Colletes solidaginis*, p. 269; Byturidae, p. 305; Nitidulidae, p. 306; Cucujidae, p. 308, 309; Clambidae, p. 313; *Cupes capitatus*, p. 315; *Leiodes assimilis*, p. 323; *Catops* sp., p. 323; *Byrrhus cyclophorus*, p. 331; *Limnichites punctatus*, p. 331; *Laricobius rubidus*, p. 339; *Strigocis opacicollis*, p. 341; *Pytho americanus*, p. 345; *Thymalus marginicollis* (both images), p. 349; *Byturus unicolor*, p. 351; *Cerylon castaneum*, p. 351; *Cucujus clavipes*, p. 357; *Brachypterus urticae*, p. 357; *Phylloporia bistrigella*, p. 391; *Eudarcia eunitariaeella*, p. 391; *Bondia crescentella*, p. 399; *Blastodacna atra*, p. 401; *Merope tuber*, p. 435; *Atherix lantha* (larva), p. 467; *Atherix lantha* (adult), p. 467; *Scenopinus fenestralis*, p. 471; *Dialysis elongata*, p. 473; *Lonchoptera* sp., p. 475; *Actenoptera* sp., p. 483; *Cordilura setosa*, p. 489.

PHOTOGRAPHIC CREDITS

Riley Nelson—*Capnia gracilaria*, p. 127; *Capnia nana*, p. 127; *Prostoia besametsa*, p. 127; *Zapada cinctipes*, p. 127; *Yoraperla brevis*, p. 127; *Doddsia occidentalis*, p. 127.

Stot Noble—*Promalactis suzukiella*, p. 403.

Harsi Parker—*Morsea californica*, p. 139.

Steve Pelikan—*Nallachius americanus*, p. 285.

Hans Pohl—first instar larva in life cycle, p. 290; *Stylops* sp., p. 291; *Xenos* sp. (both images), p. 291.

Mike Quinn—Corioxenidae, front endsheet ordinal key; *Anisembia texana* (right image), p. 151; *Aetalion nervosopunctatum*, p. 195; *Mesovelia mulsanti*, p. 202; Corioxenidae, p. 291, 293; *Haliplus* sp., p. 315; *Glaresis* sp., p 325; *Olceclostera seraphica*, p. 418.

David Reed—Epermeniidae, p. 387; *Dryadaula terpsichorella*, p. 393; *Ypsolopha dentella*, p. 393; *Argyresthia goedartella*, p. 395; *Bedellia somnulentella*, p. 395; *Epermenia albapunctella*, p. 397.

Larry Reis—*Baetisca* sp., p. 89.

Marlin Rice—*Megischus bicolor*, p. 261.

John Rosenfeld—*Aeolothrips* sp. (both images), p. 219; *Caliothrips* sp., p. 219; *Helorus* sp., p. 251; *Exallonyx* sp., p. 251; *Proctotrupes* sp., p. 251; *Tretoserphus*, p. 251; *Conura* sp., p. 253; *Euperilampus triangularis*, p. 257; *Trichogramma* sp., p. 257; Eucoilinae, p. 259; Brentidae, p. 303; Staphylinidae (left image), p. 310; Platypezidae, p. 448; *Chalarus* sp., p. 475; *Polyporivora polypori*, p. 475; *Chamaemyia* sp., p. 479; *Calycomyza* sp., p. 481; *Mumetopia occipitalis*, p. 481; *Ceratobarys eulophus*, p. 485; *Clusia lateralis*, p. 485; *Pholeomyia* sp., p. 487; *Nemopoda nitidula*, p. 487.

Edward Ruden—*Parajapyx isabellae*, p. 73.

Harvey Schmidt—*Machilinus aurantiacus*, p. 77.

Udo Schmidt—Rhysodidae (CC-BY-SA; https://en.wikipedia.org/wiki/Omoglymmius_batantae#/media/File:Omoglymmius_batantae_R.T._&_J.R._Bell,_2009_(28691620916).png), p. 300.

Nolie Schneider—*Lepisma saccharina*, p. 91; *Helophorus* sp., p. 321; *Epimartyria auricrinella*, p. 389; *Pedicia* sp., p. 459.

Chuck Sexton—*Prays atomocella*, p. 395.

Michael Skvarla—*Merope tuber* (claspers), p. 435.

Vincent Smith—*Pthiris pubis*, p. 227.

Ryszard Szczygiel—*Lyonetia prunifoliella*, p. 395.

Jan van Duinen—*Panorpa* sp. larva, p. 433; *Panorpa* sp. eggs, p. 433.

Jaco Visser—*Stylops* sp., p. 288.

Salvador Vitanza—*Brachystomella parvula*, p. 65; *Labidura riparia*, p. 119; *Calinda longistylus*, p. 191; *Diaphorina citri* (nymphs), p. 191; *Heteropsylla texana*, p. 191; *Diceroprocta* sp., p. 195; *Blissus leucopterus*, p. 209; *Homaemus parvulus*, p. 213; *Franklinothrips vespiformis*, p. 219; *Acordulecera* sp., p. 247; *Dirhinus* sp., p. 253; *Leucospis affinis*, p. 255; Trichogrammatidae (unidentified species), p. 257; Dryinidae (larva), p. 262; *Chyphotes* sp., p. 265; *Colocistis crassa*, p. 265; *Hylaeus* sp., p. 269; *Desmopachria mexicana*, p. 319;

Giant Stag Beetle
(*Lucanus elaphus*)
Lucanidae

abdomen—third (posterior) division of the insect body that contains the reproductive system and most of the digestive and respiratory systems. **Fig. 1**

aculeate—group of Hymenoptera with ovipositor modified into stinger.

ametabolous—simplest type of metamorphosis. Immature stages resemble adults, including proportions, but lack genitalia. Individuals simply increase in size as they develop. **Fig. 2**

anal area—the posterior portion of the wing, usually including anal veins. **Fig. 3**

anal stylet—an elongate projection, with 1 long apical seta and 1 or more shorter preapical setae, located dorsally on S8 posterior to sensilium; found in most female fleas (Siphonaptera). **Fig. 4**

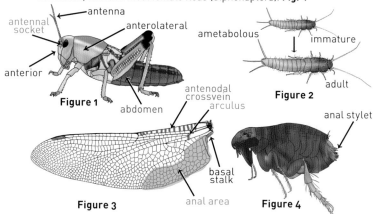

Figure 1

Figure 2

Figure 3

Figure 4

anamorphosis—type of metamorphosis in which the immature stages possess fewer abdominal segments than the adults. In Protura, 4 juvenile stages: prelarva, larva I, larva II, and maturus junior—with 9, 9, 10, and 12 abdominal segments respectively.

annulated—comprised of rings.

antemedial line—line that separates the basal and median areas of the forewing in moths (Lepidoptera). **Fig. 5**

antenna (pl. antennae)—paired, segmented, sensory appendages, usually arising in front and above on the head. **Fig. 1**

antennal socket—membranous area on head reinforced by ridge, in which the antenna is inserted. **Fig. 1**

antennomere—subunit of the antenna, including the scape, pedicel, and individual segments of the flagellum. **Fig. 6**

antenodal crossvein—crossveins between the costa and subcostal, between wing base and nodus (Odonata). **Fig. 3**

anterior—at or toward the front. **Fig. 1**

anterolateral—at or toward the front and side. **Fig. 1**

apex—usually referring to the distal-most point. **Fig.5**

apical—at or toward the apex.

arcuate—gently arched or bent like a bow.

arculus—basal crossvein between the radius and cubitus (Odonata). **Fig. 3**

arista—bristle-like distal part of antenna (made up of 3 flagellomeres) of higher Diptera. **Fig. 6**

aristate—bristle-like antenna found in higher Diptera. **Fig. 6**

aspirator—collection device used to suck up small insects.

Auchenorrhyncha—suborder of Hemiptera characterized by a beak that appears to arise from the lower part of the head; includes cicadas, leafhoppers, spittle bugs, and planthoppers.

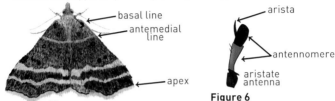

Figure 5

Figure 6

auricle—small ear-like structure on side of abdominal segment 2 in some Odonata.

basal—at or toward the base of the main body or closer to the point of attachment.

basal line—dark line cutting across basal section of wing (Lepidoptera). **Fig. 5**

basal stalk—narrowing of wing, where it attaches to thorax, in some damselflies (Odonata). **Fig. 3**

beak—elongated set of mouthpart structures, often used for sucking fluids; also known as proboscis. **Fig. 7**

beat sheet—collection device; stretched fabric placed under a substrate to catch insects beaten from vegetation.

Figure 7

Berlese funnel—collecting device used to extract living arthropods from soil, leaf litter, and similar substrate samples. It consists of a piece of screen or hardware cloth set inside a funnel, with a light mounted above and a collecting jar below.

bifurcated—a structure that is divided or forked into 2 arms.

bioluminescence—the production of light by a living organism, including insects. The process involves the oxidation of luciferin catalyzed by the enzyme luciferase.

bipectinate—having comb-like teeth on both sides, as in some antennae. **Fig. 8**

biting-chewing—general category of mouthparts wherein the mandibles are largely unmodified and used for tearing and grinding food.

bothrotricha—long slender setae that arise from depressions or pits on the body (Collembola).

brachypterous—having shortened or reduced flight wings that typically don't cover the abdomen. **Fig. 9**

brochosome—intricately structured microscopic granules secreted by leafhoppers and typically found on their body surface and, more rarely, eggs (Hemiptera). **Fig. 10**

buccula (pl. bucculae)—small, distended area consisting of elevated sclerites or ridges on the ventral part of the head and side of the rostrum in some Hemiptera. **Fig. 11**

calypter (pl. calypteres)—1 or 2 small lobes at base of wing, just above the haltere (Diptera). **Fig. 12**

campaniform sensilla—dome-shaped, usually elongated, sense organ that functions to sense stress on the surrounding cuticle. **Fig. 13**

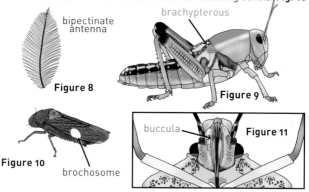

bipectinate antenna

Figure 8

brachypterous

Figure 9

Figure 10

brochosome

buccula

Figure 11

canthus—portion of cuticle that subdivides the compound eye (Coleoptera). **Fig. 14**

capitate—type of antennae that have a distinctive club at tip. **Fig. 15**

capitulum—referring to the head.

carina (pl. carinae)—ridge or keel.

caudal filament—thread-like process at the posterior end of the abdomen. **Fig. 16**

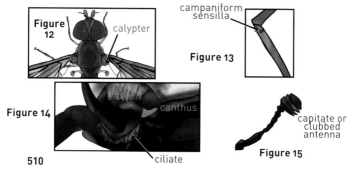

Figure 12

calypter

campaniform sensilla

Figure 13

Figure 14

canthus

ciliate

capitate or clubbed antenna

Figure 15

cephalothorax—body region consisting of the head and thorax (Strepsiptera).

cercus (pl. cerci)—1 of a pair of appendages originating from abdominal segment 11, appearing to originate on segment 10. **Fig. 16**

ciliate—fringed with a row of thin parallel hairs. **Fig. 14**

clavate—club-like or expanded at tip, usually referring to antennae. **Fig. 17**

claviform spot—a round spot positioned between the orbicular spot and the inner margin on the forewing in some moths (Lepidoptera). **Fig. 18**

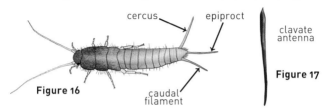

clavus—delineated area of the wing in Hemiptera, posterior and lateral to the scutellum. **Fig. 19**

clubbed—antennae with a distinct club at their tip. **Fig. 15 & Fig. 20**

clypeal—the clypeus.

clypeus—part of the insect head between the frons and labrum. **Fig. 21**

cocoon—protective silk covering for the pupa in many holometabolous insects. **Fig. 22**

collophore—ventral tube in springtails (Collembola). **Fig. 23**

complete metamorphosis—see holometabolous.

compound eye—aggregation of ommatidia, each representing a single facet of the eye. **Fig. 24**

connate—fused together or immovable (Coleoptera).

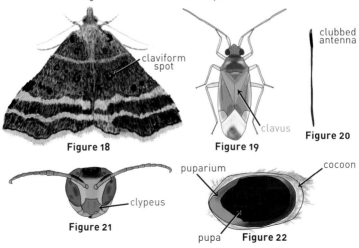

convex—outline or surface curved like the exterior of a circle.

copula—act of mating.

corium—elongated, middle portion of heteropteran hemelytron that is usually leathery (Hemiptera). **Fig. 24**

cornicle—paired tubular structures on the abdomen of aphids that discharge defensive lipids and alarm pheromones. **Fig. 25**

costa (C)—longitudinal wing vein that usually forms the anterior margin of the wing. **Fig. 26**

costal area—portion of the wing immediately posterior to the anterior margin or costa. **Fig. 26**

coxa (pl. coxae)—the basal part of the leg, part that attaches to the body.

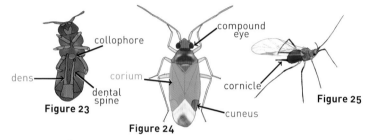

Figure 23

Figure 24

Figure 25

coxopodite—basal segment of leg, homologous to the coxa (Archaeognatha).

crenulate—having a waved or scalloped edge.

crossvein—transverse wing veins that connect the longitudinal veins. **Fig. 26**

cubital loop—area of cubitus formed by joining of Cu_{1a} and Cu_{1a} (Psocodea). **Fig. 26**

cubitus (Cu)—the sixth longitudinal vein in wings. **Fig. 26**

cuneus—distal section of the corium in the heteropteran forewing (Hemiptera). **Fig. 24**

deflexed—bent downward.

dens (pl. dentes)—central segment of furcula attached to the manubrium basally and with the mucro attached apically (Collembola). **Fig. 23**

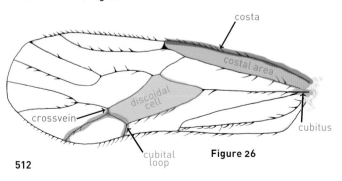

Figure 26

dental spine—one of a series of enlarged, modified setae, usually found along the inner dorsal base of the dens (Collembola). **Fig. 23**

detritus—organic matter produced by decomposition of organisms.

diapause—period of arrested development with reduced metabolic rate in which growth and metamorphosis cease, not attributable to adverse environmental conditions.

discoidal cell—median cell in wing (Psocodea). **Fig. 26**

distal—at or near the farthest end from the attachment of an appendage.

dorsal—upper surface. **Fig. 27**

dorsolateral—the top and side.

dorsoventrally—the axis joining the dorsal and ventral surfaces. **Fig. 28**

ectoparasite—a parasite that lives externally on and at the expense of another organism (host), which it does not kill.

elbowed—bent at an approximate 90° angle. **Fig. 29**

elytriform—appearing to have elytra (Sphaeropsocidae, Psocodea).

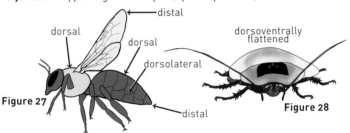

Figure 27

Figure 28

elytron (pl. elytra)—modified, hardened; forewing of a beetle that serves to protect the hindwings and abdomen (Coleoptera). **Fig. 30**

emarginate—with a notch or indentation. **Fig. 31**

endophytic—within living plant or tree tissue.

epiproct—terminal process located above the anus and appearing to arise from S10. **Fig. 32**

episternum (pl. episterna)—area of a thoracic pleuron (side of thorax) anterior to the pleural suture. **Fig. 33**

exophytic—outside of living plant or tree tissue.

exudate—secretion or modified excretory product from gland. **Fig. 34**

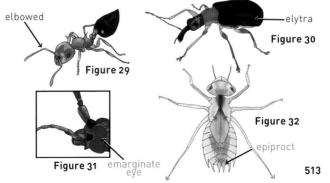

Figure 29

Figure 30

Figure 31

Figure 32

exuviae (sing. and pl.)—portion of the integument of a nymph or larva that is shed from the body during the process of molting.

facet—individual elements of the compound eye. **Fig. 35**

fastigium—extreme point or front of the vertex (Orthoptera). **Fig. 33**

femur (pl. femora)—third segment (from body) of the leg, usually largest. **Fig. 36**

filament—thread-like slender process of uniform diameter. **Fig. 36**

filamentous—long, thread-like antennae. **Fig. 36**

filiform—hair- or thread-like; type of antenna. **Fig. 36**

flabellate—having fan-like processes or projections for most segments of antennae; type of antenna. **Fig. 37**

flagellomere—segment of the flagellum. **Fig. 38**

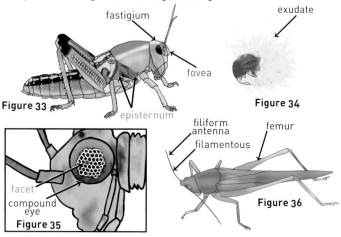

Figure 33

Figure 34

Figure 35

Figure 36

flagellum—multiple segments (flagellomeres) making up the third part of the antennae, beyond the second segment or pedicel. **Fig. 38**

foretarsi—the tarsi on the front pair of legs. **Fig. 39**

forewing—the first pair of wings attaching to the second thoracic segment. **Fig. 39**

fovea—pit or depression. **Fig. 33**

frons—large sclerite, above the clypeus, representing front of head. **Fig. 41**

frontoclypeal suture—suture or line of demarcation between the frons and clypeus. **Fig. 41**

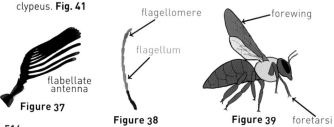

Figure 37

Figure 38

Figure 39

furcula)—springing organ of springtails found on ventral surface of abdomen (Collembola). **Fig. 40**

gaster—swollen part of the abdomen in some Hymenoptera that lies posterior to the petiole (wasp). **Fig. 42**

gena (pl. genae)—portion of the head below and behind each of the compound eyes. **Fig. 41**

genal—referring to the gena.

genal comb—row of strong spines on the lateroventral border of the head in fleas (Siphonaptera). **Fig. 43**

geniculate—strongly elbowed antennae. **Fig. 42**

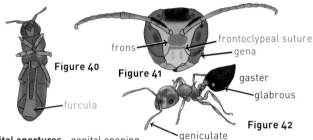

Figure 40

Figure 41

frons

frontoclypeal suture

gena

gaster

glabrous

Figure 42

furcula

geniculate

genital apertures—genital opening.

glabrous—characterized by being smooth, and without hair or punctures. **Fig. 42**

glassine envelope—envelopes made of a smooth glossy paper that is water-resistant; used when collecting Odonata and Lepidoptera.

globular—spherical in shape. **Fig. 44**

glossa (pl. glossae)—1 or a pair of lobes located at apex of labium between the paraglossae. **Fig. 45**

gonopore—external opening of the reproductive organs.

gregarious—referring to insects that congregate or live in communities, but are not social.

haltere (or halter) (pl. halteres)—modified hindwing of flies used for balancing in flight (Diptera). **Fig. 46**

hemelytron (pl. hemelytra)—forewing of heteropteran true bugs (Hemiptera). **Fig. 47**

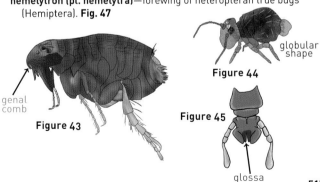

globular shape

Figure 44

genal comb

Figure 43

Figure 45

glossa

GLOSSARY

hemimetabolous—incomplete development in which the body gradually changes at each molt, wing buds growing larger. **Fig. 48**

herbivore—an animal that feeds on plants.

Hexapoda—subphylum of arthropods containing insects.

holometabolous—development in which there is a pupal stage between the larval and adult stage. **Fig. 49**

holoptic—eyes contiguous above (Diptera). **Fig. 53**

humeral angle—angle at base of anterior margin of wing. **Fig. 50**

humerus—referring to the shoulder. **Fig. 51**

hypermetamorphosis—undergoing a major change in morphology between larval instars.

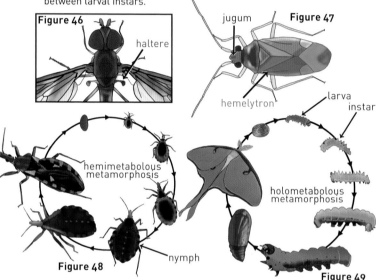

Figure 46

haltere

Figure 47

jugum

hemelytron

hemimetabolous metamorphosis

holometabolous metamorphosis

larva

instar

nymph

Figure 48

Figure 49

hyperparasitoid—a secondary parasitoid that develops upon another parasite or parasitoid.

hypognathous—head and mouthparts directed downward. **Fig. 52**

hypopharynx—centrally located mouthpart structure anterior to labium.

hypopleural—referring to the hypopleuron.

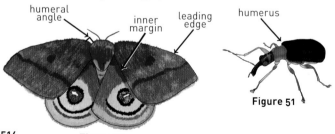

humeral angle

inner margin

leading edge

humerus

Figure 50

Figure 51

hypopleuron (pl. hypopleura)—sclerite located laterally on thorax, just above the coxa (Diptera). **Fig. 53**

inferior appendage—used by males to grasp females during copulation in Odonata. Anisoptera with 1, Zygoptera with 2.

inner margin—the inside or most medial area. **Fig. 50**

instar—stage between successive molts. **Fig. 49**

intercalary vein—short, unattached veins (Ephemeroptera). **Fig. 54**

jugal lobe—a basal, lobe-like projection on the posterior side of the hindwing (Hymenoptera). **Fig. 55**

jugum—paired lateral lobes of the head in heteropteran true bugs (Hemiptera) **Fig. 47**; also used for basal lobe of hindwing in several insect orders. **Fig. 55**

keel—elevated ridge or carina. **Fig. 56**

keeled—having a keel.

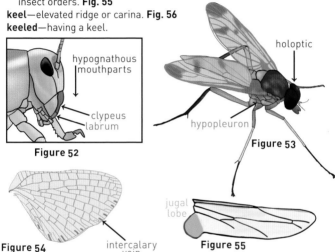

Figure 52

Figure 53

Figure 54

Figure 55

labial palp—paired appendage of the labium, 1- to 5-segmented. **Fig. 57**

labium—bottom-most mouthpart, also known as the "lower lip." **Fig. 57**

labrum—top-most mouthpart, also known as the "upper lip." **Fig. 52**

lacinia (pl. laciniae)—inner lobe of the maxilla; part of the stipes. **Fig. 58**

lamellate—antennae with apical segments forming a club of closely opposed leaf-like surfaces (Scarabaeoidea). **Fig. 59**

larva (pl. larvae)—immature insect; usually restricted to holometabolous insects. **Fig. 49**

lateroclinate—laterally directed setae (Diptera). **Fig. 60**

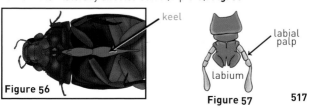

Figure 56

Figure 57

leading edge—anterior margin of wing when in flight. **Fig. 64**

littoral zone—region of a lake lying along the shore.

maculation—marks or spots of color. **Fig. 61**

mandible—paired mouthparts positioned below the labrum, can be jawed or variously modified. **Fig. 62**

manubrium (pl. manubria)—large, median base of furcula bearing the dentes (Collembola). **Fig. 63**

margin—the edge or border. **Fig. 64**

mating hook—genital structure in male Zoraptera.

maxilla (pl. maxillae)—second pair of jaws, variously modified in different groups. **Fig. 58**

maxillary palp—1- to 7-segmented structure on the maxilla. **Figs. 58 and 62**

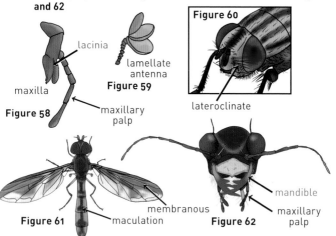

lacinia

lamellate antenna
Figure 59

maxilla

Figure 58

maxillary palp

Figure 60

lateroclinate

mandible

membranous
Figure 61 maculation

maxillary palp

Figure 62

media (M)—longitudinal wing vein between radius and cubitus. **Fig. 65**

median line—line that passes through the median area of the forewing, usually between the orbicular and reniform spots, in some moths (Lepidoptera). **Fig. 64**

median ridge—elevated area along the midline. **Fig. 66**

membranous—like a membrane; soft, thin, and sometimes more or less transparent. **Fig. 61**

Figure 64

leading edge

median line

margin

manubrium

Figure 63

meracanthus—distinct, conical, posterior projection from hind coxae (Psyllidae, Hemiptera). **Fig. 67**
mesonotum—dorsal sclerite on mesothorax. **Fig. 68**
mesopleuron—lateral sclerite(s) of the mesothorax. **Fig. 69**
mesoscutellum—scutellum of the mesothorax. **Fig. 71**
mesoscutum—scutum of the mesothorax. **Figs. 70 and 71**
mesosoma—in Apocrita, middle tagma of body consisting of 3 thoracic segments and first abdominal segment (Hymenoptera). **Fig. 72**
mesothorax—second or middle thoracic segment. **Fig. 68**

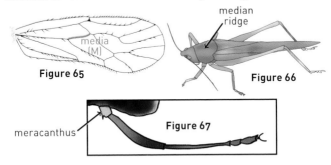

Figure 65

median ridge

Figure 66

meracanthus

Figure 67

mesotibia—tibia on the mesothorax. **Fig. 72**
metabasitarsus—basal tarsal segment on hind leg. **Figs. 69 and 72**
metacoxa—coxa on metathoracic (hind) leg. **Fig. 73**
metamorphosis—change in body form between immature and adult stages. **Figs. 49 and 50**
metasoma—in Apocrita, posterior tagma of body including all but first abdominal segment; segments posterior to propodaeum (Hymenoptera). **Fig. 73**
metasternal—referring to metasternum. **Fig. 74**

mesonotum

mesopleuron

mesothorax

metabasitarsus

meta-pleuron

Figure 68

Figure 69

pronotum

mesosoma

metacoxa

meta-basitarsus

mesoscutum

mesoscutellum

Figure 71

mesotibia

mesocoxa

Figure 70

Figure 72

metasternum—ventral sclerite of the metathorax. **Fig. 74**

metathorax—third or last thoracic segment. **Fig. 74**

metatibia—tibia on the metathorax or last pair of legs. **Fig. 74**

metatibial flange—expanded area of tibia on hind leg in some leaf-footed bugs (Hemiptera: Coreidae). **Fig. 75**

metepisternum (pl. metepisterna)—episternum of metathorax. **Fig. 76**

middorsal—middle of upper surface. **Fig. 75**

molar plate—projection of the basal lobe of mandible, directed toward the middle of the head and usually equipped with numerous rows of teeth (Collembola).

molt—process of shedding the exoskeleton.

molting—formation of new cuticle followed by shedding of old cuticle.

moniliform—bead-like antennae. **Fig. 77**

monophyletic—a group of organisms descended from a common evolutionary ancestor.

mucro—third segment of the furcula arising from the apex of the dens (Collembola). **Fig. 78**

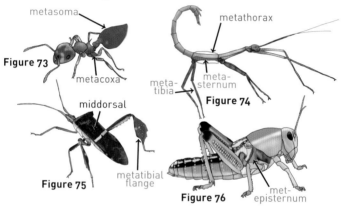

Figure 73

metasoma
metacoxa

Figure 74

metathorax
meta-sternum
meta-tibia

Figure 75

middorsal
metatibial flange

Figure 76

met-episternum

mucronal seta—setae on the mucro (Collembola). **Fig. 78**

naiad—aquatic, gill-breathing immature stage of an insect (Odonata).

nasute—individual of a termite soldier caste in which the head narrows anteriorly into a snout-like projection (Isoptera, Blattodea). **Fig. 79**

nodulus—where Cu$_2$ connects with anal vein in Psocodea. **Fig. 80**

nodus—indentation near the middle of the anterior margin of the wing in dragonflies and damselflies (Odonata). **Fig. 81**

nymph—an immature insect with gradual metamorphosis (ametabolous or hemimetabolous). **Fig. 49**

moniliform antenna

Figure 77

mucro

Figure 78

mucronal seta

ocellus (pl. ocelli)—a simple eye of adult and nymph insects. Typically, 3 in number and located on the vertex. **Fig. 82**

ommatidium (pl. ommatidia)—a single facet of the compound eye. **Fig. 83**

omnivorous—organisms that feed on both plants and animals.

ootheca (pl. oothecae)—a protective covering containing multiple eggs (Blattodea and Mantodea). **Fig. 84**

orbicular spot—a round spot or outline in the inner median part of the forewing of moths (Lepidoptera). **Fig. 85**

ornate—adorned with intricate shapes or complex patterns.

outer margin—lateral-most edge. **Fig. 85**

ovipositor—female organ used for laying eggs. **Fig. 86**

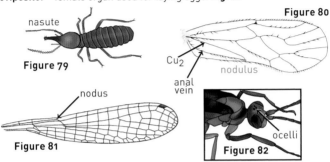

nasute

Figure 79

Figure 80

Cu$_2$

nodulus

anal vein

nodus

Figure 81

ocelli

Figure 82

palp—segmented process on the maxilla or labium. **Fig. 83**

paraprocts—1 or a pair of caudal or terminal lobes bordering anus lateroventrally. **Fig. 87**

parasitoid—a parasite that kills its host.

parthenogenetic—development from an unfertilized egg.

pectinate—a comb-like structure, including antennae. **Fig. 88**

pedicel—second segment of the antenna. **Fig. 88**

penultimate—next to last. **Fig. 88**

percussion—drumming or tapping a structure on the substrate (Plecoptera).

petiolate—having a narrow stalk or stem.

petiolate stalk—attached by narrow stem. **Fig. 89**

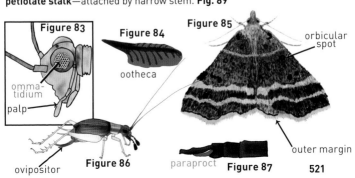

Figure 83

Figure 84

Figure 85

orbicular spot

omma-tidium

palp

ootheca

ovipositor

Figure 86

paraproct

Figure 87

outer margin

521

petiolate waist—having a narrowed waist. **Fig. 90**

petiole—a stalk or stem.

pheromone—a chemical produced by an individual that results in a specific reaction by other individuals of the same species.

phytophagous—organisms that eat plants.

piercing-sucking—mouthpart stylets used for piercing animal or plant tissue and then sucking up the resulting fluids. **Fig. 91**

pilose—covered with hair. **Fig. 92**

postantennal organ (PAO)—a ring-like sensory area located just posterior to the base of the antenna (Symphyla, Collembola).

postclypeus—posterior or upper part of the clypeus when a line of demarcation exists. **Fig. 93**

posterolateral—the side of the rearmost portion of a structure. **Fig. 91**

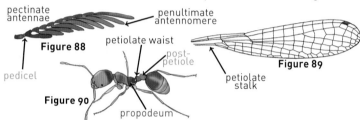

pectinate antennae

penultimate antennomere

Figure 88

pedicel

petiolate waist

post-petiole

Figure 89

petiolate stalk

Figure 90

propodeum

postmedial line—line that separates the median area from the subterminal area of the forewing in some moths (Lepidoptera). **Fig. 94**

postocular—area behind compound eyes. **Fig. 95**

postpetiole—area of abdomen in aculeate Hymenoptera posterior to petiole. **Fig. 90**

postscutellum—sclerite below the scutellum, subscutellum (Diptera). **Fig. 96**

preapical claw—claw anterior to the apex. **Fig. 97**

preapical spur—spur situated just before the apex. **Fig. 96**

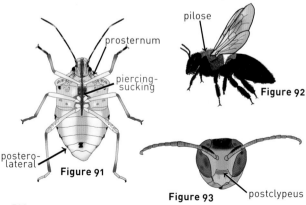

pilose

prosternum

piercing-sucking

Figure 92

postero-lateral

Figure 91

Figure 93

postclypeus

prepectus—area along anterioventral margin of the mesepisternum set off by a suture (Hymenoptera). **Fig. 98**

proboscis—elongated, tube-like mouthparts. **Fig. 99**

proclinate—forward- or downward-directed setae (Diptera). **Fig. 100**

procoxal—coxa on first leg of thoracic segment. **Fig. 99**

prolarva—newly hatched nymph (Odonata).

pronotum—upper surface of prothorax. **Fig. 99**

propodeum—in aculeate Hymenoptera, first abdominal segment, which is fused with the thorax. **Fig. 90**

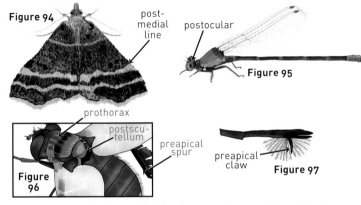

Figure 94

post-medial line

postocular

Figure 95

prothorax

postscutellum

preapical spur

Figure 96

preapical claw

Figure 97

propupa—stage in the nymphal development of some thrips and scale insects (Thysanoptera and Hemiptera).

prosternum—the anterior-most sternal sclerite between the forelegs. **Fig. 91**

protarsi—tarsal segments on the first leg (proleg). **Fig. 99**

prothorax—first thoracic segment. **Fig. 96**

protibia—tibia of first leg (proleg). **Fig. 99**

proximal—area nearest to the body. **Fig. 101**

pruinesence—waxy or powdery covering on some dragonflies and damselflies that exudes from the cuticle and turns the body light blue, gray, or white (Odonata). **Fig. 101**

pruinose—exhibiting pruinesence.

pruinosity—having pruinesence.

pseudocellus (pl. pseudocelli)—area of thin, corrugated cuticle on the dorsal part of the head (Collembola). **Fig. 102**

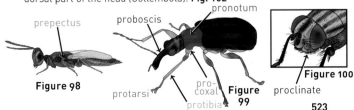

prepectus

proboscis

pronotum

Figure 98

protarsi

procoxal

protibia

Figure 99

proclinate

Figure 100

pseudopostocellar—referring to a pair of setae arising just behind the ocelli (Diptera). **Fig. 103**

pteropleural—referring to sclerite on side of thorax, below base of wing and consisting of upper part of mesepimeron. **Fig. 103**

pterostigma—pigmented spot near the anterior margin of the fore- and sometimes hindwings. **Fig. 101**

pterothorax—wing-bearing segments of thorax; meso- and metathorax. **Fig. 104**

pubescence—clothed in fine, short setae.

pubescent—covered with short, fine hairs. **Fig. 105**

punctate—surface with punctures.

punctures—pits or small depressions. **Fig. 106**

pupa—the developmental stage between the larva and adult in insects with complete development (holometabolous). **Fig. 107**

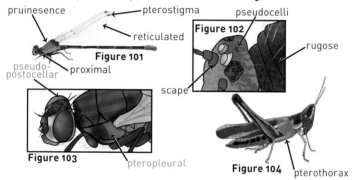

pruinesence

pterostigma

reticulated

pseudocelli

Figure 102

rugose

scape

pseudo-postocellar

proximal

Figure 101

Figure 103

pteropleural

Figure 104

pterothorax

puparium (pl. puparia)—covering formed by hardening of last larval skin, encloses pupa. **Fig. 107**

pupating—becoming a pupa.

pupation—transformation to the pupa stage.

pygidium—last dorsal segment of abdomen. **Fig. 109**

quadridentate—having 4 teeth.

radius (R)—longitudinal wing vein between the subcostal and the medial veins. **Fig. 108**

raptorial—adapted for capturing prey by grasping. **Fig. 109**

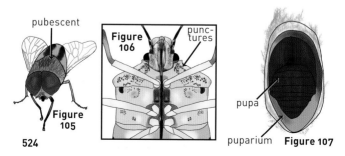

pubescent

Figure 105

Figure 106

punc-tures

pupa

puparium Figure 107

reniform spot—a spot, often kidney shaped, in the outer median part of the forewing in some moths (Lepidoptera). **Fig. 110**

retenaculum—small structure on ventral side of S3 that clasps furcula (Collembola). **Fig. 111**

reticulated—a surface that has a net-like or intermeshed appearance. **Fig. 101**

rostrum—tube-like mouthparts found in some insects. **Fig. 112**

rugose—appearing wrinkled. **Fig. 102**

scale—a flattened seta or hair. **Fig. 110**

scalloped—a series of rounded, convex projections forming an ornamental edge. **Fig. 112**

scape—the first segment of the antenna. **Fig. 102**

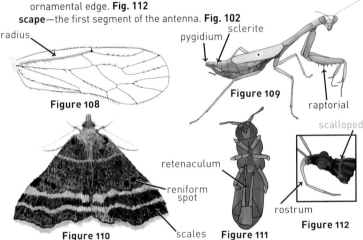

Figure 108

Figure 109

Figure 110

Figure 111

Figure 112

sclerite—a plate on the body wall surrounded by membrane or sutures. **Fig. 109**

sclerotized—hardened cuticle.

scopa (pl. scopae)—small dense tuft of hair (Hymenoptera).

scutellum—posterior portion of the meso- or metanotum, behind the scutum. **Fig. 113**

scutum—middle third of the meso- or metanotum, in front of the scutellum. **Fig. 113**

secondarily wingless—groups whose ancestors once had wings and have subsequently lost them through evolution.

secondary copulatory organs—copulatory organ on abdomen of male in which sperm is held after being produced by primary genitals (Odonata). **Fig. 114**

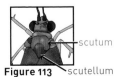

Figure 113

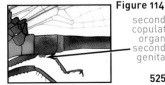

Figure 114
secondary copulatory organs/ secondary genitalia

secondary genitalia—set of structures on male dragonflies and damselflies used as an intromittent organ to transfer sperm to the female (Odonata). **Fig. 114**

sensillium (pl. sensilla)—sensory organ that can detect external stimuli. **Fig. 120**

sensorium (pl. sensoria)—slender and pointed sensory structures on antennal segments (Thysanoptera). **Fig. 115**

sensory cone—slender and pointed sensory structures on antennal segments; syn. sensorium (Thysanoptera). **Fig. 115**

serrate—toothed along edge like a saw; type of antenna. **Fig. 116**

seta (pl. setae)—a hair arising from the cuticle. **Fig. 115**

setaceous—bristle-like; type of antenna. **Fig. 117**

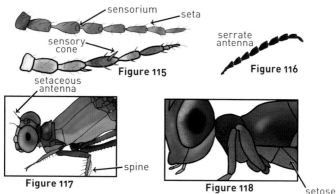

setose—covered in setae. **Fig. 118**

sexually dimorphic—male and female sexes are morphologically different beyond the differences in sexual organs.

spatulate—shaped as a spatula; typically rounded or broad at the apex and tapered at the base. **Fig. 119**

spermatophore—an encapsulated package of spermatozoa.

spine—an unjointed cuticular extension, often thorn-like. **Fig. 117**

spinelet—small spine, not much longer than wide. **Fig. 120**

spiracle—external opening of the respiratory or tracheal system. **Fig. 121**

spur—moveable spine. **Fig. 122**

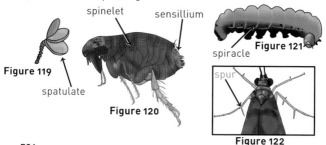

spurious vein—longitudinal wing vein between the radius and media, crossing the r-m crossvein and not reaching wing margin (Syrphidae, Diptera). **Fig. 123**

sternite—sclerotized plate on the sternum. **Fig. 124**

sternopleuron (pl. sternopleural)—lateral sclerite just above base of middle leg (Diptera). **Fig. 125**

Sternorrhyncha—suborder of Hemiptera that includes psyllids, whiteflies, aphids, scales, and mealybugs.

sternum (pl. sterna)—ventral surface of a segment. **Fig. 124**

stigmasac—enlarged area where R₁ separates from R in winged Psocodea. **Fig. 126**

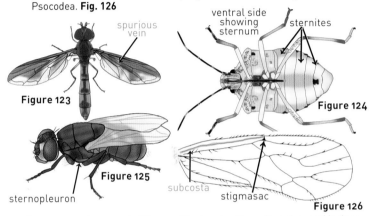

Figure 123

Figure 125

sternopleuron

ventral side showing sternum — sternites

Figure 124

subcosta

stigmasac

Figure 126

stipes—second segment of the maxilla, bearing the palp, galea, and lacinia. **Fig. 127**

stria (pl. striae)—a groove or depressed line.

striate—surface with numerous fine, parallel lines. **Fig. 128**

stridulate—making noise by rubbing 2 structures or surfaces together.

stylate—antenna type with a style, or short, cylindrical appendage (Diptera). **Fig. 129**

subapical—the area below or just before the apex. **Fig. 128**

subcosta (Sc)—longitudinal wing vein between costa and radius. **Fig. 126**

subequal—about equal.

subgenital plate—sternite that lies under the genitalia. **Fig. 130**

subimago—winged, penultimate instar; subadult (Ephemeroptera).

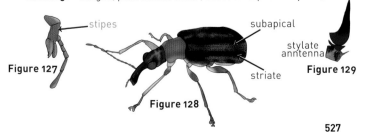

stipes

Figure 127

subapical

striate

Figure 128

stylate anntenna

Figure 129

submargin—just within the margin. **Fig. 131**

subquadrate—nearly quadrate, or having 4 equal sides. **Fig. 131**

subterminal line—line of scales often present between the postmedial and terminal lines in some moths (Lepidoptera). **Fig. 132**

sulcus (pl. sulci)—groove or furrow in the integument. **Fig. 133**

superior appendage—1 of the 2 upper appendages at the end of the abdomen; cercus (Odonata). **Fig. 134**

supertriangle—wing cell just in front of the triangle (Odonata).

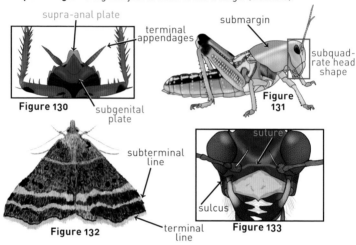

Figure 130

Figure 131

Figure 132

Figure 133

supra-anal plate—abdominal tergum 10 (Blattodea). **Fig. 130**

suture—external groove that may indicate the fusion of 2 sclerites. **Fig. 133**

symphyta—basal Hymenoptera lineages, now recognized to not have a single common ancestor.

tagma (pl. tagmata)—a group of body segments specialized for a given function; head, thorax, and abdomen in insects.

talus slope—area where debris piles up to a characteristic angle on rocky soils and slopes.

tandem—2 individuals, usually male and female, connected, but not by reproductive structures.

tarsomere—any of the individual segments of the tarsi. **Fig. 135**

tarsus (pl. tarsi)—last segment of the leg, distal to the tibia. **Fig. 135**

Figure 134

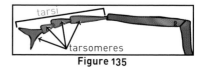

Figure 135

tegmen (pl. tegmina)—thickened, leathery, elongated forewing (Orthoptera). **Fig. 136**

tegula (pl. tegulae)—small, scalelike structure overlying base of forewing. **Fig. 137**

tergite—sclerotized, dorsal surface of a segment. **Fig. 136**

tergum—dorsal surface of an abdominal segment. **Fig. 137**

terminal appendages—appendages at the end of the abdomen. **Fig. 130**

terminal line—outermost line on the wings before the fringe in some moths (Lepidoptera). **Fig. 132**

thorax—middle of the 3 major divisions of the insect body. **Fig. 137**

tibia—fourth leg segment (from the body), following the femur. **Fig. 137**

tibiotarsus (pl. tibiotarsi)—fused tibia and tarsus in some insects. **Fig. 138**

tomentum—a type of pubescence made up of short, matted, woolly hair. **Fig. 139**

transscutal groove—transverse suture on scutum. **Fig. 140**

transverse—extending across the surface.

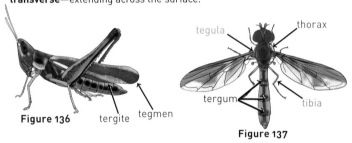

Figure 136 tergite tegmen

tegula thorax

tergum tibia

Figure 137

transverse rows—referring to rows at a right angle to the longitudinal axis. **Fig. 141**

tremulation—act of trembling or moving up and down without contacting the substrate beyond the legs.

trichobothria—minute sensory hairs (Thysanoptera). **Fig. 142**

tridentate—with 3 teeth.

trochanter—second segment of the leg, between the coxa and femur. **Fig. 143**

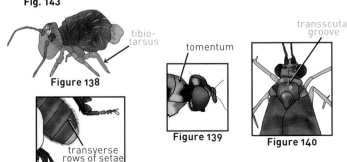

tibio-tarsus

Figure 138

tomentum

Figure 139

transscutal groove

Figure 140

transverse rows of setae

Figure 141

529

GLOSSARY

trochanteral organ—shortened, slightly expanded distal seta located in a conspicuous pit (Kitiannidae) or a series of short, differentiated setae on the inner surface of the trochanter (Entomobryidae, Collembola).

truncate—abruptly ending or cutting off.

tubercle—small, rounded protuberance. **Fig. 144**

tylus—distal part of the clypeus or anteclypeal region of the head in heteropteran Hemiptera. **Fig. 146**

tympanum (pl. tympana)—structure sensitive to vibration, comprising a membrane, air sac, and specialized sensory organ. **Fig. 147**

unguiculus—small tarsal claw (Collembola). **Fig. 145**

unguis—large tarsal claw (Collembola). **Fig. 145**

univoltine—having a single generation per year.

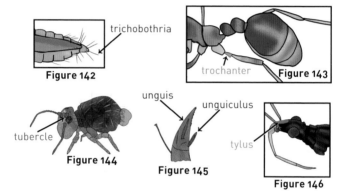

trichobothria

Figure 142

trochanter **Figure 143**

tubercle

Figure 144

unguis unguiculus

Figure 145

tylus

Figure 146

ventral—referring to the underside. **Fig. 147**

ventrolateral—referring to the outer margin of the underside.

vertex—top of the head. **Fig. 147**

vestigial—structure or process that is small or degenerate.

vibrissa (pl. vibrissae)—1 or more stout setae on the vibrissal angle (Diptera). **Fig. 148**

vibrissal angle—more or less rounded angle formed by facial ridges just above the oral margin (Diptera).

whorl—ring of hairs.

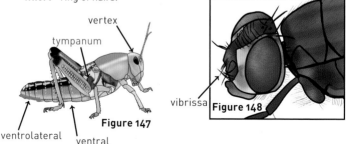

vertex

tympanum

ventrolateral

ventral

Figure 147

vibrissa **Figure 148**

Owlfly
(*Ascaloptynx appendiculata*)
Ascalaphidae

LITERATURE

There are many print and digital resources available today for the insect enthusiast and professional alike. Only a few years ago, resources for the identification of most groups were limited to the peer-reviewed literature. Especially with the advent of the internet, and various citizen science initiatives in particular, there has been resurgence in the appreciation and study of insect natural history. Below are some general references not otherwise cited in this book that you may find useful. Below that is the cited literature that appears within the text.

GENERAL REFERENCES

Arnett, R.H. 2000. *American Insects: A handbook of the Insects of America North of Mexico*, 2nd edition, CRC Press, Boca Raton. 1003 pp.

Borror, D.J. & R.E. White. 1970. *Peterson Field Guide to Insects.* Houghton Mifflin Company, Boston. 404 pp.

Crenshaw, W. 2004. *Garden Insects of North America: The ultimate guide to backyard bugs.* Princeton University Press, Princeton. 656 pp.

Eaton, E.R. & K. Kaufman. 2007. *Field Guide to Insects of North America.* Houghton Mifflin Company, Boston. 391 pp.

Evans, A.V. 2007. *Field Guide to Insects and Spiders of North America.* National Wildlife Federation, New York. 497 pp.

Grimaldi, D. & M.S. Engel. 2005. *Evolution of the Insects.* Cambridge University Press, Cambridge. 755 pp.

Marshall, S.A. 2017. *Insects: Their natural history and diversity with a photographic guide to insects of eastern North America,* 2nd ed. A Firefly Book, Buffalo. 735 pp.

McCafferty, W.P. 1998. *Aquatic Entomology: The fisherman's and ecologists' illustrated guide to insects and their relatives.* Jones and Bartlett, Sudbury. 448 pp.

Merritt, R.W., Cummins, K.W. & M.B. Berg (eds). 2008. *An Introduction to the Aquatic Insects of North America,* 4th edition. Kendall Hunt Publishing Company, Dubuque. 898 pp.

Voshell, J.R. 2002. *A Guide to Common Freshwater Invertebrates of North America.* The McDonald and Woodward Publishing Company, Blacksburg. 442 pp.

BugGuide—BugGuide.net
iNaturalist—iNaturalist.org

CITED LITERATURE

Abbott, J.C. & K.K. Abbott. 2020. *Common Insects of Texas and Surrounding States: A field guide.* The University of Texas Press, Austin. 446 pp.

Allen, R.T. 2002. "A synopsis of the Diplura of North America: Keys to higher taxa, systematics, distributions and descriptions of new taxa (Arthropoda: Insecta)." *Transactions of the American Entomological Society*, 128(4): 403–466.

Anderson, K. 2018. *Praying Mantises of the United States and Canada.* FastPencil Publishing. 291 pp.

Ando, H. 1982. *Biology of the Notoptera.* Kashiyo-Insatsu Co., Nagano. 194 pp.

Arment, C. 2006. *Stick Insects of the Continental United States and Canada: Species and early studies.* Coachwhip Publications. 204 pp.

Arnett, R.H., Jr., N.M. Downie & H.E. Jaques. 1980. *How to Know the Beetles.* William C. Brown/McGraw-Hill, New York. 416 pp.

Arnett, R.H., Jr. & M.C. Thomas. 2001. *American Beetles, vol. 1. Archostemata, Myxophaga, Adephaga, Polyphaga: Staphyliniformia.* CRC Press, Boca Raton. 443 pp.

Arnett, R.H., Jr., Thomas, M.C., Skelley, P.E. & J.H. Frank. 2002. *American Beetles, vol. 2. Polyphaga: Scarabaeoidea through Curculionoidea.* CRC Press, Boca Raton. 861 pp.

Aspöck, H. 2002. "The Biology of Raphidioptera: A review of present knowledge." *Acta Zoologica Academiae Scientiarum Hungaricae.* 48(Suppl. 2): 35–50.

Atkinson, T.H., P.G. Koehler & R.S. Patterson. 1991. "Catalog & Atlas of the Cockroaches (Dictyoptera) of North America North of Mexico." *Miscellaneous Publications of the Entomological Society of America,* no. 78. 85 pp.

Beadle, D. & S. Leckie. 2012. *Peterson Field Guide to Moths of Northeastern America.* Houghton Mifflin Harcourt, Boston. 611 pp.

Brock, J.P. & K. Kaufman. 2003. *Kaufman Field Guide to Butterflies of North America.* Houghton Mifflin Company, Boston. 391 pp.

Byers, C.W. 2002. "Scorpionflies, Hangingflies, and other Mecoptera." *The Kansas School Naturalist.* 48(1). 15pp.

Capinera, J.L., R.D. Scott & T.J. Walker. 2004. *Field Guide to Grasshoppers, Katydids, and Crickets of the United States.* Comstock Publishing Associates, Ithaca. 280 pp.

Christiansen, K. & P. Bellinger. 1998. *The Collembola of North America North of the Rio Grande: A taxonomic analysis,* vols. 1–4. Grinnell College, 1500 pp.

Corbet, P.S. 1999. *Dragonflies: Behavior and Ecology of Odonata.* Comstock Publishing Associates, Ithaca. 864 pp.

Covell, C.V., Jr. 1984. *A Field Guide to the Moths of Eastern North America: The Peterson Field Guide Series.* Houghton Mifflin Company, Boston. 496 pp.

Dindal, D.L. 1990. *Soil Biology Guide.* John Wiley & Sons, New York. 1349 pp.

Edmunds, G.F. 1976. *The Mayflies of North and Central America.* University of Minnesota Press, Minneapolis. 330 pp.

Elliott, L. 2007. *The Songs of Insects.* Houghton Mifflin Harcourt, Boston. 227 pp.

LITERATURE

Erezyilmaz, D.F., A. Hayward, Y. Huang, J. Paps, Z. Acs, J.A. Delgado, F. Collantes & J. Kathirithamby. 2014. "Expression of the Pupal Determinant Broad During Metamorphic and Neotenic Development of the Strepsipteran *Xenos vesparum* Rossi." *PLoS One,* https://doi.org/10.1371/journal.pone.0093614

Evans, A.V. 2014. *Beetles of Eastern North America.* Princeton University Press, Princeton. 560 pp.

Evans, A.V. 2021. *Beetles of Western North America.* Princeton University Press, Princeton. 624 pp.

Faust, L.F. 2017. *Fireflies, Glow-worms, and Lightning Bugs: Identification and natural history of the fireflies of the eastern and central United States and Canada.* The University of Georgia Press, Athens. 356 pp.

Gibson, G.A.P., J.T. Huber & J.B. Wooley (eds). 1997. *Annotated Keys to the Genera of Nearctic Chalcidoidea (Hymenoptera).* National Research Council of Canada, Ottawa. 794 pp.

Giribet, G. & G.D. Edgecombe. 2019. "The Phylogeny and Evolutionary History of Arthropods." *PubMed.* 29(12). doi: 10.1016/j.cub.2019.04.057.

Goulet, H. & J.T. Huber (eds). 1993. *Hymenoptera of the World: An identification guide to families. Centre for Land and Biological Resources Research,* Ottawa. 668 pp.

Hallmann, C.A., M. Sorg, E. Jongejans, H. Siepel, N. Hofland, H. Schwan, et al. 2017. "More than 75 percent decline over 27 years in total flying insect biomass in protected areas." *PLoS ONE* 12 (10): e0185809. https://doi.org/10.1371/journal.pone.0185809

Helfer, J.R. 1987. *How to Know the Grasshoppers, Crickets, Cockroaches and their Allies.* Dover Publications, New York. 363 pp.

Kim, K.C., H.D. Pratt & C. Stojanovich. 1986. *The Sucking Lice of North America: An illustrated manual for identification.* Pennsylvania State University Press, State College. 241 pp.

Knopp, M. & R. Cormier. 1997. *Mayflies: An angler's study of the trout water Ephemeroptera.* The Lyons Press, Guilford. 366 pp.

Kocárek, P., I. Horká & R. Kundrata. 2020. "Molecular Phylogeny and Infraordinal Classification of Zoraptera (Insecta)." *Insects.* 11(1), 51: https://doi.org/10.3390/insects11010051

Kosztarab, M. 1996. *Scale Insects of Northeastern North America: Identification, Biology, and Distribution.* Virginia Museum of Natural History, Special Publications Number 3. Martinsville. 650 pp.

Krasnov, B.R. 2008. *Functional and Evolutionary Ecology of Fleas: A model for ecological parasitology.* Cambridge University Press, Cambridge. 610 pp.

LaFontine, G. *Caddisflies.* 1981. The Lyons Press, New York. 336 pp.

Leckie, S. & D. Beadle. 2018. *Peterson Field Guide to Moths of Southeastern North America.* Houghton Mifflin Harcourt, Boston. 652 pp.

Marshall, S.A. 2012. *Flies: the natural history and diversity of Diptera.*

Firefly Books, Buffalo. 616 pp.

Marshall, S.A. 2018. *Beetles: the natural history and diversity of Coleoptera.* Firefly Books, Buffalo. 784 pp.

McAlpine, J.F. & D.M. Wood (eds). 1989. *Manual to Nearctic Diptera, vol. 3.* Research Branch, Agriculture Canada Monograph No. 32, Quebec. 1333–1580 pp.

McAlpine, J.F., B.V. Peterson, G.E. Shewell, H.J. Teskey, J.R. Vockeroth & D.M. Wood (eds). 1981. *Manual to Nearctic Diptera, vol. 1.* Research Branch, Agriculture Canada Monograph No. 27, Quebec. 1–1331 pp.

McAlpine, J.F., B.V. Peterson, G.E. Shewell, H.J. Teskey, J.R. Vockeroth & D.M. Wood (eds). 1987. *Manual to Nearctic Diptera, vol. 2.* Research Branch, Agriculture Canada Monograph No. 28, Quebec. 675–1332 pp.

McMahon, D.P., A. Hayward & J. Kathirithamby. 2011. "The First Molecular Phylogeny of Strepsiptera (Insecta) Reveals an Early Burst of Molecular Evolution Correlated with the Transition to Endoparasitism." *PLoS One* 6(6):e21206. DOI: 10.1371/journal. pone.0021206

Miller, D.R. & J.A. Davidson. 2005. *Armored Scale Insect Pests of Trees and Shrubs (Hemiptera: Diaspididae).* Comstock Publishing Associates, Ithaca. 442 pp.

Misof, B., S. Liu, K. Meusemann....and X. Zhou. 2014. "Phylogenomics Resolves the Timing and Pattern of Insect Evolution." *Science.* 346(763). DOI: 10.1126/science.1257570

Mound, L.A. & G. Kibby. 1998. *Thysanoptera: An identification guide,* 2nd ed. CAB International, Wallingford. 70 pp.

Otte, D. 1981. *The North American Grasshoppers, Volume I: Acrididae: Gomphocerinae and Acridinae.* Harvard University Press, Cambridge. 304 pp.

Otte, D. 1984. *The North American Grasshoppers, Volume II: Acrididae: Oedipodinae.* Harvard University Press, Cambridge. 376 pp.

Paulson, D.R. 2009. *Dragonflies and Damselflies of the West. Princeton Field Guides.* Princeton University Press, Princeton. 535 pp.

Paulson, D.R. 2011. *Dragonflies and Damselflies of the East. Princeton Field Guides.* Princeton University Press, Princeton. 538 pp.

Paulson, D.R. 2019. *Dragonflies and Damselflies: A Natural History.* Princeton University Press, Princeton. 224 pp.

Pearson, D.L., C.B. Knisley & C.J. Kazilek. 2006. *A Field Guide to the Tiger Beetles of the United States and Canada.* Oxford University Press, Oxford. 227 pp.

Penny, N.D., P.A. Adams & L.A. Stange. 1997. *Species catalog of the Neuroptera, Megaloptera, and Raphidioptera of America North of Mexico.* Proceedings of the California Academy of Sciences. 50(3): 39–114.

Price, M.A. & O.H. Graham. 1997. *Chewing and Sucking Lice as Parasites of Mammals and Birds.* U.S. Department of Agriculture, Agricultural Research Service. Vol. 1849. 256 pp.

Reddell, J. 1983. *A checklist and bibliography of the Japygoidea (Insecta: Diplura) of North America, Central America, and the West Indies.*

Pearce-Sellards Series No. 37., An Occasional Publication of the Texas Memorial Museum, The University of Texas at Austin. http://hdl.handle.net/2152/29904

Schaller, F. 1968. *Soil Animals*. University of Michigan Press, Ann Arbor. 144 pp.

Schuh, R.T. & J.A. Slater. 1995. *True Bugs of the World (Hemiptera Heteroptera): Classification and Natural History*. NCROL, Washington D.C. 416 pp.

Skevington, J.H. 2019. *Field Guide to the Flower Flies of Northeastern North America. Princeton Field Guide*. Princeton University Press, Princeton. 512 pp.

Smith, L.M. 1960. "The family Projapygidae and Anajapygidae (Diplura) in North America." *Annals of the Entomological Society of America*. 53(5): 575–583. https://doi.org/10.1093/aesa/53.5.575

Stark, B.P., S.W. Szczytko & C.R. Nelson. 1998. *American Stoneflies: A photographic guide to the Plecoptera*. The Caddis Press, Columbus. 126 pp.

Stewart, K.W. & B.P. Stark. 2002. *Nymphs of North American Stonefly Genera (Plecoptera)*, 2nd ed. The Caddis Press, Columbus. 510 pp.

Szetycki, A. 2007. "Catalogue of the World Protura." *Acta Zoologica Cracoviensia, Ser.B.*—Invertebrata. 50B(1): 210 pp.

Trietsch, C. & A.R. Deans. 2018. "The Insect Collectors' Code." *American Entomologist*. 64(3): 156–158. https://doi.org/10.1093/ae/tmy035

Tuxen, S.L. 1964. *The Protura: A revision of the species of the world with keys for determination*. Hermann, Paris. 360 pp.

Wagner, D.L. 2005. *Caterpillars of Eastern North America: A Guide to Identification and Natural History. Princeton Field Guides*. Princeton University Press, Princeton. 512 pp.

Wagner, D.L., E.M. Grames, M.L. Forister, M.R. Berenbaum & D. Stopak. 2021. "Insect Decline in the Anthropocene: Death by a thousand cuts." *Proceedings of the National Academy of Sciences of the United States of America*. 18(2), e2023989118; https://doi.org/10.1073/pnas.2023989118

Wallwork, J.A. 1970. *Ecology of Soil Animals*. McGraw-Hill, New York. 283 pp.

White, R.E. 1983. *A Field Guide to the Beetles of North America, Peterson Field Guide Series*. Houghton Mifflin Company, Boston. 368 pp.

Wiggins, G.B. 1996. *Larvae of the North American Caddisfly Genera*. University of Toronto Press, Toronto. 456 pp.

Wiggins, G.B. 2004. *Caddisflies: The Underwater Architects*. University of Toronto Press, Toronto. 292 pp.

Wilson, J.S. & O. Messinger Carrill. 2016. *The Bees in Your Backyard: A Guide to North America's Bees*. Princeton University Press, Princeton. 287 pp.

Wipfler, B., M. Bai, S. Schoville, R. Dallai, T. Uchifune, R. Machida, C. Yingying & Beutel, R. 2014. "Ice Crawlers (Grylloblattodea)—the history of the investigation of a highly unusual group of insects." *Journal of Insect Biodiversity* 2(2): 1-25. DOI: 10.12976/jib/2014.2.2

Guava Skipper
(*Phocides polybius*)
Hesperiidae

INDEX

Aaroniella badonneli 233
Abachrysa eureka 281, 285
Abaeis nicippe 409
Abbott's Sphinx 421
Abedus 205
Ablerus 252
Acalypteratae 450, 453
Acalyptratae 437, 441, 443, 476,
    477, 478, 480, 482, 484, 486
Acanalonia servillei 195
Acanaloniid Planthoppers 194,
    195
Acanaloniidae 181, 194
Acanthametropodidae 86, 94
Acanthinus argentinus 347
Acanthocephala femorata 207
Acantholyda erythrocephala 245
Acanthopteroctetidae 385, 388
Acanthosomatidae 184, 210
Acari 494
Acaricoris 206
Acariformes 493
Acartophthalmid Flies 477
Acartophthalmidae 477
Acartophthalmus 477
Acentrella 93
Acerentomata 59
Acerentomidae 56, 57, 58, 59
Acerentomids 59
Acerentomon 56, 57, 58, 59
Achemon Sphinx 420
Achilid Planthoppers 194, 195
Achilidae 181, 194, 262
Achurum sumichrasti 137
Aclerdidae 186
Aclista 253
Acmaeodera flavomarginata 329
Acordulecera 247
Acordulecera dorsalis 247
Acorn Moth 401
Acrididae 130, 135, 136
Acrobat Ant 266
Acrocera 473
Acroceridae 446, 472
Acrolepiopsis incertella 395
Acrolophus popeanella 391
Acroneuria abnormis 129

Acroneuria arenosa 129
Acronicta americana 414
Acropteroxys gracilis 357
Acrotaphus wiltii 248
Actenoptera 483
Actias luna 423
Actina viridis 467
Aculeata 237, 260
Acutalis tartarea 177, 195
Acylomus 358
Acylophorus 323
Adela caeruleella 391
Adelgidae 180, 190
Adelidae 382, 390
Adeloidea 384, 386, 387, 390
Adephaga 295, 298, 300, 314, 318
Aderidae 309, 346
Admirable Grasshopper 137
Adraneothrips 219
Aedes 465
Aenictopecheidae 184, 198
Aenictophecheids 198
Aeolothripidae 217, 218
Aeolothrips 219
Aeolus mellillus 335
Aeshnidae 100, 104, 109
Aetalion 194
Aetalionds 194, 195
Aetalionidae 180, 194
Aetalion nervosopunctatum 195
Aethina tumida 355
Aetole unipunctella 395
Agaonidae 252
Agapostemon splendens 269
Agarodes crassicornis 369
Agathidinae 249
Agraulis vanillae 378, 411
Agrilus bilineata 329
Agriphila vulgivagellus 379
Agrius cingulatus 421
Agromyzidae 453, 454, 455, 480
Agrypnus rectangularis 335
Agulla 272, 273
Agulla bicolor 273, 275
Agyrtidae 322
Ailanthus Webworm Moth 395
Akalyptoischiidae 354

Akalyptoischiid Scavenger Beetles 354
Akalyptoischion 354
Albatross Flea 429
Alderflies 279
Alderfly 276, 279
Aleuroplatus 187
Aleyrodidae 181, 186
Aleyrodoidea 186
Alkali Grasshopper 136
Alloperla 129
Alobates pensylvanica 343
Alticini 361
Alucita montana 399
Alucitidae 383, 385, 398
Alucitoidea 385, 398
Alydidae 185, 206
Alydus eurinus 207
Alypia octomaculata 414
Alysiinae 249
Amblycera 228
Amblyomma americanum 495
Amblyomma maculatum 495
Amblyptilia pica 399
Amblypygi 493, 496
Amblytropidia mysteca 136
Ambrosia Beetle 363
Ambrysus lunatus 205
Ameletidae 87, 94
Ameletus 94
American Bird Grasshopper 137
American Bumble Bee 271
American Carrion Beetle 323
American Cockroach 171
American Dagger Moth 414
American Dog Tick 495
American Emerald 107
American Ermine Moth 393
American Hover Fly 475
American Idia 417
American Lady 410
American Plum Borer 413
American Rubyspot 103
American Silkworm Moths 383, 418
American Snout 411
Ametropodidae 86, 94

Ametropus 94
Ammophila 269
Amorpha juglandis 421
Amphiboips quercusinanis 259
Amphicerus bicaudatus 339
Amphicerus cornutus 339
Amphientomidae 224
Amphion floridensis 421
Amphipoda 493, 500
Amphipods 493, 500
Amphipsocidae 232
Amphizoa 318
Amphizoa lecontei 319
Amphizoidae 301, 318
Amphotis schwarzi 355
Ampulicidae 268
Anajapygidae 72
Anajapygids 72
Anajapyx hermosa 72
Anamorphid Beetles 352
Anamorphidae 352
Anaspis 347
Anax junius 105, 109
Anaxipha 141
Anaxyelidae 246
Anaxyeloidea 241, 246
Ancient Barklice 230, 231
Ancistrocerus campestris 263
Andrena 269
Andrenidae 268
Androchirus femoralis 343
Aneugmenus flavipes 245
Angel Insects 113
Angulose Prominent 415
Anicla infecta 414
Anisembia texana 151
Anisembiid Webspinner 151
Anisembiidae 151
Anise Swallowtail 406
Anisolabididae 117, 118
Anisolabis maritima 119
Anisomorpha 155
Anisomorpha buprestoides 157
Anisopodidae 456, 460
Anisoptera 100, 104
Anisota virginiensis 423
Ankothrips 218

INDEX

Annulipalpia 365, 368, 370
Anommatus 352
Anopheles crucians 465
Anoplius 267
Anoplura 221
Anostostomatidae 135, 140
Antaeotricha humilis 403
Antaeotricha schlaegeri 403
Ant Crickets 140, 141
Anteoninae 262
Anthaxia 329
Antheraea polyphemus 422
Anthicidae 309, 346
Anthidiellum notatum 270
Anthocoridae 185, 199
Anthomyiidae 450, 488
Anthomyzid Flies 480, 481
Anthomyzidae 455, 480
Anthophoridae 266
Anthrenus verbasci 339
Anthribidae 303, 362
Antispila ampelopsifoliella 247, 391
Antlike Flower Beetles 346, 347
Ant-like Leaf Beetles 346, 347
Antlike Weevil 363
Antlions 281, 286, 287
Ants 237, 260, 266, 267
Anurapteryx crenulata 412
Anuroctonus phaiodactylus 497
Apache degeeri 197
Apataniidae 367, 376
Apatelodes torrefacta 418
Apatelodidae 383
Aphalaridae 190
Aphalarids 190
Aphelinid Wasps 252
Aphelinidae 252
Aphididae 181, 186
Aphidoidea 186
Aphids 175, 180, 186
Aphis nerii 187
Aphonopelma 499
Aphorista vittata 353
Aphrophora 193
Aphrophora cribrata 193
Aphrophora saratogensis 193

Aphrophoridae 192
Aphrophorid Froghoppers 192, 193
Aphylla angustifolia 109
Apiaceae 396
Apidae 270
Apiocera 468, 469
Apiocerid flies 468, 469
Apioceridae 468
Apiomerus californicus 201
Apis mellifera 271
Aplodontia rufa 428
Apocrita 240
Apodemia mormo 410
Apoidea 242, 268, 270, 292
Apple Pith Moth 401
Apple Twig Borer 339
Apsilocephala longistyla 468
Apsilocephalid Flies 468
Apsilocephalidae 468
Apterobittacus apterus 430, 434, 435
Apystomyia elinguis 472
Apystomyiid Flies 472
Apystomyiidae 472
Aquarius 203
Arachnida 493, 494, 496, 498
Aradidae 185, 206
Aradus 206
Araneae 493, 498
Araneidae 499
Archaeognatha 61, 71, 75, 79
Archaic Sun Moths 388
Archasia auriculata 195
Archeocrypticidae 342
Archidermaptera 115
Archilestes grandis 103
Archipsocidae 230
Archipsocus 231
Archostemata 295, 301, 314
Archotermopsidae 172
Archytas apicifer 491
Arcigera Flower Moth 414
Arctiinae 382
Arenigena 471
Arenivaga erratica 172
Arge 245

Argentine Ant 266
Argia 97
Argia fumipennis 103
Argia moesta 97
Argia plana 103
Argia translata 109
Argid Sawflies 244, 245
Argidae 244
Argiope aurantia 499
Argoporis 343
Argyresthia goedartella 395
Argyresthia oreasella 395
Argyresthiidae 394
Argyrotaenia quercifoliana 397
Arhyssus nigristernum 207
Arilus cristatus 201
Arizona Carpenter Bee 271
Arizona Tailless Whip Scorpion 497
Armadillidiidae 501
Armadillidium vulgare 501
Armored Lice 228
Armored Mayflies 88
Armored Scale Insects 186, 187
Army Ant 267
Aroga morenella 401
Arphia sulphurea 136
Arrenuridae 494
Arrenurus 494
Arrhenodes minutus 363
Arrhopalitidae 68
Arrhopalitids 68
Arrowhead Spiketail 105, 109
Artace cribrarius 419
Artematopodidae 332
Artheneidae 206
Artheneids 206
Arthroplea bipunctata 92
Arthropleidae 86, 92
Asbolus verrucosus 343
Ascalapha odorata 417
Ascalaphidae 286
Ascaloptynx appendiculata 280, 287
Aschiza 437, 441, 443, 446, 474
Asellidae 501
Ash Bark Beetle 363

Ash-gray Leaf Bugs 210
Ashy Clubtail 99
Ashy Gray Lady Beetle 353
Asian Ambrosia Beetle 363
Asian Citrus Liviid 191
Asian Needle Ant 266
Asilidae 442, 448, 470
Asiloidea 446
Asilomorpha 440, 468, 470, 472, 474
Asiopsocidae 229
Asphondylia auripila 461
Assassin Bugs 200, 201
Asteiid Flies 477
Asteiidae 477
Asterocampa clyton 411
Asterolecaniidae 186
Astrotischeria helianthi 389
Atalantycha bilineata 337
Atanycolus 249
Atelestid Flies 472
Atelestidae 472
Athericidae 447, 466
Atherix lantha 467
Atlanticus pachymerus 142
Atomosia puella 471
Atopomacer ites 360
Atopsyche 374
Attalus scincetus 351
Attelabidae 303, 362
Atteva aurea 395
Attevidae 394
Auchenorrhyncha 175, 177, 179, 180, 192, 262
Aulacid Wasps 260, 261
Aulacidae 260
Aulacigastrid Flies 477
Aulacigastridae 477
Austrotinodes texensis 370
Automeris io 422
Autostichid Moths 400, 401
Autostichidae 400
Axymyia furcata 461
Axymyiid Flies 460, 461
Axymyiidae 460
Azotid Wasps 252
Azotidae 252

Aztec Grasshopper 137
Aztec Pygmy Grasshopper 139

Backswimmers 204, 205
Baetidae 87, 92
Baetis 93
Baetisca 88
Baetiscidae 87, 88
Bagrada Bug 211
Bagrada hilaris 211
Bagworm Moth 390, 391
Bagworms 390, 391
Baileya ophthalmica 416
Bald Cypress Sphinx 420
Bald-faced Hornet 263
Banded Cucumber Beetle 361
Banded Hairstreak 408
Banded Hickory Borer 359
Banded Sphinx 420
Banksiola 377
Banksiola concatenata 377
Barber's Brown Lacewing 285
Barber's Lady Beetle 353
Barce 201
Bark-gnawing Beetles 350, 351
Barklice 221
Bat Bugs 199
Bat Fleas 427, 428
Bat Flies 490
Batrachedrid Moths 400
Batrachedridae 400
Battigrassiella wheeleri 79, 81
Battus philenor 407
Beach flies 477
Beaded Lacewings 284, 285
Beautiful Banded Lebia 317
Bed Bugs 199
Bedelliidae 394
Bee Flies 470, 471
Bee Killer 442
Bee Lice 476
Beech Blight Aphid 187
Bees 237, 260
Beetles 295
Behningiidae 86, 88
Belidae 360
Belonocnema treatae 259

Belostoma 205
Belostomatidae 183, 204
Belotus bicolor 337
Belyta 253
Beraea 374
Beraea gorteba 369
Beraeid Caddisflies 374
Beraeidae 366, 369, 374
Berosus ordinatus 321
Berothidae 284
Berytidae 185, 208
Bess Beetles 324, 325
Bethylid Wasps 260, 261
Bethylidae 260, 264
Bibio articulatus 461
Bibionidae 456, 460
Bibionomorpha 441, 460, 462
Bicolored-Crown-of-thorns Wasp 261
Big-eyed Bugs 208, 209
Big-headed Ant 267
Big-headed Flies 474, 475
Biphyllidae 350
Birch Catkin Bug 209
Birch Shield Bug 211
Bird and Rodent Fleas 427, 428, 429
Bird Lice 226, 227
Bird Nest Barklice 224, 225
Bistanta mexicana 161, 162
Biting Midges 462, 463
Bitoma quadricollisus 345
Bittacidae 430, 433, 434
Bittacomorpha clavipes 459
Bittacus texanus 435
Bizarre Caddisflies 377
Blaberid Cockroaches 170, 171
Blaberidae 170
Blaberoidea 167, 168, 170
Black & Yellow Argiope 499
Black Blow Fly 489
Black Caterpillar Hunter 317
Black Flies 464, 465
Black Larder Beetle 339
Black Saddlebags 108
Black Scavenger Flies 486, 487
Black Soldier Fly 467

Black Swallowtail 406
Black Treehopper 177
Black Turpentine Beetle 363
Black Webspinner 151
Black Witch 417
Blackberry Skeletonizer 397
Black-etched Prominent 415
Black-patched Glaphyria 413
Black-waved Flannel Moth 405
Blapstinus fortis 343
Blaste posticata 221
Blastobasidae 400
Blastobasis glandulell 401
Blastodacna atra 401
Blatta orientalis 171
Blattaria 167
Blattella germanica 171
Blattidae 167, 170
Blattodea 111, 115, 145, 149,
        167, 221
Blattoidea 167, 168, 169, 170, 172
Blephariceridae 466
Blind Click Beetle 335
Blind Silverfish 79
Blissidae 208
Blissids 208, 209
Blissus leucopterus 209
Blister Beetles 344, 345
Blow Flies 488, 489
Blue Death Feigning Beetle 343
Blue Ghost Firefly 337
Blue Metalmark 410
Blue Mud Wasp 269
Blueberry Digger Bee 271
Blueberry Mason Bee 270
Blues 408
Body Lice 226, 227
Boisea trivittata 207
Bolbocerosoma farctum 325
Bolbomyia 466
Bolbomyiid Flies 466
Bolbomyiidae 466
Bolitophilid Flies 460
Bolitophilidae 460
Bolitotherus cornutus 343
Boll's Sandroach 172
Bombus griseocollis 271

Bombus pensylvanicus 271
Bombycidae 418
Bombycoidea 387, 418, 420, 422
Bombyliidae 449, 470
Bondia crescentella 399
Booklice 221, 224, 225
Boopiidae 228
Bordered Plant Bugs 212, 213
Boreidae 425, 434
Boreothrinax maculipennis 483
Boreus brumalis 433, 435
Boridae 344
Bostrichidae 308, 310, 338
Bostrichoidea 299, 338
Bot Flies 488, 489
Bothrideridae 352
Bothrotes canaliculatus 343
Bourletiellidae 68
Bourletiellids 68, 69
Brachiacantha barberi 353
Brachycentridae 367, 369, 377
Brachycentrus numerosus 369
Brachycera 437, 443, 444, 466
Brachygastra mellifica 263
Brachymeria 253
Brachymeria tegularis 253
Brachynemurus sackeni 287
Brachypanorpa oregonensis 435
Brachyponera chinensis 266
Brachypsectra fulva 332
Brachypsectridae 332
Brachypterus urticae 357
Brachystola magna 139
Brachystomatid Flies 474
Brachystomatidae 474
Brachystomella parvula 65
Brachystomellidae 64
Brachystomellids 64, 65
Braconid Wasps 248, 249
Braconidae 248
Branchiopoda 493
Brasema 255
Braula coeca 476
Braulidae 476
Brentidae 303, 362
Brephidium exilis 408
Brine Shrimp 493

Bristle-legged Moths 396, 397
Bristletails 71, 75, 79
Bristly Millipede 501
Broad-headed Bugs 206, 207
Broad-headed Sharpshooter 192
Broad-hipped Flower Beetles
    346
Broad-striped Forceptail 109
Broad-tipped Conehead 143
Broad-winged Damsels 102
Bronze Alder Moth 395
Bronze Blister Beetle 345
Bronzed River Cruiser 109
Brown Harvestman 498
Brown Lacewings 284, 285
Brown Marmorated Stink Bug
    211
Brown Recluse 499
Brown Scoopwing 413
Brown Stink Bugs 210
Brown Trig 141
Brown Winter Grasshopper 136
Brown-banded Cockroach 171
Brown-belted Bumble Bee 271
Brown-hooded Cockroaches 170,
    171
Bruchomorpha tristis 197
Brunneria borealis 161, 163
Brunner's Mantis 161, 163
Brush-footed Butterflies 410, 411
Bucculatricidae 392
Bucculatrix coronatella 393
Bumble Bee Scarab Beetles 326
Bumble Bee Scarabs 327
Bumble Bees 271
Bumble Flower Beetle 327
Bumelia 396
Bumelia Borer 359
Bumelia Webworm Moth 397
Buprestidae 306, 328
Buprestis decora 329
Buprestoidea 298, 328
Burnet Moths 404, 405
Burrowing Bugs 210, 211
Burrowing Water Beetles 318,
    319
Bush Katydid 131

Bushtailed Caddisflies 376
Buthidae 497
Buttercup Oil Beetle 345
Butterflies 378, 379
Byrrhidae 312, 330
Byrrhoidea 298, 330, 332
Byrrhus cyclophorus 331
Byturidae 305, 350
Byturus unicolor 351

Cabbage White 409
Cactus Coreid 207
Cactus Flies 478, 479
Cactus Lady Beetle 353
Caddisflies 365, 373
Caecidotea 501
Caenia dimidiata 334
Caenida 90
Caenidae 86, 88
Caenis 89
Caenocholax 292
Calamoceratidae 367, 369, 372
Calamocertid Caddisflies 372,
    373
Calephelis nemesis 410
Calico Pennant 107
California Bee Assassin 201
Calinda longistylus 191
Caliothrips 219
Caliscelidae 181, 196
Calledapteryx dryopterata 413
Calleta Silkmoth 423
Callibaetis 93
Callibaetis pretiosus 93
Calligrapha serpentina 361
Calliphoridae 437, 451, 488
Callirhipid Cedar Beetles 332
Callirhipidae 332
Callophrys niphon 408
Callosamia angulifera 422
Callosobruchus maculatus 361
Calobatina geometra 481
Calocedrus 246
Calophya nigripennis 191
Calophyidae 190
Calophyids 190, 191
Calopteron 334

Calopteron discrepans 334
Calopteron terminale 334
Calopterygidae 99, 100, 102, 109
Calopteryx maculata 99, 103, 109
Caloptilia 393
Calosoma sayi 317
Calosoma scrutator 317
Calycomyza 481
Calycopis cecrops 408
Calypteratae 490
Calyptratae 441, 443, 488
Cambaridae 501
Cambarus diogenes 501
Camel Cricket 131, 142, 143
Camel Spiders 493
Cameraria quercivorella 393
Camillid Flies 476
Camillidae 476
Campodeidae 71, 73
Camponotus castaneus 267
Campopleginae 249
Camptonotus carolinensis 141
Canacidae 477
Cantharidae 307, 310, 336
Cantharis rufa 337
Canthon viridis 326
Canthyloscelid Flies 460
Canthyloscelidae 460
Capnia anna 127
Capnia gracilaria 127
Capniidae 124, 126
Capperia 399
Carabidae 295, 300, 316
Carausius morosus 156
Carnid Flies 477
Carnidae 477
Carnivore Fleas 427, 428
Carolina Grasshopper 137
Carolina Leaf-roller 141
Carolina Mantis 158, 160, 161, 165
Carolina Metallic Tiger Beetle 316
Carolina Satyr 410
Carolina Sphinx 420
Carolina Walkingstick 156

Carpenter and Leopard Moths 404, 405
Carpenter Ant 267
Carpenter-worm Moth 405
Carpet Beetles 338, 339
Carpophilus melanopterus 355
Carposinidae 398
Carrion Beetles 322, 323
Carrionflower Moth 395
Carrot Wasps 260, 261
Carthasis decoratus 201
Casebearer Moths 382, 400, 401
Cassini Periodical Cicada 195
Castianeira 499
Cat Flea 424, 425, 429
Catocala epione 417
Catogenus rufus 357
Catonia nava 195
Catops 323
Cattail Toothpick Grasshopper 130
Cauchas simpliciella 391
Caurinus dectes 435
Cave Barklice 229
Cecidomyiidae 437, 445, 456, 458, 460
Cecropia Moth 423
Cedusa 197
Celithemis elisa 107
Celithemis eponina 99
Centipedes 493, 500
Centris rhodopus 271
Centrozoros snyderi 113
Centruroides gracilis 497
Centruroides vittatus 497
Cephalocarida 493
Cephidae 241, 246
Cephoidea 241, 246
Ceraclea 368, 375
Ceraclea punctata 375
Ceraclea tarsipunctata 369
Cerambycidae 304, 311, 358
Ceraphron 251
Ceraphronid Wasps 250, 251
Ceraphronidae 250
Ceraphronoidea 243, 250

Cerastipsocus venosus 231
Ceratobarys eulophus 485
Ceratocombidae 184, 198
Ceratocombids 198, 199
Ceratocombus 199
Ceratonyx satanaria 413
Ceratophyllidae 428
Ceratopogonidae 445, 462
Cercopid Froghoppers 192, 193
Cercopidae 192
Cercopoidea 180, 192
Cercyonis pegala 410
Ceresa 195
Cerobasis guestfalica 225
Cerococcidae 188
Cerococcus 188
Cerophytidae 333
Cerophytum 333
Ceroplastes 189
Ceroptres 259
Cerura scitiscripta 415
Cerylon castaneum 351
Cerylonidae 350
Ceuthophilus divergens 143
Ceuthophilus pallidus 143
Chaetopsis massyla 483
Chalarus 475
Chalcid Wasps 253
Chalcidid Wasps 252
Chalcididae 252
Chalcidoidea 241, 243, 252, 254, 256
Chalcophora virginiensis 329
Chalk-fronted Corporal 108
Chalybion californicum 269
Chamaemyia 479
Chamaemyiid Flies 478, 479
Chamaemyiidae 457, 478
Chaoboridae 445, 464
Chaoborus 465
Chaparral Monkey Grasshopper 139
Charcoal Seed Bug 209
Chariessa pilosa 347
Chauliodes rastricornis 279
Chauliognathus pensylvanicus 337
Checkered Beetles 346, 347

Checkered White 409
Cheese Skipper 482, 483
Chelicerata 493, 494
Chelifer 497
Chelinidea vittiger 207
Chelonariidae 332
Chelonarium lecontei 332, 333
Cherry Fruit Moth 395
Cherry Lace Bug 200
Chestnut Short-wing Katydid 142
Cheumatopsyche 371
Chewing Lice 226, 228
Chickweed Geometer 413
Chihuahuanus coahuilae 497
Chilocorus 353
Chilocorus cacti 353
Chilopoda 493, 500
Chimarra 373
Chimoptesis pennsylvaniana 397
Chinavia hilaris 211
Chinese Mantis 159, 161, 165
Chinquapin Leaf-miner Moth 389
Chironomidae 445, 464
Chironomini 465
Chironomus 465
Chironomus decorus 465
Chlaenius emarginatus 317
Chloroperlidae 123, 124, 128
Chloropidae 454, 484
Chlorotabanus crepuscularis 469
Choreutidae 383, 384, 396
Choreutoidea 384, 396
Chorisoneura 170
Chorisoneura texensis 166, 168
Choristoneura pinus 397
Choristoneura rosaceana 397
Chortophaga viridifasciata 137
Chrysanthrax cypris 471
Chrysididae 262
Chrysidoidea 241, 243, 260, 262
Chrysina gloriosa 327
Chrysomelidae 294, 297, 305, 360
Chrysomeloidea 299, 358, 360
Chrysopa chi 285
Chrysoperla rufilabris 285
Chrysopidae 284
Chrysopilus thoracicus 469

Chrysops geminatus 469
Chrysops separatus 469
Chrysops virgulatus 469
Chymomyza amoena 479
Chyphotes 265
Chyphotid Wasps 264, 265
Chyphotidae 264
Chyromyid Flies 484, 485
Chyromyidae 484
Cicada Parasite Beetles 328, 329
Cicadas 175, 177, 194, 195
Cicadellidae 174, 180, 192, 292
Cicadidae 180, 194
Cicadoidea 192, 194
Cicindela oregona 316
Cicindela sexguttata 316
Cicindelidia punctulata 316
Cicindelinae 264
Ciidae 313, 340
Cimberid weevils 360
Cimberidae 360
Cimbicid Sawflies 244, 245
Cimbicidae 244
Cimex lectularius 199
Cimicidae 185, 199
Cimicomorpha 199, 200
Cinara 187
Cinteoti Vinegaroon 496
Citheronia regalis 381, 423
Citheronia sepulcralis 423
Cixiid Planthoppers 196, 197
Cixiidae 181, 196
Clambidae 313, 328
Clambus howdeni 329
Clamp-tipped Emerald 107
Clastoptera achatina 193
Clastoptera octonotata 193
Clastoptera xanthocephala 193
Clastopterid Froghoppers 192, 193
Clastopteridae 192
Clearlake Clubtail 104
Clearwing Moths 383, 404, 405
Cleftfooted Minnows 94
Clemens' Grass Tubeworm Moth 391

Clepsis peritana 397
Cleridae 305, 306, 311, 346
Cleroidea 299, 346, 348, 350
Click Beetles 334, 335
Climacia areolaris 287
Climaciella brunnea 287
Clio Stripetail Stonefly 120, 129
Clioperla clio 120, 129
Clip-wing Grasshopper 137
Clogmia albipunctata 467
Clothes Moths 390, 391
Clover Stem Borer 357
Clover Thrip 219
Clown Beetles 322, 323
Club Fleas 427, 428, 429
Clubtails 104
Clusia lateralis 485
Clusiid Flies 484, 485
Clusiidae 453, 484
Cluster Flies 488, 489
Clymene Moth 417
Cnemotetti 141
Coastal Plain Meganola Moth 416
Coboldia 461
Coccidae 176, 188
Coccinella septempunctata 353
Coccinellidae 307, 352
Coccinelloidea 299, 350, 351, 352, 354
Coccoidea 180, 186, 188
Cochineal Insects 186, 187
Cockroach Egg Parasitoid Wasp 261
Cockroach Wasps 268
Cockroaches 167, 168
Cocoa Clubtail 105
Coecobrya tenebricosa 67
Coelichneumon 248
Coelodasys unicornis 415
Coelopa frigida 479
Coelopidae 457, 478
Coenagrionidae 97, 101, 102, 109
Colaspis favosa 361
Coleomegilla maculata 353
Coleophora 401
Coleophora tiliaefoliella 401

Coleophoridae 382, 400
Coleoptera 115, 131, 175, 260, 264, 289, 295
Coleorrhyncha 175
Colias eurytheme 409
Colias interior 409
Colladonus clitellarius 192
Collembola 57, 61, 71, 75, 79
Colletes solidaginis 269
Colletidae 268
Collophora quadrioculata 68
Collophoridae 68
Collophorids 68
Collops vittatus 351
Colocistis crassa 265
Colomychus talis 413
Colopterus unicolor 355
Combmouthed Minnows 94
Common Barklice 230, 231, 234
Common Buckeye 410
Common Burrower Mayfly 82
Common Burrowers 90
Common Checkered-skipper 407
Common Earwigs 118, 119
Common Eastern Firefly 337
Common Fleas 427, 428, 429
Common Green Bottle Fly 489
Common Green Darner 105, 109
Common Peafowl Louse 227
Common Pill Bug 501
Common Sanddragon 109
Common Sawflies 244, 245
Common Scorpionflies 434, 435
Common Scorpionfly 431, 432, 433
Common Silverfish 81
Common Stoneflies 128, 129
Common Thrips 218, 219
Common True Katydid 143
Common Walkingsticks 156, 157
Common Wood Nymph 410
Compsocryptus 248
Concealer Moths 402, 403
Condylostylus 436, 473
Condylostylus sipho 473
Coneheads 57
Conifer Bark Beetles 344

Conifer Sawflies 246, 247
Coniopterygidae 284
Coniopteryx 285
Conocephalus fasciatus 142
Conopidae 452, 480
Conotrachelus nenuphar 363
Contacyphon 329
Conura 253
Copidosoma floridanum 253
Coppers 408
Copromorphidae 398
Copromorphoidea 387, 398
Copromyza equina 487
Coproporus 323
Coptopterygid Mantises 162, 163
Coptopterygidae 162
Coptotermes formosanus 173
Coptotomus 319
Coptotriche citrinipennella 389
Cordilura setosa 489
Cordulegaster obliqua 105, 109
Cordulegastridae 101, 104, 109
Cordulia shurtleffi 107
Corduliidae 101, 106, 109
Coreidae 185, 206, 292
Corethrella 464
Corethrellidae 464
Corimelaena 213
Corinnidae 499
Corioxenidae 291, 292, 293
Corixidae 182, 204
Corydalidae 277, 279
Corydalus cornutus 277, 278, 279
Corydiidae 170, 172
Corydioidea 167, 168, 172
Corylophidae 354
Coryphaeschna ingens 105
Corythucha associata 200
Cosmet Moths 400, 401
Cosmetidae 498
Cosmopterigidae 400
Cossidae 404
Cossoidea 386, 404
Cossula magnifica 405
Cotesia empretiae 249
Cotinis nitida 326
Cotton Stainer 213

Cottonwood Borer 359
Cottony Cushion Scale 189
Cow Killer 265
Cowpea Weevil 361
Crab Louse 227
Crab Spider 499
Crabonid Wasps 268, 269
Crabronidae 268, 292
Crabs 493
Crabwalkers 94
Crambid Snout Moths 412
Crambidae 379, 382, 412
Cramptonomyia spenceri 460
Cratichneumon 249
Crawling Water Beetles 314
Crayfish 493, 500, 501
Crayfishes 493
Cream-bordered Dichomeris 401
Crematogaster laeviuscula 266
Creosote Gall Midge 461
Crescent-marked Bondia 399
Crescentia cujete 398
Crickets 131
Crown Wasps 260, 261
Cruisers 106, 107
Crumomyia 487
Crustacea 493, 500
Cryphocricos hungerfordi 205
Cryptarcha ampla 355
Cryptic Fungus Beetles 342
Cryptic Slime Mold Beetles 354, 355
Cryptocephala binominis 361
Cryptocercidae 167, 170
Cryptocercus darwini 171
Cryptocercus punctulatus 171
Cryptochetidae 476
Cryptochetum iceryae 476
Cryptolaemus montrouzieri 353
Cryptolestes 357
Cryptophagidae 311, 358
Cryptophagus 358
Cryptorhopalum 339
Cryptostemma 198
Cryptothelea 391
Ctenocephalides canis 429
Ctenocephalides felis 424, 425, 429

Ctenolepisma longicaudata 78, 81
Ctenophthalmidae 428
Cuckoo Wasps 262
Cucujidae 308, 309, 356
Cucujoidea 299, 354, 356, 358
Cucujus clavipes 357
Cuerna costalis 174
Culicidae 445, 464
Culicoides 463
Culicomorpha 440, 462, 464
Cupedidae 314
Cupes capitatus 315
Cupido comyntas 408
Curaliidae 184, 200
Curaliids 200
Curalium cronini 200
Curculio 363
Curculionidae 302, 303, 362
Curculionoidea 299, 360, 362
Curicta scorpio 205
Curtonotid Flies 476
Curtonotidae 476
Curtonotum 476
Cuterebra 489
Cybister fimbriolatus 318, 319
Cybocephalid Beetles 358
Cybocephalidae 358
Cybocephalus 358
Cycad Weevils 360
Cycloneda munda 353
Cycloptilum trigonipalpum 141
Cyclorrhapha 443
Cyclotrachelus unicolor 317
Cydia deshaisiana 396
Cydnidae 184, 210, 292
Cylindera lemniscata 316
Cylindromyia intermedia 491
Cylindrotomid Crane Flies 458
Cylindrotomidae 458
Cymatodera undulata 347
Cymidae 208
Cymids 208
Cymoninus notabilis 208
Cynipidae 258
Cynipoidea 243, 258
Cynthia 422
Cyphoderris monstrosa 143

Cyphoderus similis 67
Cypress Twig Gall Midge 461
Cypselosomatidae 476
Cypselosomatids 476
Cyrpoptus belfrage 197
Cyrtacanthacridinae 136
Cyrtolobus tuberosus 195
Cyrtophorus verrucosus 359
Cysteodemus armatus 345

Dacoderus striaticeps 347
Dactylopiidae 186
Dactylopius 187
Dainty Sulphur 409
Dalcerid Moths 404
Dalceridae 404
Dalcerides ingenita 404
Dalquestia formosa 498
Damaeidae 494
Damaeus grossmani 494
Damp Barklice 229
Damsel Bugs 200, 201
Damselflies 97
Danaus gilippus 411
Danaus plexippus 411
Dance Flies 472, 473
Dancer 97
Dancing Moths 392, 393
Darapsa myron 420
Dark Fishfly 279
Dark Marathyssa 416
Dark-collared Tinea 391
Darkling Beetles 342, 343
Dark-spotted Palthis 417
Dark-winged Fungus Gnats 462, 463
Dark-winged Striped-Sweat Bee 269
Darners 104
Dascillidae 328
Dascilloidea 298
Dasydemellidae 232, 235
Dasymutilla bioculata 265
Dasymutilla occidentalis 265
Dasymutilla quadriguttata 265
Datana 415

Davis' Southeastern Dog-Day Cicada 195
Dead Log Beetles 344, 345
Death-Watch Beetles 338, 339
Decapoda 493, 500
Deep Yellow Euchlaena 413
Deer Flies 468, 469
Deer Louse 227
Deileater atlanticus 335
Deloyala guttata 361
Delphacid Planthoppers 196, 197
Delphacidae 181, 196, 292
Delphinia picta 483
Delta Flower Scarab 326
Deltophora sella 401
Dendroctonus terebrans 363
Depressariid Moths 402, 403
Depressariidae 402
Derbid Planthoppers 196, 197
Derbidae 181, 196
Dermacentor variabilis 495
Dermaptera 115, 131, 145, 153, 295
Dermestes ater 339
Dermestidae 312, 313, 338
Derodontidae 338
Derodontoidea 299, 336, 338
Derolathrus 336
Desert Cockroaches 172
Desert Firetail 103
Desert Long-horned Grasshoppers 138
Desert Shadowdamsel 103
Desmia funeralis 413
Desmopachria mexicana 319
Deuterophlebiidae 458
Deuterosminthurus 69
Devil Crawfish 501
Diabrotica balteata 361
Diadocidia 462
Diadocidiid Flies 462
Diadocidiidae 462
Dialysis elongata 473
Diamondback Moths 392, 393
Diamondback Spittlebug 193
Diaperis maculata 343
Diapheromera 155

Diapheromera carolina 156
Diapheromera femorata 157
Diapheromera velii 153
Diapheromeridae 153, 156
Diaphorina citri 191
Diapriid Wasps 252, 253
Diapriidae 252
Diaprioidea 241, 252
Diaspididae 186
Diaspis echinocacti 187
Diastatid Flies 476
Diastatidae 476
Dicellurata 73
Diceroprocta 195
Dichomeris flavocostella 401
Dicosmoecus 365, 377
Dicrepidius palmatus 335
Dicromantispa sayi 287
Dictyopharid Planthoppers 196, 197
Dictyopharidae 181, 196, 292
Dictyoptera 159, 167
Dicyrtomidae 68
Dicyrtomids 68, 69
Didymops transversa 107
Dielis plumipes 267
Differential Grasshopper 136
Dilaridae 284
Dilophus 461
Dilophus orbatus 461
Dimorphic Macalla 413
Dineutus 315
Dinky Light Summer Sedges 371
Diogmites 471
Diopsidae 452, 484
Diphyllostomata 324
Diphyllostomatidae 324
Diplocentridae 497
Diplocentrus lindo 497
Diplopod 493
Diplopoda 500
Diploschizia impigritella 395
Diplura 57, 61, 71, 75, 79
Diprionidae 246
Dipseudid Caddisflies 370
Dipseudopsidae 366, 370

Dipsocoridae 184, 198
Dipsocorids 198
Dipsocoromorpha 198
Diptera 83, 97, 237, 252, 289, 292, 431, 437
Diradius vandykei 151
Dirhinus 253
Dirt-colored Seed Bugs 210, 211
Dissomphalus 261
Dissosteira carolina 137
Disteniid Beetles 358, 359
Disteniidae 358
Distinguished Colomychus 413
Distremocephalus texanus 336
Ditomyiid Flies 462
Ditomyiidae 462
Dixidae 464
Dobsonflies 279
Doddsia occidentalis 127
Dog Biting Louse 222
Dog Flea 429
Doid Moths 412
Doidae 412
Dolania americana 88
Dolba hyloeus 421
Dolichomitus irritator 249
Dolichopodidae 436, , 448, 472
Dolichovespula maculata 263
Dolophilodes distinctus 372
Donacia 361
Doratopsylla blarinae 429
Doru taeniatum 117, 119
Dot-lined White 419
Dot-tailed Whiteface 108
Dotted Anteotricha Moth 403
Double-striped Bluet 103
Douglas moths 394
Douglasiidae 386, 394
Downy Leather-winged Beetle 337
Dracotettix monstrosus 139
Dragonflies 97
Dragonhunter 105
Drepanidae 383
Drepanoidea 387, 412
Drosophila 479
Drosophila repleta 479

Drosophilidae 455, 476, 478
Drugstore Beetle 339
Dry Bark Beetles 352
Dryadaula terpsichorella 393
Dryadaulidae 392
Dryinidae 262
Dryinids 262
Dryocampa rubicunda 423
Dryomyzid Flies 476
Dryomyzidae 476, 477
Dryopidae 312, 330
Drywood Termites 172, 173
Dubious Checkered Beetle 347
Dun Skipper 407
Dung Flies 488, 489
Dusky Dancer 109
Dusky Lady Beetle 353
Dusky-winged Hover Fly 475
Dustywings 284, 285
Dynastes tityus 327
Dysdercus concinnus 213
Dysdercus mimulus 213
Dysdercus suturellus 213
Dyseriocrania griseocapitella 389
Dysmicoccus wistariae 187
Dytiscidae 301, 318

Eacles imperialis 379, 423
Early Smoky Wing Sedges 376
Earth-boring Scarab Beetles 324, 325
Earwigflies 434, 435
Earwigs 115
Eastern Amberwing 96
Eastern Ant Cricket 141
Eastern Black-legged Tick 495
Eastern Blood-sucking Conenose 201
Eastern Boxelder Bug 207
Eastern Carpenter Bee 271
Eastern Cicada Killer 269
Eastern Dobsonfly 277, 278, 279
Eastern Green June Beetle 326
Eastern Hercules Beetle 327
Eastern Lubber Grasshopper 139
Eastern Pine Elfin 408

Eastern Pondhawk 106, 107
Eastern Subterranean Termite 173
Eastern Swiftwing 475
Eastern Tailed-blue 408
Eastern Tent Caterpillar & Moth 419
Eastern Tiger Swallowtail 407
Eastern Yellowjacket 263
Ebony Bugs 212, 213
Ebony Jewelwing 99, 103, 109
Eccoptura xanthenes 129
Echinophthiriidae 228
Echmepteryx hageni 225
Ecnomidae 366, 370
Ecnomidae Caddisflies 370
Ectobiidae 166, 170
Ectoedemia rubifoliella 389
Ectopsocidae 223, 230, 234
Ectopsocopsis cryptomeriae 231
Ectopsocus californicus 223
Ectopsocus meridionalis 231
Efferia 471
Eidoreus politus 351
Eight-spotted Forester 414
Elachistidae 400
Elasmostethus interstinctus 211
Elasmus polistis 255
Elateridae 306, 334
Elateroidea 299, 332, 333, 334, 336
Elegant Sphinx 421
Elenchidae 292, 293
Eleodes 343
Elephant Mosquito 465
Elfin Skimmer 106
Elipsocidae 229
Ellipsoptera nevadica 316
Elliptical Barklice 229
Ellychnia 337
Elm Leaf Beetle 361
Elmidae 312, 330
Elongate Twig Ant 267
Elongate-bodied Springtails 66
Elonus basalis 347
Elytrimitatrix undata 358, 359

Embioptera 111, 115, 121, 145, 149, 153, 167, 221
Embolemid wasps 262
Embolemidae 262
Emeralds 106
Emmelina monodactyla 399
Empididae 449, 472, 474
Empidoidea 446
Empis 473
Empis spectabilis 473
Empoasca 192
Empria 245
Enallagma basidens 103
Enallagma civile 103
Encyrtid Wasps 252, 253
Encyrtidae 252
Endecatomid Beetles 338
Endecatomidae 338
Endecatomus 338
Enderleinellidae 228
Endomychidae 307, 313, 352
Endomychus biguttatus 353
Enicocepalomorpha 198
Enicocephalidae 184, 198
Enigmatic Scarab Beetles 324, 325
Enoclerus nigripes 347
Ensifera 140, 142
Ensign scales 189
Ensign Wasps 260, 261
Entomobrya clitellaria 67
Entomobryidae 66
Entomobryids 66, 67
Entomobryomorpha 66
Entylia carinata 195
Eodermaptera 115
Eosentomidae 59
Eosentomata 59
Eosentomids 59
Epalpus signifer 491
Epaphroditid Mantises 162, 163
Epaphroditidae 162
Epargyreus clarus 407
Epeorus 93
Epeorus longimanus 93
Epermenia albapunctella 397

Epermeniidae 383, 387, 396
Epermenioidea 387, 396
Ephemerella 91
Ephemerella dorothea infrequens 91
Ephemerellidae 87, 90
Ephemeridae 82, 86, 90
Ephemeroptera 83, 97, 121
Ephydra hians 479
Ephydridae 454, 455, 478
Epiaeschna heros 105
Epicallima argenticinctella 403
Epicauta atrivittata 345
Epicauta polingi 345
Epimartyria auricrinella 389
Epimecis hortaria 413
Epimetopid Beetles 320, 321
Epimetopidae 320
Epimetopus 320, 321
Epione Underwing 417
Epipaschia superatalis 413
Epipsocidae 229
Epipyropida 402
Epitheca costalis 107
Epitheca princeps 107
Epuraea 355
Eralea albalineella 401
Erebid Moths 416, 417
Erebidae 382, 416
Eremiaphilid Mantises 164
Eremiaphilidae 164
Eremobatidae 497
Eremochrysa punctinervis 285
Eremocoris depressus 211
Eremoneura 443
Eriaporid Wasps 252
Eriaporidae 252
Eriococcidae 188
Eriocrania semipurpurella 389
Eriocraniid Moths 388, 389
Eriocraniidae 385, 388
Eriocranioidea 385, 388
Erioptera caliptera 459
Eriosomatinae 187
Ermine Moth 392, 393
Ernobius 339

Eros humeralis 334
Erotylidae 305, 311, 356
Erynnis horatius 407
Erythemis simplicicollis 107
Erythraeidae 494
Etainia sericopeza 389
Euborellia 119
Euborellia annulipes 115, 119
Eucalantica polita 393
Euceratia securella 393
Eucharitid Wasps 254
Eucharitidae 254
Euchlaena amoenaria 413
Eucinetidae 328
Euclea delphinii 405
Euclemensia bassettella 401
Eucnemidae 333
Eucoilinae 259
Eudarcia eunitariaeella 393
Eudryas unio 414
Euetheola rugiceps 326
Eugnamptus angustatus 363
Eugnophomyia luctuosa 459
Euholognatha 121, 126
Eulichadidae 332
Eulophid Wasps 254, 255
Eulophidae 254
Eumastacidae 134, 138
Eumenes fraternus 263
Eumesosoma 498
Eumorpha achemon 420
Eumorpha fasciatus 420
Eumorpha pandorus 420
Eumorsea balli 139
Eupackardia calleta 423
Euparius marmoreus 362
Eupelmid Wasps 254, 255
Eupelmidae 254
Eupelmus 255
Eupeodes 475
Euperilampus triangularis 257
Euphoria inda 327
Euphyes vestris 407
Eupleurida 346
Eupompha elegans 345
Eupsilobiidae 351

European Earwig 114, 119
European Hornet 263
Europs pallipennis 357
Eurybrachidae 292
Eurylophella doris 91
Eurytides marcellus 406
Eurytomid Wasps 254, 255
Eurytomidae 254
Euschistus servus 210
Eustrophopsis bicolor 341
Euteliid Moths 383, 416
Euteliidae 383, 416
Euthycera arcuata 481
Euxestidae 350
Euzophera semifuneralis 413
Evallijapyx 73
Evania appendigaster 261
Evaniidae 260
Evanioidea 242, 260
Evergreen Bagworm Moth 391
Everlasting Tebenna Moth 397
Exallonyx 251
Exclamation Moth 403
Excultanus excultus 192
Exema neglecta 361
Exoprosopa fasciata 471
Eyed Baileya 416
Eyed Paectes 416

Fairy Moths 382, 390, 391
Fairy Shrimp 493
Fairyflies 254
False Blister Beetles 344, 345
False Bombardier Beetle 295, 317
False Burnet Moths 396, 397
False Click Beetles 333
False Clown Beetles 320
False Darkling Beetles 340, 341
False Ermine moths 394, 395
False Fairy Wasps 258
False Flower Beetles 346, 347
False Ground Beetles 318, 319
False Jewel Beetles 328
False Lizard Barklice 229
False Longhorn Beetles 342

False Mealworm Beetle 343
False Metallic Wood-boring
    Beetles 333
False Owlet Moths 412, 413
False Pit Scales 189
False Potato Beetle 361
False Skin Beetles 350
False Soldier Beetles 333
False Stag Beetles 324
Familiar Bluet 103
Fannia 489
Fanniid Flies 488, 489
Fanniidae 451, 488
Fatal Metalmark 410
Fateful Barklice 230, 231, 234
Featherwinged Beetles 322, 323
Felt scales 188
Ficus 190
Field Cricket 141
Fiery Searcher 317
Fiery Skipper 407
Fig Wasps 252
Figitid Wasp 258, 259
Figitidae 258
Filigreed Chimoptesis 397
Filter Fly 467
Fingernet Caddisflies 372, 373
Firebrat 81
Fire-Colored Beetles 344, 345
Fireflies 336, 337
Fishflies 279
Five-banded Thynnid Wasp 265
Flame Skimmer 108
Flannel Moths 404, 405
Flat Bark Beetles 356, 357
Flat Bugs 206
Flat Wireworm 335
Flataloides scabrosa 197
Flat-footed Flies 474, 475
Flatheaded Mayflies 92, 93
Flatid Planthoppers 196, 197
Flatidae 181, 196, 292
Flatormenis proxima 197
Fleas 425
Flesh Flies 490, 491
Flies 437

Florida Bark Scorpion 497
Flower Moths 402, 403
Flutter Flies 482, 483
Folsomia candida 61, 67
Forcepstails 73
Forest Scaly Cricket 141
Forest Silverfish 80
Forest Stream Beetles 332
Forest Tent Caterpillar & Moth
    419
Forficula auricularia 114, 119
Forficulidae 117, 118
Forked Fungus Beetle 343
Formicidae 237, 240, 266
Formicoidea 240, 242, 266
Formosan Subterranean Termite
    173
Four-spotted Fungus Beetle 357
Four-spotted Ghost Moth 389
Four-spotted Skimmer 107
Four-spotted Velvet Ant 265
Four-toothed Mason Wasp 263
Frankliniella 219
Franklinothrips vespiformis 219
Free-living Caddisflies 374
Freeloader Flies 486, 487
Fringe-tufted Moths 383, 396,
    397
Frit Flies 484, 485
Frog-biting Midges 464
Froghoppers 177
Fruit Flies 482, 483
Fruitworm Beetles 350, 351
Fruitworm Moths 388, 399
Fulgoraecia exigua 402, 403
Fulgorid Planthoppers 196, 197
Fulgoridae 181, 196, 292
Fulgoroidea 180, 194, 196
Fungus Gnats 462, 463
Fungus Weevils 362
Furcula borealis 415

Galacticid Moths 396
Galacticidae 396
Galacticoidea 396
Galerita 295

INDEX

Galerita bicolor 317
Galgula partita 414
Galgupha 213
Gall Midges 460, 461
Gall Wasps 258, 259
Gall-inducing Aphids 187
Gall-like Scales 188
Gambrinus griseus 335
Gammarotettix 131, 143
Garden Tortrix 397
Gargaphia iridescens 200
Gasteracantha cancriformis 499
Gasteruptiidae 260
Gasteruption 261
Gelastocoridae 182, 204
Gelastocoris rotundatus 205
Gelechiidae 382, 400
Gelechioidea 384, 385, 400, 402
Geocoridae 208
Geocoris 209
Geometer Moths 382, 412, 413
Geometridae 382, 412
Geometroidea 385, 387, 412
Geomyza 487
Geophilidae 501
Georissidae 320
Georissus 320
Geotrupes splendidus 325
Geotrupidae 303, 324
Geranium Plume Moth 399
Gerdana caritella 401
German Cockroach 171
Germarostes globosus 327
Gerridae 183, 202
Gerromorpha 202
Ghost Moths 388, 389
Giant Bark Aphid 187
Giant Casemakers 377
Giant Conifer Aphid 187
Giant Ichneumon 248
Giant Lacewings 286, 287
Giant Leopard Moth 417
Giant Mealybugs 188
Giant Red Sedge 377
Giant Redheaded Centipede 501
Giant Scale Insects 189
Giant Silkworm Moths 422

Giant Stag Beetle 325
Giant Stoneflies 126, 127
Giant Swallowtail 407
Giant Walkingstick 157
Giant Water Bugs 204, 205
Gibbium aequinoctiale 339
Glaphyria fulminalis 413
Glaphyridae 303, 326
Glaresidae 324
Glaresis 325
Glassy-winged Sharpshooter 193
Glenurus gratus 287
Globemallow Leaf Beetle 361
Globipedidae 498
Globular Springtails 68
Glorious Habrosyne 413
Glorious Scarab 327
Glossosomatidae 366, 372
Glossosomatoidea 372
Glowworm Beetles 336
Glyphidocera flosella 401
Glyphipterigidae 394
Gnat Bugs 198
Gnathamitermes 169, 173
Goera calcarata 369
Goerid Caddisflies 376
Goeridae 366, 369, 376
Goldcap Moss-eater Moth 389
Golden Silk Orbweaver 492
Golden Snow-flea 60, 65
Golden-backed Snipe Fly 469
Goldenrod Cellophane Bee 269
Goldenrod Soldier Beetle 337
Gomphidae 99, 101, 104, 109
Gomphurus hybridus 105
Gonatista grisea 162, 163
Goniodes pavonis 227
Gracillariidae 383, 392
Gracillarioidea 386, 392
Granary Booklice 224, 225
Grape Leaffolder 413
Grape Tube Gallmaker 437, 461
Grapeleaf Skeletonizer 405
Grapevine Beetle 327
Graphopsocus cruciatus 233
Grass Miner Moths 400, 401
Grass scales 186

Grass Shrimp 501
Grasshoppers 131
Grass-like Mantid 162
Gray Hairstreak 408
Gray Petaltail 105, 109
Gray Silverfish 78, 81
Gray Walkingstick 153
Great Black Wasp 269
Great Grig 143
Great Spreadwing 103
Green Banana Cockroach 171
Green Cloverworm 417
Green Cutworm Moth 414
Green Lacewings 281, 284, 285
Green Mantisfly 287
Green Stink Bug 211
Green Stoneflies 128, 129
Greenhouse Millipede 501
Green-striped Grasshopper 137
Grizzled Mantis 163
Ground Beetles 316
Ground Mantid 165
Ground Pearls 189
Gryllacrididae 135, 140
Gryllidae 135, 140
Grylloblatta 145
Grylloblatta campodeiformis 144, 146, 147
Grylloblattidae 147
Grylloblattodea 145
Grylloprociphilus imbricator 187
Gryllotalpidae 134, 140
Gryllus 141
Grynocharis quadrilineata 349
Guinea Pig Lice 228
Gulf Coast Tick 495
Gulf Fritillary 378, 411
Gymnochiromyia 485
Gymnopternus 473
Gyponana octolineata 193
Gyrinidae 301, 314
Gyrinus 315
Gyropidae 228

Habropoda laboriosa 271
Habrosyne gloriosa 413
Hackberry Psyllid 191

Hacklegilled Burrowers 90
Haematopinidae 226
Haematopinus suis 227
Haematopis grataria 413
Hagenius brevistylus 105
Hairstreaks 408
Hairy Chinch Bug 209
Hairy Fungus Beetles 340, 341
Hairy-eyed Crane Flies 458, 459
Hairy-veined Barklice 229
Hairy-winged Barklice 232, 233
Halictidae 268
Halictophagidae 292, 293
Halictophagus 293
Halictoxenos 293
Halictus ligatus 269
Haliplidae 300, 314
Haliplus 315
Halloween Pennant 99
Halyomorpha halys 211
Handsome Fungus Beetles 352, 353
Hangingflies 431, 433, 434, 435
Haploa clymene 417
Haploembia 151
Haplopus mayeri 156, 157
Haplothrips 219
Haplothrips leucanthemi 219
Hardwood Stump Borer 359
Harlequin Bug 210
Harmonia axyridis 353
Harrisina americana 405
Harvesters 408
Harvestmen 493, 498
Hawaiian Dancing Moth 393
Hawk Lice 226, 227
Head Louse 227
Hebrew Moth 414
Hebridae 183, 202
Hebrus 203
Hedychrum 262
Helcomyza mirabilis 477
Helcomyzid Flies 477
Helcomyzidae 477
Heleomyzid Flies 484, 485
Heleomyzidae 453, 455, 484
Heleomyzinae 485

INDEX

Helichus lithophilus 331
Helicopsyche 374, 375
Helicopsyche borealis 369
Helicopsychidae 367, 369, 374
Heliocis repanda 345
Heliodinidae 394
Heliozelidae 390
Helluomorphoides nigripennis 317
Helophorid Beetles 320, 321
Helophoridae 320
Helophorus 320, 321
Helorid Wasps 250, 251
Heloridae 250
Helorus 250, 251
Hemerobiidae 284
Hemerobiiformia 284
Hemerobius stigma 285
Hemipsocidae 230, 234
Hemipsocus chloroticus 231
Hemiptera 131, 153, 159, 175,
    292, 293, 295
Hemisphaerota cyanea 361
Hepialidae 381, 385, 388
Hepialoidea 385, 388
Heptagenia 93
Heptageniidae 87, 92
Hermetia illucens 467
Hermeuptychia sosybius 410
Hesperapis 270
Hesperentomidae 57
Hesperentomon macswaini 57
Hesperiidae 406
Hesperoctenes 199
Hetaerina americana 103
Hetaerina titia 99, 103
Heterocampa obliqua 415
Heterocampa pulverea 415
Heteroceridae 313, 330
Heterocerus 331
Heterocheila hannai 477
Heterocheilid Flies 477
Heterocheilidae 477
Heterodoxus spiniger 228
Heterogaster 208
Heterogastridae 208
Heterogastrids 208

Heteronemiidae 156
Heteroplectron americanum 369
Heteropsylla texana 191
Heteroptera 175, 178, 179, 198
Heterothripidae 217, 218
Hexagenia 91
Hexagenia bilineata 91
Hexagenia limbata 82, 91
Hibiscus 396
Hidden Barklice 229
Hide Beetles 324, 325
Hieroglyphic Cicada 195
Higher Termites 172, 173
Hilara 473
Hilarimorpha 470
Hilarimorphid Flies 470
Hilarimorphidae 470
Hill-Prairie Spittlebug 193
Hippiscus ocelote 136
Hippoboscidae 444, 490
Hister 323
Histeridae 304, 322
Hog Louse 227
Hololepta yucateca 323
Holopsis marginicollis 355
Holorusia hespera 459
Homadaula anisocentra 396
Homaemus parvulus 213
Homaeotarsus 323
Homaledra heptathalama 403
Homalodisca vitripennis 193
Homidia sauteri 67
Homoeolabus analis 363
Homoneura 479
Homosetia 391
Homotomidae 190
Homotomids 190
Honey Bees 270, 271
Honeysuckle Moth 393
Hood Casemakers 374, 375
Hooded Grouse Locust 139
Hooded Tickspiders 493
Hooktip Moths 383, 412, 413
Hoplopleuridae 228
Hoplothrips 214
Hoppers 175, 180

Hop-tree Borer Moth 395
Horace's Duskywing 407
Horned Passalus 325
Horned Powder-post Beetles 338, 339
Hornets 262, 263
Horntails 246, 247
Horse Flies 468, 469
Horsefly-like Carpenter Bee 271
Horseshoe Crabs 493
Hoshihananomia octopunctata 341
House Centipede 501
House Flies 488, 489
Household Cockroaches 170, 171
Hover Flies 474, 475
Huachuca Monkey Grasshopper 139
Hubbard's Angel Insect 110, 111, 113
Human Flea 429
Hummingbird Lice 228
Hummingbird Moths 420
Humpless Casemaker Caddisflies 377
Hump-winged Crickets 142, 143
Hyalella 501
Hyalellidae 501
Hyalophora cecropia 423
Hyalymenus tarsatus 207
Hyblaea puera 398
Hyblaeidae 387, 398
Hyblaeoidea 387, 398
Hybosoridae 302, 326
Hybosorus illigeri 327
Hybotid Dance Flies 472, 473
Hybotidae 448
Hydaticus bimarginatus 319
Hydraenidae 322
Hydrellia 479
Hydrobiosid Caddisflies 374
Hydrobiosidae 366, 374
Hydrocanthus 319
Hydrocanthus atripennis 319
Hydrochara soror 321
Hydrochid Beetles 320, 321
Hydrochidae 320

Hydrochus 320, 321
Hydrometra 203
Hydrometridae 182, 202
Hydrophilidae 307, 320
Hydrophiloidea 298, 320, 322
Hydrophilus triangularis 321
Hydropsyche 371
Hydropsyche simulans 369
Hydropsychidae 366, 368, 369, 370
Hydropsychoidea 370
Hydroptila 373
Hydroptilidae 366, 372
Hydroptiloidea 372
Hydroscapha natans 315
Hydroscaphidae 314
Hylephila phyleus 407
Hyles lineata 420
Hylesinus mexicanus 363
Hymenaphorura cocklei 60, 65
Hymenoptera 83, 97, 167, 237, 240, 289, 292, 293, 365, 379, 431
Hypena scabra 417
Hypercompe scribonia 417
Hypogastrura 65
Hypogastruridae 64
Hypogastrurids 64, 65
Hypoprepia fucosa 417
Hyporhagus 345
Hypselonotus punctiventris 207
Hystrichopsylla schefferi 428
Hystrichopsyllidae 428

Ianassa lignicolor 415
Ibalia anceps 259
Ibaliid Wasps 258, 259
Ibaliidae 258
Ice Crawlers 145, 147
Icerya purchasi 189
Ichneumon Wasps 248, 249
Ichneumonid Wasps 260
Ichneumonidae 248
Ichneumonoidea 240, 242, 248
Icosta americana 491
Idia americalis 417

INDEX

Idris 259
Imperial Moth 423
Imperial Moth Caterpillar 379
Incense Cedar Wood Wasp 246
Incurvariidae 390
Inflated Beetle 345
Inocelliidae 273, 275
Integripalpia 365, 368, 372
Io Moth 422
Iris oratoria 161, 164, 507
Ironclad Beetles 344, 345
Ironweed Curculio 363
Isabella Tiger Moth 417
Ischaliidae 346
Ischnocera 226
Ischnopsyllidae 428
Ischnoptera bilunata 171
Ischnura ramburii 103
Ischyrus quadripunctatus 357
Isonychia 83, 92
Isonychiidae 83, 87, 92
Isoparce cupressi 420
Isoperla 122, 129
Isoperla davisi 129
Isophrictis magnella 401
Isopoda 493, 500
Isopods 493
Isoptera 167
Isotoma viridis 67
Isotomidae 61, 66
Isotomids 66, 67
Isotomurus bimus 67
Issid Planthoppers 196, 197
Issidae 181, 196, 292
Ithonidae 284
Iuridae 497
Ixodes 495
Ixodes scapularis 495
Ixodidae 494, 495

Jack Pine Budworm 397
Jacobsoniidae 336
Jacobson's Beetles 336
Jagged Ambush Bug 201
Jalysus 209
Jalysus spinosus 209
Janusius sylvestris 69

Japanese Beetle 327
Japygidae 70, 73
Jerusalem Crickets 142, 143
Jugular-horned Beetles 342
Jumping Bristletails 77
Jumping Plant Lice 176, 190, 191
Jumping Soil Bugs 198
Jumping Spider 499
Jumping Thrips 218
Juniperus 246
Junonia coenia 410

Kalotermitidae 172
Kateretidae 356
Katianna 69
Katiannidae 68
Katiannids 68, 69
Katydids 131, 142, 143
Kazacholthrips triassicus 215
Kelp Flies 478, 479
Kennedy's Emerald 109
Kermes Scale Moth 401
Kermesidae 188
Keroplatidae 462
Kerriidae 181, 188
King Crickets 140, 141
Kinnarid Planthoppers 194
Kleidocerys resedae 209
Kleptochthonius griscomanus 497
Knulliana cincta 359
Kudzu Bug 212
Kuschelina thoracica 361
Kuwaniidae 188
Kuwaniids 188

Labia minor 116, 118
Labidomera clivicollis 361
Labidura riparia 119
Labiduridae 117, 118
Labidus coecus 267
Lac scales 188
Laccophilus fasciatus 319
Lace Bugs 200
Lacewing 281
Lachesilla floridana 231
Lachesilla nubilis 231
Lachesilla penta 231

Lachesillidae 230, 234
Lacosoma chiridota 418
Lactista azteca 137
Lactura subfervens 405
Lacturidae 404
Ladona julia 108
Lady Beetles 352, 353
Laemobothriidae 226
Laemobothrion 227
Laemophloeidae 356
Lagynodes 251
Lampetis drummondi 329
Lampyridae 307, 310, 336
Lance Flies 480, 481
Lancet Clubtail 105
Lance-wing Moths 402, 403
Languria mozardi 357
Laphria grossa 471
Lappet Moths 382
Large Crane Flies 458, 459
Large Milkweed Bug 209
Large Paectes Moth 416
Large Squaregilled Mayflies 88
Larger Empty Oak Apple Wasp
     259
Larger Pygmy Mole Grasshopper
     139
Large-winged Psocids 229
Largidae 185, 212
Largus 213
Largus Bug 213
Largus succinctus 213
Laricobius rubidus 339
Lasaia sula 410
Lash-Faced Psocids 230, 231
Lasiocampidae 382, 418
Lasiocampoidea 387, 418
Lasiochilidae 199
Lasiochilids 199
Lasioglossum 269
Lasochilidae 184
Latheticomyia 476
Latridiidae 354
Latrodectus 499
Lauxaniid Flies 478, 479
Lauxaniidae 453, 457, 478
Leaf Beetles 360, 361

Leaf Blotch Miner Moths 383
Leaf Blotch Miners 392, 393
Leaf Litter Barklice 230, 231, 234
Leaf Miner Flies 480, 481
Leaf Rolling Weevils 362, 363
Leaf Skeletonizer 383
Leaf-cutter Bee 270
Leafcutter Bees 270
Leafcutter Moths 390, 391
Leaf-footed Bugs 206, 207
Leafhoppers 177, 192, 193
Leafroller Moths 396, 397
Leafrolling Sawflies 244
Lebia pulchella 317
Lecanodiaspididae 189
Lecanodiaspis 189
Lecithocerid Moths 402
Lecithoceridae 402
LeConte's Seedcorn Beetle 317
Leiobunum 498
Leiodes assimilis 323
Leiodidae 322
Leiomyza 477
Lepidopsocidae 224
Lepidoptera 237, 260, 289, 365,
     379
Lepidostoma 376
Lepidostomatidae 367, 369, 377
Lepinotus reticulatus 225
Lepisma saccharina 81
Lepismatidae 81
Leptinotarsa juncta 361
Leptoceridae 367, 368, 369, 374
Leptoceroidea 372
Leptogastrinae 471
Leptoglossus phyllopus 207
Leptohyphidae 86, 90
Leptomantispa pulchella 287
Leptophlebia intermedia 89, 95
Leptophlebiidae 87, 88, 95
Leptopodidae 182, 198
Leptopodomorpha 198
Leptopsyllidae 428
Lepturobosca chrysocoma 359
Leptysma marginicollis 130
Lepyronia gibbosa 193
Lepyronia quadrangularis 193

Lesser Angle-winged Katydid 142
Lesser Dung Flies 486, 487
Lesser Earwig 116, 118
Lesser Stripetail Scorpion 497
Lestes australis 103
Lestes disjunctus 109
Lestidae 100, 102, 109
Lethocerus uhleri 205
Leucochrysa insularis 285
Leucorrhinia intacta 108
Leucospid Wasps 254, 255
Leucospidae 254
Leucospis 254
Leucospis affinis 255
Leuctra 127, 128
Leuctridae 124, 126
Libellula flavida 109
Libellula incesta 108
Libellula luctuosa 108
Libellula quadrimaculata 107
Libellula saturata 108
Libellulidae 96, 99, 101, 106, 109
Liburniella ornata 197
Libytheana carinenta 411
Lichen Moths 382
Lichenomima sparsa 231
Lichenophanes bicornis 339
Lichnanthe 326
Lichnanthe rathvoni 327
Ligated Sweat Bee 269
Light Western Drywood Termite 173
Ligia 501
Ligiidae 501
Limacodidae 382, 404
Limenitis archippus 411
Limenitis arthemis astyanax 411
Limnephilidae 364, 365, 367, 376
Limnephiloidea 376
Limnephilus 377
Limnia conica 481
Limnichidae 330
Limnichites punctatus 331
Limnophila macrocera 459
Limnoporus 203
Limoniid Crane Flies 458, 459
Limoniidae 444, 458

Lined Earwig 117, 119
Lined Flat Bark Beetles 356, 357
Lined June Beetle 327
Linepithema humile 266
Linognathidae 228
Linyphilidae 499
Liopterid wasps 258
Liopteridae 258
Lipoptena mazamae 491
Liposcelididae 224
Liposcelis bostrychophila 225
Litaneutria minor 165
Lithariapteryx jubarella 395
Lithoseopsis hellmani 225
Litter Bug 199
Little Black Caddisflies 372, 373
Little Earwigs 118
Little Stout Crawlers 90
Little Yellow 409
Little Yucatán Mantis 165
Liturgusa maya 161, 162, 163
Liturgusid Mantises 162, 163
Liturgusidae 162
Livia saltatrix 191
Liviidae 190, 191
Liviids 190, 191
Lizard Barklice 232, 233
Lobiopa undulata 355
Lobsters 493
Loensia 231
Lomamyia 285
Lonchaeidae 452, 480
Lonchoptera 474, 475
Lonchopteridae 449, 474
Lone Star Tick 495
Long-beaked Fungus Gnats 462
Long-horned Bee 271
Long-horned Beetles 358, 359
Long-horned Caddisflies 374, 375
Longistigma caryae 187
Long-jointed Beetle 343
Long-legged Velvet Mite 494
Long-necked Seed Bug 211
Long-tailed Skipper 407
Long-toed Water Beetles 330, 331

Longlegged Flies 436, 472, 473
Lonicera 398
Lophocaterid Beetles 348, 349
Lophocateridae 348
Lophognathella 64
Lotisma trigonana 399
Louse Flies 490, 491
Lovebugs 461
Loving Barklice 232, 233, 235
Loxosceles reclusa 499
Lubber Grasshoppers 138, 139
Lucanidae 302, 324
Lucanus capreolus 325
Lucanus elaphus 325
Lucidota atra 337
*Lucilia sericata*489
Luna Moth 423
Lutrochidae 330
Lutrochus 330
Lutrochus laticeps 331
Lycaenidae 408
Lycidae 310, 334
Lycosidae 499
Lyctocoridae 184, 199
Lyctocorids 199
Lyctocoris 199
Lycus sanguineus 334
Lygaeidae 208, 292
Lygaeoidea 185
Lygaeus kalmii 209
Lygistorrhina sanctaecatharinae 462
Lygistorrhinidae 462
Lygus hesperus 201
Lymantriinae 382
Lymexylidae 340
Lyonet Moths 394, 395
Lyonetia prunifoliella 395
Lyonetiidae 394
Lytta magister 345
Lytta polita 345

Machilidae 76, 77
Machilinus 75
Machilinus aurantiacus 77
Machiloides banksi 74, 77
Macrocera formosa 463

Macrolepidoptera 379
Macromia annulata 109
Macromia taeniolata 107
Macromiidae 101, 106, 109
Macropis 270
Macrosiagon limbata 295, 341
Macrostemum carolina 371
Macrovelia hornii 203
Macroveliid Shore Bugs 202, 203
Macroveliidae 182, 202
Macrurocampa marthesia 415
Magicicada 194
Magicicada cassini 195
Malacosoma americana 419
Malacosoma californicum 419
Malacosoma disstria 419
Malacostraca 493, 500
Mallodon dasystomus 359
Mallophaga 221
Mallophora fautrix 442
Malthodes 337
Mammal Chewing Lice 226, 227
Mandibulate Archaic Moths 388, 389
Manduca sexta 420
Mantidae 158, 159, 164
Mantidflies 286, 287
Mantids 164, 165
Mantises 159
Mantispidae 159, 286
Mantodea 153, 159, 293
Mantoid Mantises 164, 165
Mantoida maya 164, 165
Mantoididae 164
Mantophasmatodea 145
Many-plume Moths 383, 398, 399
Maple Callus Borer 405
Maple Leafcutter Moth 391
Maple Webworm Moth 413
Marathyssa inficita 416
March Flies 460, 461
Margarodidae 189
Marginitermes hubbardi 173
Marilia flexuosa 373
Maritime Earwig 119
Marsh Beetles 328, 329
Marsh Flies 480, 481

Marsupial Chewing Lice 228
Marsupial Lice 228
Masked Bees 268, 269
Mason Bees 270
Mason Wasps 262, 263
Master Blister Beetle 345
Mastigoproctus cinteoti 496
Matsucoccidae 189
Mauroniscid Beetles 348
Mauroniscidae 348
Maxillopoda 493
May Beetle 326
Mayan Lichen Mantis 161, 163
Mayer's Walkingstick 157
Mayflies 83
Mealybug Destroyer 353
Mealybugs 180, 186
Mecaphesa 499
Mecoptera 273, 281, 425, 431
Mediterranean Mantis 161, 164
Meessiid Moths 392, 393
Meessiidae 392
Megachile 270
Megachilidae 266, 270
Megacopta cribraria 212
Megalodacne heros 357
Megalomus fidelis 285
Megalopodid Leaf Beetles 360
Megalopodidae 360
Megaloptera 83, 97, 273, 281
Megalopyge crispata 405
Megalopygidae 404
Meganola phylla 416
Megaphasma denticrus 155, 157
Megarhyssa macrurus 248
Megaselia scalaris 475
Megaspilid Wasps 250, 251
Megaspilidae 250
Megastigmid Wasps 254
Megastigmidae 254
Megathymus yucca 407
Megatibicen auletes 195
Meghyperus 472
Megischus bicolor 261
Megisthanidae 495
Megisthanus 495
Meinertellidae 76, 77

Melacoryphus lateralis 209
Melanactes piceus 335
Melandryidae 309, 340
Melanis pixe 410
Melanoliarus 197
Melanophthalma 355
Melanoplus bivittatus 136
Melanoplus differentialis 136
Melanoplus femmurrubrum 136
Melanthripid Thrips 218
Melanthripidae 218
Melissodes tepaneca 271
Melittid Bees 270
Melittidae 270
Melittobia 255
Meloe americanus 345
Meloidae 306, 309, 344
Melyridae 307, 311, 350
Melyrodes cribrata 351
Membracidae 180, 292
Menacanthus stramineus 220, 227
Meniscus Midges 464
Menoponidae 220, 226
Meracantha contracta 343
Merope tuber 434, 435
Meropeidae 434
Merostomata 493
Merothripidae 217, 218
Meskea dyspteraria 399
Mesopsocidae 229
Mesovelia 202
Mesovelia mulsanti 202
Mesoveliidae 183, 202
Mesquite Psyllid 191
Metajapyx 70, 73
Metaleptea brevicornis 137
Metallic Wood-boring Beetles 328, 329
Metalmark Moths 383, 396, 397
Metalmarks 410
Metopius 249
Metretopodidae 87, 94
Metylophorus novaescotiae 232
Metylophorus purus 231
Mexican Honey Wasp 263
Mexican Jumping Bean 396
Mezira 206

Michaelophorus identatus 399
Micracanthia humilis 199
Micro Bee Flies 470, 471
Micro Velvet Mite 494
Microcaddisflies 372, 373
Microcentrum retinerve 142
Micromalthidae 314
Micromalthus debilis 314
Micropezidae 452, 480
Microphysidae 184, 200
Micropterigidae 385, 388
Micropterigoidea 385
Microrhopala excavata 361
Microtonus sericans 341
Microtrombidiidae 494
Microvelia 203
Micro-Whipscorpions 493
Middle Barklice 229
Midges 464, 465
Mid-winter Boreus 433, 435
Milesia virginiensis 475
Milichiidae 453, 455, 486
Milkweed Aphid 187
Millipedes 493, 500, 501
Mimallonidae 383, 387
Mimallonoidea 387, 418
Mimosa webworm 396
Minettia 479
Miniature Trap-jaw Ant 267
Mining Bees 268, 269
Minute Bark Beetles 350, 351
Minute Beetles 328, 329
Minute Black Scavengers Flies 460, 461
Minute Bladder Bugs 200
Minute Bog Beetles 314
Minute Brown Scavenger Beetles 354, 355
Minute Clubbed Beetles 356, 357
Minute Hooded Beetles 354, 355
Minute Marsh-loving Beetles 330, 331
Minute Moss Beetles 322
Minute Mud-loving Beetles 320
Minute Pirate Bugs 199
Minute Tree-Fungus Beetles 340, 341

Miridae 185, 200
Mischievous Bird Grasshopper 137
Mites 493, 494
Mogoplistidae 140
Molanna 375
Molanna tryphena 375
Molannidae 367, 374
Mole Crickets 140, 141
Mompha 403
Momphid Moths 402, 403
Momphidae 402
Monarch 411
Monkey Grasshoppers 138, 139
Mono Lake Alkali Fly 479
Monobia quadridens 263
Monoceros Beetle 347
Monophlebidae 189
Monotomidae 356
Montana Six-plume Moth 399
Moonseed Moth 417
Mordella 341
Mordella cervicalis 341
Mordella melaena 341
Mordellidae 304, 340
Mordellistena comata 341
Mormon Metalmark 410
Morning-glory Plume Moth 399
Morsea californica 139
Mortarjoint Casemakers 372, 373
Mosquitoes 464, 465
Moss Bugs 175
Moth Flies 466, 467
Moth Lacewings 284
Moths 379
Mottled Prominent 415
Mottled Tortoise Beetle 361
Mountain Midges 458
Mourning Cloak 411
Mourning Scorpionfly 435
Mouse-like Barklice 230, 231
Mozena obtusa 207
Multicolored Asian Lady Beetle 353
Mumetopia occipitalis 481
Murgantia histrionica 210
Murmidiidae 351

INDEX

Musca domestica 489
Muscidae 451, 488
Mutillidae 264
Mycetaea subterranea 352
Mycetaeid Beetles 352
Mycetaeidae 352
Mycetophagidae 313
Mycetophagus serrulatus 341
Mycetophilidae 456, 462
Mycomya 463
Mycotrupes cartwrighti 325
Mycteridae 342
Mydas chrysostomus 471
Mydas clavatus 471
Mydas Flies 470, 471
Mydidae 449, 470
Myiocnema comperei 252
Mymaridae 254
Mymarommatidae 258
Mymarommatoidea 243, 258
Myodocha serripes 211
Myoplatypus flavicornis 363
Myopsocidae 230
Myriapoda 493, 500
Myrmecolacidae 289, 292, 293
Myrmecophilidae 134, 140
Myrmecophilus pergandei 141
Myrmeleon 287
Myrmeleontidae 286
Myrmeleontiformia 286
Myrmex 363
Myrmilloides grandiceps 265
Myrmosid Wasps 266
Myrmosidae 266
Mystacides 375
Mythicomyia 471
Mythicomyiidae 449, 470
Myxophaga 295, 301, 314
Myzia pullata 353
Myzinum quinquecinctum 265

Nabidae 185, 200
Nabis 201
Nabis subcoleoptratus 201
Nadata gibbosa 415
Nallachius americanus 284, 285

Nallachius pulchellus 284
Nannothemis bella 106
Nanopsocus oceanicus 225
Narceus 501
Narrow Barklice 232, 233, 235
Narrow-waisted Bark Beetles 346, 347
Narrow-winged Moths 394
Nathalis iole 409
Naucoridae 183, 204
Neacoryphus bicrucis 209
Neanuridae 64
Neanurids 64, 65
Nearctopsylla brooksi 429
Necrodes surinamensis 323
Necrophila americana 323
Nectopsyche 375
Nectopsyche candida 375
Necydalis mellita 359
Neelidae 66
Neelids 66
Neelipleona 66
Nehalennia gracilis 103
Nematocera 437, 443, 444, 445, 458
Nemestrinidae 472
Nemognatha nemorensis 345
Nemonychidae 360
Nemopoda nitidula 487
Nemoria bistriaria 413
Nemouridae 124, 126
Neobarrettia spinosa 143
Neochodaeus praesidii 327
Neochoroterpes 89
Neocicada hieroglyphica 195
Neoclytus mucronatus 359
Neoconocephalus triops 143
Neodermaptera 115
Neodiprion 247
Neoephemera 88
Neoephemeridae 86, 88
Neoharmonia venusta 353
Neolepolepis occidentalis 225
Neopamera bilobata 211
Neoperla clymene 129
Neoporus arizonicus 319

Neopseustoidea 385, 388
Neopyrochroa femoralis 345
Neorthopleura thoracica 347
Neoscapteriscus borellii 141
Neoscona crucifera 499
Neotermes castaneus 173
Neotibicen davisi 195
Neotridactylus apicialis 139
Neotropical Deer Ked 491
Neoxabea bipunctata 141
Nephrotoma 459
Nepidae 183, 204
Nepomorpha 204
Nepticulidae 388
Nepticuloidea 386, 388
Neriene radiata 499
Neriidae 452, 478
Nersia florida 197
Nerthra 205
Nessus Sphinx 421
Net Tube Caddisflies 370
Netspinning Caddisflies 370, 371
Nettle Pollen Beetle 357
Net-winged Beetles 334
Net-winged Midges 466
Neuroptera 83, 97, 121, 159, 167, 221, 273, 277, 281, 365, 431
Nevada Tiger Beetle 316
Nicoletia wheeleri 81
Nicoletiidae 81
Nicrophorus orbicollis 323
Niesthrea louisianica 207
Nigronia fasciata 279
Ninidae 208
Ninids 208
Nitidulidae 305, 306, 311, 354, 358
Noctuidae 382, 414
Noctuoidea 385, 387, 414, 416
Nocturnal Velvet Ant 265
Nola pustulata 416
Nola triquestrana 416
Nolid Moths 416
Nolidae 416
Nomad Bee 271
Nomada 271

Non-insect Arthropods 493
Norrbomia frigipennis 487
Northern Apple Sphinx 421
Northern Caddisflies 376, 377
Northern Caddisfly 364
Northern Dusk Singing Cicada 195
Northern Flatid Planthopper 197
Northern Oak Hairstreak 408
Northern Rotund-Resin Bee 270
Northern Spreadwing 109
Northern Walkingstick 157
No-see-ums 462, 463
Nosodendridae 338
Nosodendroidea 338
Nosodendron 338
Nosodendron unicolor 339
Noteridae 301, 318
Notiophilus aeneus 316
Notodontidae 382, 414
Notonecta irrorata 205
Notonectidae 182, 204
Notoptera 145, 153
Notoxus monodon 347
Notoxus murinipennis 347
Nycteribiid Flies 490
Nycteribiidae 490
Nyctiophylax 371
Nylanderia fulva 266
Nymphalidae 378, 410
Nymphalis antiopa 411
Nymphomyia 458
Nymphomyiidae 458
Nymphomyiids 458
Nysius 209

Oak Leafroller 397
Oak Timberworm 363
Oak Treehopper 195
Oblique Heterocampa 415
Oblique-banded Leafroller 397
Obolopteryx castanea 142
Ochodaeidae 303, 326
Ochteridae 2182, 04
Ochterus 204
October Caddis 365

Ocyptamus fuscipennis 475
Odiniid Flies 486
Odiniidae 486
Odonata 83, 97, 277, 281, 494
Odontellidae 64
Odontellids 64
Odonteus thoracicornis 325
Odontoceridae 367, 369, 372
Odontoloxozus longicornis 479
Odontomyia rufipes 467
Odontosciara nigra 463
Odontotaenius disjunctus 325
Odontotaenius floridanus 325
Oebalus pugnax 210
Oecanthus californicus 141
Oecetis 375
Oeclidius 194
Oecophoridae 402
Oedemasia semirufescens 415
Oedemeridae 309, 344
Oestridae 450, 488
Oestrophasia calva 491
Ofatulena duodecemstriata 397
Oiceoptoma inaequale 323
Olbiogaster 461
Olceclostera seraphica 418
Oliarces clara 284
Oligoneuridae 86
Oligoneuriid Brushlegged
    Mayflies 92
Oligoneuriidae 92
Oligonicella scudderi 162
Oligotoma nigra 151
Oligotoma saundersii 151
Oligotomid Webspinner 151
Oligotomidae 151
Olla v-nigrum 353
Omethidae 333
Omoglymmius hamatus 315
Omorgus monachus 325
Oncerometopus nigriclavus 201
Oncometopia orbona 192
Oncopeltus fasciatus 209
Oncopeltus sexmaculatus 209
Oncopoduridae 66
Oncopodurids 66
Onthophagus hecate 326

Onychiuridae 60, 64
Onychiurids 64, 65
Ophion 249
Opiliones 493
Opomyzid Flies 486, 487
Opomyzidae 454, 486
Opostegidae 388
Orange Sulphur 409
Orange-barred Sulphur 409
Orange-headed Epicallima 403
Orchesella celsa 67
Orchopeas howardi 429
Oreogeton 474
Oreogetonid Flies 474
Oreogetonidae 474
Oreoleptid Flies 468
Oreoleptidae 468
Oreoleptis torrenticola 468
Oreta rosea 413
Orfelia fultoni 463
Oriental Cockroach 171
Ormia ochracea 491
Ormyrid Wasps 254, 255
Ormyridae 254
Ormyrus 254, 255
Ornate Checkered Beetle 347
Ornate Pit Scales 188
Ornate Pygmy Grasshopper 139
Ornidia obesa 475
Oropsylla hirsuta 429
Orsodacne atra 361
Orsodacnidae 360
Ortheziidae 189
Orthoptera 131, 153, 159, 167,
    175, 292, 293, 295
Orthorrhapha 443
Orussidae 240, 246
Orussoidea 240, 246
Orussus 247
Oscinellinae 485
Osmia ribifloris 270
Osphya varians 341
Ostracoda 493
Outer Barklice 230, 231, 234
Owlet Moths 382, 414
Owlflies 280, 281, 286, 287
Ox Beetle 327

Oxidus gracillis 501
Oxycarenidae 206
Oxycarenids 206
Oxycopis thoracica 345
Ozodiceromyia 471
Ozophora picturata 211

Pachnaeus litus 363
Pachygronthidae 208
Pachygronthids 208
Pachyneurid Flies 460
Pachyneuridae 460
Pachypsylla 191
Pachytroctidae 224
Pachytullbergiidae 64
Pachytullbergiids 64
Pacific Coast Dampwood Termite 173
Paectes abrostoloides 416
Paectes oculatrix 416
Painted Lichen Moth 417
Palaegapetus 372
Palaemnema domina 102, 103
Palaemon 501
Palaemonidae 501
Pale Bordered Field Cockroach 171
Pale Lice 228
Pale Red Bug 213
Palingeniidae 86, 90
Pallid-winged Grasshopper 137
Pallodes pallidus 355
Pallopteridae 457, 482
Palm and Flower Beetles 342
Palm Scales 189
Palmetto Beetles 354
Palmetto Tortoise Beetle 361
Palpheaded Mayflies 92
Palpigradi 493
Palthis angulalis 417
Pamphiliidae 241, 244
Pamphilioidea 241, 244
Panchlora nivea 171
Pandora Sphinx 420
Panorpa 433
Panorpa helena 435
Panorpa lugubris 435

Panorpa nuptialis 431, 432, 435
Panorpa rupeculana 585
Panorpa vernalis 435
Panorpidae 431, 432, 433, 434, 585
Panorpodidae 434
Pantala flavescens 97, 108, 109
Pantopoda 493
Paonias myops 421
Paper Wasps 236, 262, 263
Papilio cresphontes 407
Papilio glaucus 407
Papilio polyxenes 406
Papilio zelicaon 406
Papilionidae 406
Papilionoidea 384, 406, 408, 410
Parabacillus 157
Parabacillus coloradus 156
Paraclemensia acerifoliella 391
Paracotalpa ursina 326
Paradoxosomatidae 501
Parajapygidae 73
Parajapyx isabellae 73
Paraleptophlebia 89
Paralobesia cyclopiana 397
Paramysidia mississippiensis 197
Parandra polita 359
Paraphrynus carolynae 497
Parapsyllus longicornis 429
Parasitic Apocrita 248
Parasitic Flat Bark Beetles 356, 357
Parasitic Flies 490, 491
Parasitic Lice 221
Parasitic Wood Wasps 246, 247
Parasitidae 495
Parasitiformes 493
Paratettix aztecus 139
Paratettix cucullatus 139
Paraxenetus guttulatus 201
Parcoblatta 170
Parcoblatta divisa 171
Parnassians 406
Parnassius smintheus 406
Paronellidae 66
Paronellids 66, 67
Paronychiurus ramosus 65

INDEX

Parornix 393
Pasimachus 316
Passalidae 302, 324, 495
Passandridae 308, 356
Patapius spinosus 198
Pauropoda 493
Pauropods 493
Paw Paw Sphinx 421
Peachtree Borer 405
Pearly Wood-Nymph 414
Pecan Carpenterworm Moth 405
Pecan Spittlebug 193
Pecaroecidae 228
Pecaroecus javalii 228
Peccary lice 228
Pedetontus submutans 77
Pedicia 459
Pediciidae 444, 458
Pediculidae 226
Pediculus humanus capitis 227
Pegomya winthemi 489
Pelecinid Wasps 250, 251
Pelecinidae 250
Pelecinus polyturator 250, 251
Pelecorhynchid Flies 468
Pelecorhynchidae 468
Pelidnota punctata 327
Pelochrista robinsonana 397
Pelonides quadripunctata 347
Peltidae 348
Peltis 348
Peltis pippingskoeld 349
Peltodytes sexmaculatus 315
Peltoperla 127
Peltoperlidae 124, 126
Penepissonotus bicolor 197
Pentacora 199
Pentagenia vittigera 90, 91
Pentatomidae 175, 184, 210, 292
Pentatomomorpha 206, 208, 210, 212
Penthe pimelia 341
Pepsis 267
Pergid Sawflies 246, 247
Pergidae 246
Periclista 245

Peridea angulosa 415
Perilampid Wasps 256, 257
Perilampidae 256
Perilampis 257
Periplaneta americana 171
Peripsocidae 232
Peripsocus 233
Peripsocus madidus 233
Periscelidid Flies 484
Periscelididae 484
Perithemis tenera 96
Perlesta 121, 129
Perlesta decipiens 129
Perlidae 121, 123, 124, 125, 128
Perlodidae 120, 122, 124, 128
Perplexing Barklice 229
Persea 396
Petaltails 104
Petaluridae 101, 104, 109
Petitia domingensis 398
Petrobius 76
Phacelia purshii 394
Phacopteronidae 190
Phacopteronids 190
Phalacridae 305, 311, 358
Phalangium opilio 498
Phanaeus difformis 327
Phanogomphus australis 104
Phanogomphus exilis 105
Phanogomphus lividus 99
Phantom Crane Flies 458, 459
Phantom Midges 464, 465
Phasmatidae 156
Phasmatodea 115, 145, 153, 159
Phassus giganteus 381
Phausis reticulata 337
Pheidole 267
Phengodes 336
Phengodes fusciceps 336
Phengodidae 307, 336
Pherhimius fascicularis 335
Phidippus 499
Philonix 259
Philopotamidae 366, 372
Philopotamoidea 372
Philopteridae 226

Philotarsidae 232, 235
Phlaeothripidae 214, 215, 216, 218
Phoebis philea 409
Phoenicococcidae 189
Phoenicococcus marlatti 189
Pholeomyia 487
Phoridae 448, 474
Phormia regina 489
Photinus pyralis 337
Photuris 337
Phryganeidae 366, 377
Phryganeoidea 377
Phrynidae 497
Phthiracaridae 494
Phthiracarus 494
Phthiraptera 221
Phylloicus aeneus 373
Phyllopalpus pulchellus 141
Phyllophaga 326, 482
Phylloporia bistrigella 391
Phyllovates chlorophaea 165
Phylloxera 191
Phylloxeridae 180, 190
Phylloxerids 190, 191
Phylloxeroidea 190
Phylocentropus 370
Phymata 201
Physocephala tibialis 481
Physonota alutacea 294
Picture-winged Flies 482, 483
Pieridae 408
Pieris rapae 409
Piesmatidae 184, 210
Pigeon Tremex 247
Piglet Bugs 196, 197
Pigmy Moths 388, 389
Pill Beetles 330, 331
Pill Scarab Beetle 327
Pine & Spruce Aphids 190
Pine Bast Scales 189
Pine Devil Moth 423
Pine Flower Snout Beetles 360
Pine Spittlebug 193
Pink-edged Sulphur 409
Pink-spotted Hawk Moth 421

Pinophilus 323
Pintalia vibex 197
Piophilidae 482
Pipe Organ Mud Dauber 269
Pipevine Swallowtail 407
Pipunculidae 474
Pistol Casebearer 401
Pit Scales 186
Pitch Pine Tip Moth 397
Pityococcid Scales 189
Pityococcidae 189
Plains Camel Cricket 143
Plains Lubber Grasshopper 139
Plant Bugs 200, 201
Planthopper Parasite Moth 403
Planthopper Parasites 402, 403
Planthoppers 177
Plant-parasitic Hemipterans 176
Plasterer Bees 268, 269
Plataspidae 184, 212
Plataspids 184, 212
Plateros 334
Plate-thigh Beetles 328
Platycentropus 377
Platycotis vittata 195
Platydema micans 343
Platydracus femoratus 323
Platydracus fossator 323
Platygastrid Wasps 258, 259
Platygastridae 258
Platygastroidea 241, 258
Platypalpus 473
Platypezidae 448, 474
Platystictidae 101, 102
Platystomatidae 457, 482
Pleasing Fungus Beetles 356, 357
Pleasing Lacewings 284, 285
Plecia nearctica 461
Plecoptera 115, 121, 145, 277, 281
Plectrodera scalator 359
Pleidae 183, 204
Pleocoma dubitabilis 325
Pleocomidae 303, 324
Pleuropasta mirabilis 345

Pleurota albastrigulella 403
Plinthocoelium suaveolens 359
Plum Curculio 363
Plume Moths 383, 398, 399
Plusiodonta compressipalpis 417
Plutella xylostella 393
Plutellidae 392
Pococera asperatella 413
Podabrus pruinosus 337
Podisus maculiventris 211
Podura aquatica 65
Poduridae 64
Podurids 64, 65
Poduromorpha 64
Poecilochirus 495
Poecilognathus 471
Poecilus chalcites 317
Pogonognathellus 67
Pogonomyrmex rugosus 267
Pointed-winged Flies 474, 475
Polished Lady Beetle 353
Polistes 264
Polistes carolina 263
Polistes dorsalis 263
Polistes exclamans 236
Polistes metricus 293
Pollen Wasps 262, 263
Pollenia 489
Polleniidae 451, 488
Polycaon stouti 339
Polycentropodidae 366, 368, 370
Polycentropus 368
Polyctenidae 184, 199
Polygonia interogations 411
Polygrammate hebraeicum 414
Polymitarcidae 86, 90
Polyphaga 295, 300, 302, 304, 306, 307, 308, 312, 320
Polyphemus Moth 422
Polyphylla occidentalis 327
Polyplacidae 226
Polyplax spinulosa 227
Polypleurus perforatus 343
Polypore Fungus Beetles 340, 341
Polyporivora polypori 475
Polypsocus corruptus 233

Polystoechotes punctatus 286, 287
Polystoechotidae 286
Polyxenidae 501
Polyxenus lagurus 501
Pompilidae 266
Ponana 192
Pond Damsels 102
Pontia protodice 409
Popillia japonica 327
Poronota 494
Postelichus 331
Potamanthidae 86, 90
Potamyia flava 371
Potter Wasps 262, 263
Poultry Body Louse 220, 227
Poultry Lice 226, 227
Powdered Dancer 97
Powder-post Bostrichid 339
Prairie Dog Flea 429
Prairie Walkingstick 153
Praydidae 394
Prays atomocella 395
Predaceous Diving Beetles 318, 319
Predatory Fungus Gnats 462, 463
Predatory Stoneflies 128, 129
Predatory Thrips 218, 219
Prepops insitivus 201
Primitive Carrion Beetle 322
Primitive Crane Flies 466, 467
Primitive minnows 94
Prince Baskettail 107
Prionocyphon discoideus 329
Prionoglaridae 229
Prionoxystus robiniae 405
Pristaulacus niger 261
Procampodea macswaini 72
Procampodeidae 72
Procampodeids 72
Proctotrupes 251
Proctotrupid Wasps 250, 251
Proctotrupidae 250
Proctotrupoidea 243, 250
Prodoxidae 390
Prodoxinae 391
Prodoxus decipiens 391
Progomphus borealis 109

Proleucoptera smilaciella 395
Promachus bastardii 471
Promalactis suzukiella 403
Prominent Moths 382, 414, 415
Pronggilled Mayflies 88
Prophalangopsidae 135, 142
Prosapia bicincta 193
Prostoia besametsa 127
Prostomidae 342
Prostomis americanus 342
Protentomidae 57, 59
Protentomids 59
Protoplasa fitchii 467
Protura 57, 61, 71, 75, 79
Proxys punctulatus 210
Psectrotanypus dyari 465
Psephenidae 312, 330
Psephenus texanus 331
Pseudacteon tricuspis 475
Pseudanostirus tigrinus 335
Pseudexentera 397
Pseudiron centralis 94
Pseudironidae 86, 94
Pseudocaeciliidae 229
Pseudococcidae 186
Pseudogaurotina cressoni 359
Pseudomallada 285
Pseudomethoca frigida 265
Pseudomops septentrionalis 171
Pseudomyrmex gracilis 267
Pseudophacopteron 190
Pseudophasmatidae 156
Pseudophilippia quaintancii 176
Pseudopostega cretea 389
Pseudoscorpiones 493, 496, 497
Pseudoscorpions 493, 496, 497
Pseudosermyle straminea 153
Pseudotephritis vau 483
Psila 487
Psilidae 454, 486
Psiloreta frontalis 369
Psocidae 221, 230, 232, 234
Psocodea 111, 149, 167, 175,
        221, 425
Psocomorpha 221, 223, 229, 230,
        232
Psocoptera 221

Psoquillidae 224
Psorophora howardii 465
Psychidae 390
Psychoda 467
Psychodidae 444, 466
Psychodomorpha 440, 466
Psychomyia flavida 369
Psychomyiidae 369, 370
Psyllidae 190
Psyllipsocidae 229
Psyllobora vigintimaculata 353
Psylloidea 180, 190
Ptecticus trivittatus 467
Ptenidium 323
Ptenothrix renateae 69
Pterocolus ovatus 363
Pterolonchidae 402
Pteromalid Wasps 256, 257
Pteromalidae 256
Pteromalus puparum 257
Pteronarcyidae 124, 126
Pteronarcys californica 127
Pteronarcys dorsata 127
Pterophoridae 383, 385, 398
Pterophoroidea 385, 398
Pterophylla camellifolia 143
Pthiridae 226
Pthirus pubis 227
Ptiliidae 322
Ptilocolepid Caddisflies 372
Ptilocolepidae 366, 372
Ptilodactyla 333
Ptilodactylidae 304, 312, 332
Ptiloneuridae 229
Ptilostomis postica 377
Ptinidae 310, 312, 338
Ptosima gibbicollis 329
Ptychoptera 459
Ptychopteridae 444, 458
Pubic Lice 226, 227
Pulex irritans 426, 429
Pulicidae 424, 426, 428
Punctured Tiger Beetle 316
Purplish Birch-miner Moth 389
Puto 188
Putoidae 188
Pycnogonida 493

Pycnopsyche 364
Pycnoscelus surinamensis 171
Pycnospyche scabripennis 377
Pygidicranid Earwigs 118
Pygidicranidae 118
Pygmy Backswimmers 204
Pygmy Grasshoppers 138, 139
Pygmy Mole Crickets 138, 139
Pyractomena 337
Pyractomena borealis 337
Pyragropsis buscki 118
Pyralid Moths 383
Pyralid Snout Moths 412, 413
Pyralidae 383, 412
Pyraloidea 384, 385, 412
Pyrgota undata 483
Pyrgotid Flies 482, 483
Pyrgotidae 457, 482
Pyrgus communis 407
Pyrisitia lisa 409
Pyrochroidae 309, 344
Pyromorpha dimidiata 405
Pyropyga minuta 337
Pyrota bilineata 345
Pyrrharctia isabella 417
Pyrrhocoridae 185, 212
Pythidae 344
Pytho americanus 345

Question Mark 411

Rabid Wolf Spider 499
Rabidosa rabida 499
Rain Beetles 324, 325
Rainbow Scarab 327
Rainieria antennaepes 481
Rambid Snout Moths 382
Rambur's Forktail 103
Ranatra 205
Raphidiidae 272, 273, 275
Raphidioptera 273, 277, 281
Rare Click Beetles 333
Raspberry Fruitworm 351
Raspy Crickets 140, 141
Rattus rattus 426
Ravenous Leaf Beetles 360, 361
Red Admiral 410

Red Aphid 187
Red Bay Psyllid 191
Red Bugs 212, 213
Red Flat Bark Beetle 357
Red Imported Fire Ant 266
Red Milkweed Beetle 359
Red Wasp 263
Red-banded Hairstreak 408
Red-bordered Pixie 410
Redbud Borer 329
Reddish-brown Stag Beetle 325
Red-eyed Devil 143
Red-fringed Emerald 413
Red-headed Bush Cricket 141
Red-legged Grasshopper 136
Red-legged Oil Digger 271
Red-shouldered Bostrichid 339
Red-spotted Purple 411
Red-spotted Rove Beetle 323
Reduviidae 184, 200
Red-washed Prominent 415
Regal Darner 105
Regal Moth 423
Remipedia 493
Resin Bees 270
Reticulated Beetles 314, 315
Reticulitermes flavipes 167, 169, 173
Retocomus wildii 347
Rhabdura 72, 73
Rhadalid Beetles 348, 349
Rhadalidae 348
Rhadalus testaceus 349
Rhagio 469
Rhagionidae 446, 447, 468
Rhagonycha 337
Rhagovelia 203
Rhamphomyia 473
Rhantus atricolor 319
Rhaphidophoridae 131, 135, 142
Rhigognostis interrupta 393
Rhinophoridae 451, 490
Rhinotermitidae 167, 172
Rhipiceridae 328
Rhodobaenus tredecimpunctatus 363
Rhopalidae 185, 206

Rhopalocera 379
Rhopalopsyllidae 428
Rhopalosomatid Wasps 264
Rhopalosomatidae 264
Rhopalotria 360
Rhyacionia rigidana 397
Rhyacophilidae 366, 374
Rhyacophiloidea 374
Rhyopsocus eclipticus 225
Rhyparochromidae 185, 210
Rhysodidae 300, 314
Rhytidops floridensis 486
Ribbed Cocoon-maker Moths 392, 393
Rice Bug 207
Rice Stink Bug 210
Rice Weevil 363
Richardiid Flies 482
Richardiidae 482
Ricinidae 228
Ricinulei 493
Ridgeback Grasshopper 137
Ridged Carrion Beetle 323
Riffle Beetles 330, 331
Riffle Bugs 202
Ring-legged Earwig 115, 118, 119
Riodinidae 410
Ripiphoridae 295, 309, 340
Rivellia steyskali 483
Roach-like Stoneflies 126, 127
Robber Flies 470, 471
Robinson's Pelochrista Moth 397
Rock Bristletails 74, 75, 77
Rock Crawlers 145
Rock Slater 501
Rocky Mountain Parnassian 406
Rodent Fleas 427, 428, 429
Rolled-winged Stoneflies 126, 127
Romalea microptera 138, 139
Romaleidae 135, 138
Root-maggot Flies 488, 489
Ropalomerid Flies 486
Ropalomeridae 486
Ropronia 250
Roproniid Wasps 250
Roproniidae 250

Rose Hooktip 413
Rossianid Caddisflies 376
Rossianidae 366, 376
Rosy Maple Moth 423
Rottenwood Termites 172, 173
Rough Harvester Ant 267
Round Fungus Beetles 322, 323
Roundneck Sexton Beetle 323
Rove Beetles 322, 323
Royal Moths 422, 423
Royal River Cruiser 107
Rugosana querci 192
Rust Flies 486, 487

Sack-bearer Moths 383, 418
Sacodes pulchella 329
Saddleback Leafhopper 192
Saffron Plum 404
Saldidae 182, 198
Salpingidae 346
Salticidae 499
Samia cynthia 422
Sand Flies 466
Sand Minnows 94
Sandalus 328, 329
Sand-Burrowing Mayflies 88
Sand-loving Scarab Beetles 326, 327
Sap-feeding Beetles 354, 355
Sapyga nevadica 267
Sapygid Wasps 266, 267
Sapygidae 266
Saratoga Spittlebug 193
Sarcophaga 491
Sarcophagidae 451, 490
Sargus fasciatus 467
Saturniidae 379, 381, 383
Satyrium calanus 408
Satyrium favonius 408
Satyrium liparops 408
Saucer Bugs 204
Saunders' Embiid 151
Scale Insects 176
Scale Parasite Flies 476
Scaled Fleas 427, 428
Scales 175, 180
Scalloped Sack-bearer 418

Scaly Crickets 140, 141
Scaly Winged Barklice 224, 225
Scaphinotus angusticollis 317
Scaphisomatini 323
Scaptomyza 479
Scarab Beetles 326, 327
Scarabaeidae 303, 326
Scarabaeoidea 264, 298, 324, 326
Scarites vicinus 316
Scathophagidae 488, 450
Scatopsidae 456, 460
Scavenger Moths 400, 401
Scavenger Scarab Beetles 326, 327
Sceliphron caementarium 269
Scenopinidae 470
Scenopinus fenestralis 470, 471
Scentless Plant Bugs 206, 207
Scerogibbid Wasps 262
Schinia arcigera 414
Schistocerca americana 137
Schistocerca damnifica 137
Schistocerca lineata 137
Schizomida 493
Schizomyia viticola 437, 461
Schizophora 443, 446, 476, 477, 478, 480, 482, 484, 486, 488, 490
Schizopodidae 328
Schizopteridae 184, 198
Schlaeger's Fruitworm Moth 403
Schreckensteinia festaliella 397
Schreckensteiniidae 396
Schreckensteinioidea 387, 396
Sciaridae 456, 462
Sciomyzidae 457, 480
Scirtes orbiculatus 329
Scirtidae 313, 328
Scirtoidea 298, 328
Sclerogibbidae 262
Sclerosomatidae 498
Scoliid Wasps 266, 267
Scoliidae 266
Scolopendra heros 501
Scolopendridae 501
Scooped Scarab 326
Scoopwing Moths 412, 413

Scorpiones 493, 496
Scorpionflies 431
Scorpions 493, 496, 497
Scotoleon 287
Scraptiidae 346
Scrub Cicada 195
Scud 501
Scudderia 131
Sculptured Pine Borer 329
Scutelleridae 184, 212, 292
Scutigera coleoptrat 501
Scutigeridae 501
Scuttle Flies 474, 475
Scymnus louisianae 353
Scythrididae 402
Scythris 403
Sea Spiders 493
Seal Lice 228
Seaside Grasshopper 136
Sedge Moths 394, 395
Seed Bugs 208, 209
Seed Shrimp 493
Sehirus cinctus 211
Selonodon 335
Sematurid Moths 412
Sematuridae 412
Semijulistus flavipes 349
Senopterina foxleei 483
Sensillanura caecaa 65
Sepsidae 453, 457, 486
Sepsis 487
Sericophanes heidemanni 201
Sericostomatidae 366, 369, 376
Sericostomatoidea 374
Sesiidae 383, 385, 404
Sesioidea 385, 404
Seven-spotted Lady Beetle 353
Shadowdamsels 102
Shaggy Psocids 232, 233, 235
Sharp-blotched Nola Moth 416
Sharpshooter 174
Shield Bearers 390, 391
Shield Beetles 348 349
Shield Bugs 210, 211
Shield-backed Bugs 212, 213
Shield-backed Pine Seed Bug 213

Shining Flower Beetles 358
Shiny Head-Standing Moths 394, 395
Ship-timber Beetles 340
Shore Bugs 198, 199
Shore Earwig 119
Shore Flies 478, 479
Short-faced Scorpionflies 434, 435
Short-horn Walkingsticks 156, 157
Short-horned Grasshoppers 136
Short-tailed Whipscorpions 493
Short-winged Flower Beetles 356, 357
Shrew Flea 429
Shrimp 493, 500, 501
Sialidae 276, 277, 279
Sialis 276, 279
Sicariidae 499
Sideroxylon celastrinum 404
Sierolomorphid Wasps 264
Sierolomorphidae 264
Signal Flies 482, 483
Signiphorid Wasps 256
Signiphoridae 256
Silk Moths 383
Silken Fungus Beetles 358
Silk-spinning Cricket 141
Silphidae 306, 313, 322, 495
Silvanid Flat Bark Beetles 356, 357
Silvanidae 305, 308, 356
Silvanus muticus 357
Silverfish 71, 75, 79, 81
Silver-spotted Skipper 407
Simuliidae 445, 464
Simulium 465
Siphlonuridae 87, 94
Siphonaptera 425
Siricidae 246
Siricoidea 240, 246
Sirthenea carinata 201
Sisyridae 286
Sitophilus oryzae 363
Six-spotted Flower Strangalia 359

Six-spotted Milkweed Bug 209
Six-spotted Tiger Beetle 316
Skiff Beetles 314, 315
Skimmers 106, 108
Skippers 406, 407
Slaty Skimmer 108
Sleepy Orange 409
Slender Baskettail 107
Slender Diplurans 71, 73
Slender Lizard Beetle 357
Slender Meadow Katydid 142
Slim Mantises 162
Slim Mexican Mantis 161, 162
Slime Flux Beetle 339
Slug Caterpillar Moths 382, 404, 405
Small Hive Beetle 355
Small Mammal Fleas 427, 428, 429
Small Milkweed Bug 209
Small Squaregilled Mayflies 88
Small Tolype 419
Small Winter Stoneflies 126, 127
Small Yellow Texas Cockroach 166
Small-eyed Sphinx 421
Small-headed Flies 472, 473
Small Water Striders 203
Smerinthus jamaicensis 421
Smicripidae 354
Smicrips 354
Sminthuridae 62, 68
Sminthurides 62
Sminthurididae 68
Sminthuridids 68
Sminthurids 68, 69
Sminthurus 69
Sminthurus packardi 69
Smoky Rubyspot 99, 103
Smooth Spider Beetle 339
Snail-Case Caddisflies 374, 375
Snail-killer Carabid 317
Snakeflies 273, 275
Snipe Flies 468, 469
Snow Scorpionflies 434, 435
Snowberry Clearwing 421
Soft Scales 188, 189

Soft-bodied Plant Beetles 328, 332
Soft-winged Flower Beetles 350, 351
Soil Centipede 501
Soldier Beetles 336, 337
Soldier Flies 466, 467
Solenopsis invicta 266
Solifugae 493, 496
Somatochlora kennedyi 109
Somatochlora tenebrosa 107
Southern Blue-green Citrus Root Weevil 363
Southern Dampwood Termite 173
Southern Dogface 409
Southern Longhorn Moth 391
Southern Mole Cricket 141
Southern Pink-striped Oakworm Moth 423
Southern Protean Shieldback 142
Southern Spreadwing 103
Southern Two-striped Walkingstick 157
Southern Unstriped Scorpion 497
Southern Wood Cockroach 171
Southern Yellowjacket 263
Spatodea campanulata 398
Speckled Lactura 405
Sphaerites politus 320
Sphaeritidae 320
Sphaeriusidae 314
Sphaeroceridae 454, 486
Sphaeropsocidae 229
Sphaeropthalma 265
Sphagnum Sprite 103
Spharagemon cristatum 137
Sphecidae 268, 292
Sphecius speciosus 269
Sphecodina abbottii 421
Sphecoidea 292
Sphecomyiella valida 483
Sphex lucae 269
Sphex pensylvanicus 269
Sphindidae 354
Sphindus americanus 355
Sphingidae 383, 420
Sphinx Moths 383, 420, 421

Sphinx perelegans 421
Sphinx poecila 421
Sphyracephala 484
Sphyracephala subbifasciata 485
Spider Beetles 338, 339
Spider Wasps 266, 267
Spiders 493, 498
Spiketails 104
Spilomyia longicornis 475
Spined Soldier Bug 211
Spiny Crawlers 90, 91
Spiny Oak-Slug Moth 405
Spiny Rat Louse 226, 227
Spinybacked Orbweaver 499
Spiny-Legged Bugs 198
Spiralizoridae 113
Spirobolidae 501
Spittlebug 193
Spodoptera ornithogalli 414
Spongillaflies 286, 287
Spongiphoridae 117, 118
Spot-sided Coreid 207
Spotted Apatelodes 418
Spotted Bird Grasshopper 137
Spotted Lady Beetle 353
Spotted Thyris Moth 399
Spreadwings 102
Spring Fishfly 279
Spring Stoneflies 126, 127
Spring Treetop Flasher 337
Springtails 61, 75 ,79
Springwater Dancer 103
Square-headed Snakeflies 275
Squirrel Flea 429
Squirrel Lice 228
Stag Beetles 324, 325
Stagmomantis carolina 158, 160, 161, 165
Stalk-eyed Flies 484, 485
Staphylinidae 304, 306, 307, 310, 322
Staphylinoidea 298, 322
Stathmopodid Moths 402
Stathmopodidae 402
Statira basalis 343
Stegobium paniceum 339
Steingeliidae 188

Steingeliids 188
Stelidota octomaculata 355
Stem Sawflies 246, 247
Stenacron 93
Stenelmis 331
Stenocaecilius casarum 233
Stenocolus scutellaris 332
Stenocoris 207
Stenocranus 197
Stenolophus lecontei 317
Stenomorpha opaca 343
Stenonema femoratum 93
Stenopelmatidae 135, 142
Stenopelmatus 143
Stenopogon rufibarbis 471
Stenopsocidae 232, 235
Stenotrachelidae 342
Stephanidae 260
Stephanoidea 260
Sternorrhyncha 175, 176, 179, 180, 186
Sthenopis purpurascens 389
Stiletto Flies 470, 471
Stilt Bugs 208, 209
Stilt-Legged Flies 480, 481
Stinging Wasps 260
Stink Bugs 175, 210, 211
Stonecase Caddisflies 376
Stoneflies 121
Stout Barklice 232, 233
Stout's Hardwood Borer 339
Straight-faced Windscorpion 497
Straight-snouted Weevils 362, 363
Strangalia sexnotata 359
Strategus 327
Strategus antaeus 327
Stratiomyidae 446, 447, 466
Stratiomyomorpha 441
Streaked Lady Beetle 353
Stream Cruiser 107
Streblidae 490
Strepsiptera 289, 437
Strigocis opacicollis 341
Striped Bark Scorpion 497
Striped Earwigs 118, 119
Striped Hairstreak 408
Striped Walkingsticks 156, 157

Strongylium crenatum 343
Strongylogaster 245
Strongylophthalmyia 477
Strongylophthalmyiid Flies 477
Strongylophthalmyiidae 477
Strumigenys hexamera 267
Strymon melinus 408
Stylopidae 292, 293
Stylops 291
Subterranean Silverfish 81
Subterranean Termite 167
Subterranean Termites 172, 173
Sucking Lice 226
Sugarcane Beetle 326
Suillia 485
Sulphur-winged Grasshopper 136
Sulphurs 408, 409
Sumichrast Toothpick Grasshopper 137
Sun Moths 394, 395
Sunburst Diving Beetle 319
Sunflower Spittlebug 193
Supella longipalpa 171
Surface Bug 182
Surinam Cockroach 171
Suzuki's Promalactis Moth 403
Swallowtail Moths 383, 412
Swallowtails 406
Swamp Darner 105
Swamp Milkweed Leaf Beetle 361
Sweat Bees 268, 269
Swollenstinger Scorpion 497
Sycophila 255
Sympetrum corruptum 108
Sympherobius barberi 285
Sympherobius occidentalis 285
Symphoricarpos 398
Symphyla 493, 500
Symphypleona 68
Symphyta 237, 240, 244
Symplecta cana 459
Synanthedon acerni 405
Synanthedon exitiosa 405
Synchroa Bark Beetles 342
Synchroidae 342
Synecdoche impunctata 195

INDEX

Syneches 473
Syneches thoracicus 473
Synolabus bipustulatus 363
Syntexis libocedrii 246
Syrbula admirabilis 137
Syritta pipiens 475
Syrphidae 448, 474
Systellognatha 121, 126, 128

Tabanidae 447, 468
Tabanomorpha 441, 444, 466, 468
Tabanus americanus 469
Tabanus pallidescens 469
Tabanus sulcifrons 469
Tabebuia heterophylla 398
Tabuda planiceps 471
Tachardiella 188
Tachinid Flies 260
Tachinidae 451, 490
Tachopteryx thoreyi 105, 109
Tachypompilus ferrugineus 267
Tachytes distinctus 269
Tadpole Shrimp 493
Taeniaptera trivittata 480, 481
Taeniogonalos gundlachii 261
Taenionema 127
Taeniopterygidae 124, 126
Tailless Whipscorpions 493, 496, 497
Tanaoceridae 134, 138
Tanaocerus koebelei 138
Tanaostigmatid Wasps 256, 257
Tanaostigmatidae 256
Tanaostigmodes albiclavus 257
Tangle-veined Flies 472
Tanyderidae 466
Tanypeza 484, 485
Tanypezid Flies 484, 485
Tanypezidae 452, 484
Tapinella maculata 225
Tarantula 499
Tarpela micans 343
Tawny Crazy Ant 266
Tawny Emperor 411
Taxodiomyia cupressiananassa 461
Taxus Mealybug 187

Teak Moths 398
Tebenna carduiella 397
Tebenna gnaphaliella 397
Tebenna Moth 397
Telamona concava 195
Telebasis salva 103
Telephone-pole Beetle 314
Teliapsocus conterminus 233
Temnoscheila 351
Tenebrionidae 308, 342
Tenebrionoidea 299, 340, 342, 344, 346
Tenebroides laticollis 351
Tenodera sinensis 159, 161, 165
Tenomerga cinerea 315
Tent Caterpillar 382, 418
Tenthredinidae 244
Tenthredinoidea 241, 244, 246
Tenthredo xantha 245
Tenuirostritermes 169
Tenuirostritermes cinereus 173
Tephritidae 292, 452, 482
Teratembiid Webspinner 151
Teratembiidae 151
Terebrantia 215, 218
Teredid Beetles 352
Teredidae 352
Termites 111, 167, 169
Termitidae 172
Tersa Sphinx 421
Tetracampid Wasps 256
Tetracampidae 256
Tetracha carolina 316
Tetracha virginica 316
Tetragnatha 499
Tetragnathidae 499
Tetraopes tetrophthalmus 359
Tetratomidae 309, 340
Tetrigidae 134, 138
Tetrix ornata 139
Tettigoniidae 135, 142
Tetyra bipunctata 213
Texas Beetles 332
Texas Bowlegged Bug 207
Texas Unicorn Mantis 165
Thalycra 355
Thanasimus dubius 347

Thaneroclerid Beetles 348, 349
Thaneroclerdae 348
Thaneroclerus buquet 348, 349
Thaumaleidae 464
Thaumastocoridae 184, 200
Thaumastocorids 200
Thaumastocoris peregrinus 200
The Seraph 418
Thecostraca 493
Theilopsyche 369
Theisoa constrictella 401
Thelyphonida 493, 496
Thelyphonidae 496
Theraphosidae 499
Therevidae 449, 468, 470
Theridiidae 499
Therion 249
Thermobia domestica 81
Thermonectus basillaris 319
Thermonectus marmoratus 319
Thermonectus nigrofasciatus 319
Thespidae 162
Thesprotia graminis 162
Thick Barklice 224, 225
Thick-Headed Flies 480, 481
Thionia bullata 197
Thomisidae 499
Thread-legged Bugs 201
Thread-waisted Wasps 268, 269
Three-spotted Nola 416
Thremmatid Caddisflies 376
Thremmatidae 367, 376
Thripidae 217, 218
Thrips 215
Throscidae 333
Thuja 246
Thymalid Beetles 348, 349
Thymalidae 348
Thymalus marginicollis 348, 349
Thynnid Wasps 264
Thynnidae 264
Thyreocoridae 184, 212
Thyrididae 387, 398
Thyridoidea 387, 398
Thyridopteryx ephemeraeformis 391
Thyris maculata 399

Thysanoptera 57, 215
Thysanura 71, 75, 79
Ticks 493, 494
Tiger Bee Fly 471
Tiger Crane Fly 459
Tiger Moths 382
Timema californicum 152, 157
Timema Walkingsticks 156, 157
Timematidae 152, 156
Timulla vagans 265
Tinagma gaedikei 394
Tinea apicimaculella 391
Tineidae 390
Tineoidea 384, 387, 390, 392
Tinoidea 387
Tingidae 184, 200
Tiphiid Wasps 264, 265
Tiphiidae 264
Tipula 459
Tipulidae 444, 458
Tipulomorpha 440, 458
Tischeriidae 388
Tischerioidea 384, 388
Toad Bugs 204, 205
Toe-winged Beetles 332, 333
Tolype notialis 419
Tomoceridae 66
Tomocerids 66, 67
Tooth-necked Beetles 338, 339
Tortricid Moths 383
Tortricidae 383, 396
Tortricidia testacea 405
Tortricoidea 385, 386, 396
Torymid Wasps 256, 257
Torymidae 256
Torymus 257
Toxomerus geminatus 475
Toxonevra superba 483
Toxonotus cornutus 362
Toxorhynchites rutilus 465
Trachypachidae 300, 318
Trachypachus 318
Trachypachus gibbsii 319
Trachythrips watsoni 219
Tramea lacerata 108
Travertine Beetles 330, 331
Tree Thrips 218

Treehoppers 177, 195
Tremex columba 247
Trepobates 203
Tretoserphus 251
Triaenodes 375
Triassothrips virginicus 215
Triatoma sanguisuga 201
Trichiosoma triangulum 245
Trichocera 459
Trichoceridae 458
Trichodectes canis 222
Trichodectidae 222, 226
Trichodes ornatus 347
Trichogramma 257
Trichogrammatid Wasps 256, 257
Trichogrammatidae 256
Tricholepidiidae 80
Tricholepidion gertschi 80
Tricholipeurus parallelus 227
Trichonephila clavipes 492
Trichopoda lanipes 491
Trichopsocidae 230
Trichopsocus dalii 231
Trichoptera 121, 281, 365, 369, 379
Trickle Midges 464
Tridactylidae 134, 138, 292
Tridentaforma fuscoleuca 390
Tridentaformid Moths 390
Tridentaformidae 390
Trielis octomaculata 267
Trigonalid Wasps 260, 261
Trigonalidae 260
Trigonaloidea 242, 260
Trigoniophthalmus alternatus 77
Trigonopeltastes delta 326
Trigonotylus caelestialium 201
Trimenopon hispidum 228
Trimenoponidae 228
Trimerotropis maritima 136
Trimerotropis pallidipennis 137
Trimerotropis salina 136
Trioza magnoliae 191
Triozidae 190
Triozids 190, 191
Triozocera 293
Triplax thoracica 357

Tritoxa flexa 483
Tritoxa incurva 483
Troctomorpha 221, 222, 223, 224, 226, 228, 229
Trogidae 303, 324
Trogiidae 224
Trogiomorpha 221, 223, 224, 229
Trogomorpha arrogans 248
Trogossitidae 309, 311, 313, 348, 350
Tropical Barklice 224, 225
Tropical Burnet Moths 404, 405
Tropical Ermine Moths 294, 395
Tropical Fruitworm Moths 398, 399
Tropiduchid Planthoppers 194
Tropiduchidae 181, 194
Tropisternus collaris 321
Trout-stream Beetles 318, 319
Trox foveicollis 325
Trox variolatus 325
True Bugs 175, 178
True Crickets 140, 141
True Hoppers 177
Trumpet Leaf Miner Moths 388, 389
Trupanea 483
Trypoxylon politum 269
Tube Maker Caddisflies 370, 371
Tube-tailed Thrips 215, 218
Tubulifera 215, 218
Tulip-tree Beauty 413
Tulip-tree Silkmoth 422
Tullbergiidae 64
Tullbergiids 64
Tumbling Flower Beetles 340, 341
Turtle Beetles 332, 333
Tussock Moths 382
Twenty-spotted Lady Beetle 353
Twice-stabbed Lady Beetle 353
Twin-spotted Sphinx 421
Twirler Moths 382, 400, 401
Twisted-winged Insects 289
Two-lined Spittlebug 193
Two-pronged Bristletails 71
Two-spotted Tree Cricket 141

Two-Striped Grasshopper 136

Uenoidae 366, 376
Uhler's Water Bug 205
Ulidiidae 457, 482
Uloma punctulata 343
Ululodes 287
Ungulate Lice 226, 227
Unicorn Caterpillar Moth 415
Uraniidae 383, 412
Urbanus proteus 407
Urodidae 387, 396
Urodoidea 387, 396
Uroleucon 187
Uropygi 493, 496
Usazoros hubbardi 110, 111, 113

Vaejovidae 497
Vaejovis carolinianus 497
Vagabond Crambus 379
Valenzuela subflavus 233
Vanessa atalanta 410
Vanessa virginiensis 410
Vanhornia eucnemidarum 250
Vanhorniid Wasps 250
Vanhorniidae 250
Variable Dancer 103
Varied Carpet Beetle 339
Variegated Meadowhawk 108
Variegated Mud-loving Beetles
    330, 331
Veliidae 183, 202
Velvet Ants 264, 265
Velvet Water Bugs 202, 203
Velvety Bark Beetle 341
Velvety Shore Bugs 204
Vermileonidae 468
Vermipsyllidae 428
Vesicephalus 68, 69
Vesicephalus occidentalis 69
Vespa crabro 263
Vespidae 236, 262, 266, 292
Vespoidea 241, 262, 264, 292
Vespula maculifrons 263
Vespula squamosa 263
Viceroy 411
Vinegar Flies 478, 479

Vinegaroons 493
Virginia Creeper Sphinx 420
Virginia Metallic Tiger Beetle 316
Vivid Metallic Ground Beetle 317
V-marked Lady Beetle 353
Volucella evecta 475
Vonones 498
Vostox brunneipennis 118

Walkingsticks 153
Walnut Sphinx 421
Wandering Glider 97, 108, 109
Warm-chevroned Moth 405
Wasp Mantidfly 287
Wasps 237
Water Boatman 204, 205
Water Fleas 493
Water Measurers 202, 203
Water Penny Beetles 330, 331
Water Scavenger Beetle 320, 321
Water Snipe Flies 466, 467
Water Striders 202, 203
Water Treaders 202
Waterscorpions 204, 205
Waved Light Fly 483
Wax Scales 189
Weasel Flea 429
Webbing Barklouse 231
Webspinners 148, 149
Web-spinning Sawflies 244, 245
Wedge-shaped Beetles 295, 340,
    341
Wedgling Moth 414
Well Polished Beetles 350
Western Honey Bee 271
Western Pygmy-Blue 408
Western Tent Caterpillar 419
Western Tiger Beetle 316
Western Tree Cricket 141
Western Tarnished Plant Bug 201
Wetas 140, 141
Whip Spiders 493
Whipscorpions 493, 496
Whirligig Beetles 314, 315
White Eyecap Moths 388, 389
White Furcula 415
White Miller 375

INDEX

White-blotched Heterocampa 415
Whitecrossed Seed Bug 209
White-dotted Prominent 415
White-lined Sphinx 420
White-margined Burrower Bug 211
Whiteflies 176, 186, 187
White-streaked Prominent 415
White-striped Tiger Beetle 316
Whites 408, 409
Widow 499
Widow Skimmer 108
Wild Olive Tortoise Beetle 294
Window Flies 470, 471
Window-Winged Moth 398, 399
Windscorpions 493, 496, 497
Winged Walkingsticks 156, 157
Wingless Scorpionfly 434, 435
Winter Crane Flies 458, 459
Winter Stoneflies 126, 127
Wood Cockroaches 170, 171
Wood Gnats 460, 461
Wood Midges 460, 461
Wood Soldier Flies 466, 467
Woodland Ground Beetle 317
Woodland Lucy 337
Woodlice 500
Woodlouse Flies 490
Woolly Aphids 187
Woolly Pine Scale 176
Wormlions 468
Wounded-tree Beetles 338, 339
Wrinkled Bark Beetles 314, 315
Wrinkled Grasshopper 136

Xanthogaleruca luteola 361
Xenochaetina 479
Xenos 291, 293
Xenox tigrinus 471
Xiphocentronid Caddisflies 370
Xiphocentronidae 366, 370
Xiphosura 493
Xiphydria mellipes 247
Xiphydria tibialis 247
Xiphydriid Wood Wasps 246, 247
Xiphydriidae 246
Xyela 245

Xyelid Sawflies 244, 245
Xyelidae 240
Xyeloidea 240, 244
Xylastodoris luteolus 200
Xylobiops basilaris 339
Xylococcidae 188
Xylococcids 188
Xylocopa californica arizonensis 271
Xylocopa tabaniformis 271
Xylocopa virginica 271
Xylomya 467
Xylomyidae 446, 466
Xylophagid Flies 472, 473
Xylophagidae 447, 472
Xylophanes tersa 421
Xylopinus saperdioides 343
Xylosandrus crassiusculus 363

Yelicones delicatus 249
Yelicones nigromarginatus 249
Yellow Nutsedge Moth 395
Yellowjacket Hover Fly 475
Yellowjackets 262, 263
Yellow-legged Mud-dauber 269
Yellows 408, 409
Yellow-sided Skimmer 109
Yellow-striped Armyworm Moth 414
Yersenia pestis 426
Yoraperla brevis 127
Yponomeuta multipunctella 393
Yponomeutidae 392
Yponomeutoidea 386, 387, 392, 394
Ypsolopha dentella 393
Ypsolophid Moths 392, 393
Ypsolophidae 392
Yucca Giant-Skipper 407
Yucca Moths 390, 391

Zamia 360
Zapada cinctipes 127
Zealeuctra hitei 127
Zebra Swallowtail 406
Zelleria retiniella 393
Zelus longipes 1
Zendosus sanguineus 348

Zenoa picea 332
Zenodosus sanguineus 349
Zerene cesonia 409
Zeugomantispa minuta 1, 287
Zootermopsis angusticollis 167, 173
Zopheridae 308, 313, 344
Zopherus concolor 345
Zopherus nodulosus 345

Zoraptera 57, 61, 71, 112, 149, 167, 221
Zorapterans 113
Zorotypidae 111, 113
Zorotypus 111
Zygaenidae 383, 404
Zygaenoidea 386, 402, 404
Zygentoma 57, 61, 71, 75, 79
Zygoptera 100, 102

Scorpionfly
(*Panorpa rupeculana*)
Panorpidae